THE INVENTION OF PREHISTORY

Figure 0.0. Gabriel von Max, *Affe vor Skelett* (Monkey before Skeleton, 1900).

The
Invention of
Prehistory

Empire, Violence, and
Our Obsession with
Human Origins

Stefanos Geroulanos

Liveright Publishing Corporation
A Division of W. W. Norton & Company
Independent Publishers Since 1923

For information about permission to reproduce selections from
this book, write to Permissions, Liveright Publishing Corporation,
a division of W. W. Norton & Company, Inc., 500 Fifth Avenue,
New York, NY 10110

For information about special discounts for bulk purchases,
please contact W. W. Norton Special Sales at
specialsales@wwnorton.com or 800-233-4830

Manufacturing by Lakeside Book Company
Book design by Daniel Lagin
Production manager: Lauren Abbate

ISBN 978-1-324-09145-5

Liveright Publishing Corporation
500 Fifth Avenue, New York, N.Y. 10110
www.wwnorton.com

W. W. Norton & Company Ltd., 15 Carlisle Street, London W1D
3BS

1 2 3 4 5 6 7 8 9 0

For Isabelle and Leon
with all my love

Contents

Part III
THE HORROR, PART I
(FROM 1900 TO THE 1960s)

Part IV
THE NEW SCIENTIFIC IDEOLOGIES;
OR THE HORROR, PART II
(SINCE 1930, AND STILL ONGOING)

THE INVENTION OF PREHISTORY

Introduction

THE HUMAN EPIC

It is not self-evident that humanity has a past, known or unknown.

—PAUL VEYNE,

DID THE GREEKS BELIEVE IN THEIR MYTHS?

W hen I was growing up in the 1980s and 1990s, Neanderthals were hulking creatures, evolutionary failures, savage beasts just a step above the gorilla. Illustrations in books showed that their dark skin was rough, like a hide covered in scraggly hair; their foreheads were aggressive, their eyes not so bright. Our teachers walked us through typical Neanderthal scenes: the terrible cold of the glaciers, the hunched efforts to hunt, the fires they must have built, their manifest inferiority compared to the *Homo sapiens* arriving in Europe. The stories all converged at an evolutionary dead end. Neanderthals were harsh, disappointing. Few—certainly not me—recognized that this imagery was really a set of claims about how to think about intelligence, beauty, and race.

Today, just thirty years later, the Neanderthal has changed. In newer life-size reconstructions, Neanderthal families figure as pensive-looking redheads and even blonds, blue-eyed, light-skinned, often with tools in hand and dressed in hides. We say less about the jutting forehead and more about the Neanderthal's brain cavity being larger than that of *Homo sapiens*. Their dull gaze is gone. A few scientists even consider them the creators of the earliest paleolithic art.

This seems to represent progress, at least in historical and genetic accuracy. Once more, the current image is declared true. But the new, more optimistic account of Neanderthals leads to hard questions. Are Neanderthals now smart because they are no longer depicted as dark-skinned? Or, conversely, have they become blond and white because they are now believed to have been smart, able, quintessentially human? Don't they resemble just a bit too closely those who study their fossils at sites in Europe? One theory about their extinction around 40,000 years ago proposes that the two species mated and *sapiens* "genetically swamped" Neanderthals. Another, that *sapiens* conveyed diseases to which Neanderthals had no resistance. Yet another, that the (African) *sapiens* swept in and raped and slaughtered the (European) Neanderthals. This last notion finds adherents on the political Far Right today, for whom Neanderthals sometimes figure in grotesque "white genocide" and "great replacement" claims. In such schemes, Neanderthals are the original white Europeans, who suffered after stupidly welcoming African migrants.

The Neanderthals themselves say nothing. We arrange them into whatever position we need them to take.

Neanderthals play a minor and rather benign role in modern thought and self-knowledge. Other figures from the deep past are far more troubling. For at least two centuries now, the very word "primitive" has identified living human beings, usually Indigenous peoples, with the beginning of the species, pushing them almost out of humanity and back in time. Since the eighteenth century, the word "barbarian" has been used to describe peoples deserving of scorn. The barbarian is other, it is a brute, an enemy, it is *it* and not *he or she*. But maybe barbarians are also our past, they must be understood better, they must be tamed and domesticated.

Human origins are not mere abstractions. Nor are they simple prompts for thought experiments and pure scientific inquiry. Promises and violence have regularly been unleashed in their name. Theories of our past have shaped history and the world we live in today. That, at least, is the contention of this book.

THERE IS PERHAPS NO GRANDER STORY THAN HUMANITY'S EMERgence out of nature. Scientists recount that our bodies, our posture, our food, our toil and suffering, our emotions and intelligence, our tools and our cities are all outcomes of our long and humble evolution. No epic is more stirring, none more seductive. No glowing CT scan can compare with it, no photographs of the cosmos taken by satellites we've dispatched into the void, no magnification of a virus particle, no moonwalk. The story of human origins tells us who we are, how we came to dominate this planet and each other, how we invented religion and then discarded it in favor of the gods of progress and technology. It supposedly reveals a million little things about human life, like why we desire and whom, how our emotions work, or how we love and care for others.

These grandiose claims prompt far more questions than they answer. When was humanity born? Where? How has it changed since that time? What in particular made humans human? What was language like at first? Are our emotions hard-wired? If so, since when? Which origin point matters the most—that of the body, that of culture, that of the patriarchy, or another? *Where* did culture emerge? Was it "primitive," or did it have some intense complexity we today fail to understand? Were the "early times" a kind of utopia—long halcyon millennia when our warm-hearted humanity developed? Or a gruesome hellhole from which we barely escaped? How is our singular origin as a species to be reconciled with our present-day differences? With racial violence? How do Indigenous peoples and their histories fit into the story of this past?

Many impressive names are affixed to the grand story of us: "the human adventure," the "origin of culture," the "dawn of Man," the "ascent of Man," the "descent of Man," the threshold, the quantum leap, the humanization of *Homo sapiens*, the civilizing of the beast. We speak of our "ancestral forefathers," of the cradle of humanity, of the exodus from Africa. Origins are so marvelous. We speculate about humanity's emergence, yet it feels *real*. We treat the story of our origins as the obvious triumph of modern knowledge over religious superstition, of truth over myth, of fact over ideology.

The story of human origins offers as good an answer as any we have to the fundamental question: what, after all, *is* the human?

FOR SOME SIXTY YEARS NOW, ROUGHLY SINCE AFRICA "BECAME" humanity's birthplace, these questions have been answered more or less as follows. From a small and insignificant australopithecine, a descendant of early apes, new groups of animal species evolved. Over millions of years, some of these animals diverged enough to form a new genus: *Homo*. Species in this group developed particular advantages and weaknesses. Sometimes gradually, sometimes quickly, they established quite complex technologies, a larger and more efficient brain, language, the domestication of fire and animals, hierarchy, traditions, complex social organization, gendered divisions of labor, and eventually trade, culture, religion, agriculture, and the state. Each of these things has its own history. Most are associated with the last and only surviving species of *Homo*, namely *Homo sapiens*, "anatomically modern humans." Humans founded societies that, however different from one another, were capable of enduring.

That is "how it all began." Over time, we like to think, scientists will fill in the details and breach through the remaining dead ends. After all, over the last two-and-a-half centuries, we have managed to piece together a sweeping account, one among the most enthralling and contested in the world. Ever since the Enlightenment in the late eighteenth century, the story of humanity's emergence out of nature has changed as a result of the expansion of European empires, the debates sparked by scholars such as Charles Darwin, the advent of "primitivism" in art, the discovery of ever more human fossils, religious anger, Nazi racism, Cold War competitions, decolonization, and much else.

Today, the "human adventure" is taught all over the world. News outlets from *Scientific American* to the *New York Times* (to say nothing of unreliable internet sources) report "breaking news" about the earliest past. Teams of skilled and dedicated researchers discover hundred-thousand-year fossils regularly. Some discoveries "upend" our understanding and "reveal" a new truth. Other findings "confirm what scientists have long suspected." A skull in a ravine on an Indonesian island. Some hominid footprints on a riverbed in Mexico or near a beach in Crete. (Whose footprints? Why did the people who left them live there?) A deep cave with fossilized human remains in South Africa. (Were they slaughtered? Or buried in a ritual?) Another cave, in France, full of "forgotten dreams," as Werner Herzog called its paleolithic paintings.

The newer theories are sometimes as wild as ones from decades, even centuries, ago. Some popular ideas speak of caveman strength and paleo diets: you too can be *Neander-Thin!*[1] Other, more obviously scientific ideas concern "genetic signatures" that tell us that humans have a "ghost ancestor."[2] There's the cooking-made-humanity theory, popularized by (among others) Yuval Noah Harari's deceptive hodgepodge *Sapiens*. Or the Stoned Ape theory, which declares that humanity's early "quantum leap" was due to mind-expanding hallucinogens in mushrooms. Or the neuroscientists' promise, with its bravado that we're the same as the first humans—only more complex thanks to neuroplasticity. Or anarcho-primitivism, the dream of pure freedom in a nature before the rise of states. For a while now, whenever I tell people about this book, someone invariably comes out as a convert to one or another of these theories. But who can know what to take seriously? As readers, we constantly encounter headlines promising new revelations and truths. The grand story is largely accepted, but it is also constantly under revision. The effect is dizzying.

Behind all of these questions, though, looms another that we ponder much less frequently: *Why* do we need to understand human origins? The answer is that the story of human origins has never really been about the past. It has never really been concerned with an exact, precise depiction of humanity's emergence out of nature. Prehistory is about the present day; it always has been. Over the 250 or so years that human origins have been pursued, studied, and taught, the countless stories and theories proposed have said a lot more about the current moment than the distant past. That past does not exist independently, suspended in amber, waiting for us. It doesn't simply get sharper, like an image coming slowly into focus. Rather, every natural philosopher and, later, scientist who has sought to "breathe life into the past" exhales with their own lungs.[3] Every time we find old bones, we dream up a primal scene and flesh it out with details from our own time.

Now, much the same could be said about any historical claims. But the deep past so exceeds our grasp, and at the same time it matters so much to "who we are," that the normal protections historians employ against "presentism"—like the rejection of the notion that presentism is a bad thing, or that it is avoidable— just aren't germane. Unlike, say, the history of the First World War, prehistory is often more a narcissistic fantasy than a field of inquiry. Its study is always contemporary at the same time that it is pre-modern, even pre-antique. After all, religion offered the first origin stories.

Like some religious stories, modern claims about prehistory can offer a soothing picture. Consider a primeval hominid on the savannah. It has just descended from trees and stood up. It might mate or craft tools, it might feel for others, hunt, paint on rocks, even build a sort of home. That creature speaks to us. It quenches our thirst for self-knowledge. Entire sciences have been built to tell us how "we" came to be. Careers have been made, curricula drawn up, knowledge pushed forward, policies rewritten. Thinking about human origins has been one of the most generative intellectual endeavors in modern history.[4]

It has also been one of the most ruinous. The Euromodern search for origins began in and then contributed to a long, brutal history of conquest and empire. It has been drunk on hierarchy. It is rooted in illusions—often murderous ones. It has served ferocious power. Its beautiful ideas have justified force against those deemed weak, different, ugly. It has rationalized colonial domination and eugenics. It has contributed to the destruction of Indigenous peoples. The sinister dimension of prehistory is easily disavowed and forgotten—after all, the archaeologists who dig up old bones and the biologists who study hominid genes are seldom the vectors of violence. But prehistory is, at its core, a device for creating meaning—for celebrating those who practice a particular idea of humanity and for demonizing those who don't.

Today, we might reject violence and racism, yet we still fail to recognize the blinders we wear when we look at "humanity." And we might as well admit it: without the grand story of our origins, we simply lack a good definition for humanity. That was not always the case, and I firmly believe that we *should* have a definition of humanity that does not rely at all on an origin story. But instead we spend our time pining for an origin story because it allows us to admire our grandeur. We call on it to give shape to our fears, to declare enemies to be subhuman, and then to say it was all done in the search for truth.

THE MODERN STORY OF ORIGINS WAS BORN IN THE EIGHTEENTH century, though "born" is a misnomer: concepts are not delivered, nor do they burst out fully armored like Athena out of some Zeus's brain. Even so, by around 1750, the Christian version of Creation had long lost its stranglehold, and notions of a world before civilization and history were coming into focus. This was a consequence of two major developments that shaped Western intellectual life.

First, Europe's colonization of the "New World." How could humans exist in America and not be accounted for in the biblical account of Creation? At first, the Native peoples who were "discovered" in the Americas were deemed simply subhuman. In 1537, Pope Paul III anointed them as human and rational beings in the bull *Sublimis Deus*. European empires expanded rapidly, and so did a certain awareness of other peoples' customs and a need to understand who one was ruling or trading with. Paul III's bull hardly ended the debates on Native humanity, but it led to a shift in the concepts used to reject Indigenous peoples: they became "cannibals," then "savages." Later, they got retrofitted as "primitives," meaning peoples living as the original humans had before civilization arose.

Second, the appearance of a European discourse of the earliest past that had serious empirical commitments. This was a new, and also specifically European, approach. In the sixteenth century, humanist-scientists like Johannes Kepler and Joseph Scaliger were mapping astronomy onto ancient texts. Isaac Newton and his contemporaries obsessed over how to reconcile events in the Hebrew Bible with Greek texts and Assyrian steles.[5] As these ancient peoples all belonged to the same history, shouldn't scholarship be able to show how they related to each other? By the eighteenth century, educated Europeans were confidently proposing a rise of humanity out of nature. How this had happened was very much up for debate. But no longer would they pretend, as Archbishop James Ussher had insisted in his well-known 1648 chronology, that the Creation could be pinpointed to 4004 BCE.

In other words, by around 1750 European thinkers and scientists were looking to a history before Creation, a time before history as it was known, and they were trying to understand their own "civilization" by contrasting it with those newly "discovered" peoples they dehumanized and consigned to the past. They were not reworking the Christian account of Creation, just as they were not turning to the ancients (Empedocles, Anaximander, or Lucretius). Instead, they scrutinized language, biology, material culture, and even violence in their own time for clues about the original humans. Within less than a century, all leading definitions of humanity would come to rely on theories of its origins.

THIS BOOK TELLS THE HISTORY OF PREHISTORY SINCE ITS INVEN-tion. It offers no answer to "What really happened at the beginnings of humanity?" Instead, it examines the most prominent and notable answers given over the last two and a half centuries. It is not a history of paleontology; rather, it is a history of how we and other moderns talk and think about the deepest past. I do not much care if particular theories are true: I ask what work they do, and at whose expense. The book does not obsess over the individual scientists who contributed to knowledge of human origins but rather focuses on how human origins came to saturate modern life. To do so, it identifies and tracks particular concepts, expressions, and images that have played star-ring roles in the history of prehistory. These include:

> the violent state of nature • barbarians • the infancy of humanity • the "Earth before Man" • dinosaurs • ruins and vestiges • the Sanskrit language and Indo-European conquer-ors • stadial theories • evolution and evolutionary anthropol-ogy • matriarchy • "the savage beneath the thin veneer of civilization" • primitive communism • Natives who . . . "dis-appear" • "flooding" hordes • cave art and "primitive" reli-gion • totemic feasts and Freudian oedipal murder • Nazi aryanism and antisemitism • killer apes • Neanderthals • Man the Tool-Maker • aquatic apes • "primitive warfare" • Mitochondrial Eves • lizard brains and creative explosions

I have tried to select those concepts that have been the most influential: the most controversial and also those that gained in acceptance and became almost, so to speak, second nature (at least for a period of time). To the people who subscribe to them, they evoke something primal. They draw a rich picture of the deep past and its dramas. They show up in scientific treatises but also in politicians' speeches, in novels, in news reports, in pedagogy. *The Invention of Prehistory* is their genealogy. It shows how profoundly they have influenced not just scientific research and popular ideas about humanity, but state violence—how terribly they have determined what qualifies as human. This is a book about science and speculation, about the space where each loses itself in the other, the great gray zone where rigorous research meets with righteous belief.

There are reasons why I have told this story through concepts rather than individual scientists or treatises or books. First, I do not write to cast aspersions on particular researchers, their good will, or their ability (though on occasion I cannot help it). Nor is scientific inquiry my target here. I am much more interested in how these concepts shaped modern humanity. Concepts are intertwined with the knowledge they sustain, with the way we experience the world, with the social conditions that make them possible, with the human lives they help form. Concepts don't float in mental clouds; rather they thread together the discrete aspects of the world for us. To understand them is to realize how they escape the meaning we want them to have, how they often tell a very different story than what those who devised them wanted to tell.

PERHAPS EVERY STORY ABOUT HUMANITY IS A TALE OF GRANDEUR and dehumanization. The stories of human origins certainly are. They have offered those who wield them the extraordinary capacity to decide who is human, who is advanced, and who has not changed for tens of thousands of years and can be dispossessed. What is more, when we link our politics to the past, we do more than continue an ongoing violence: we shackle what is actually politically possible.

Bracket the deepest past, stop trying to answer the grand, grandiose question—Where did humanity come from?—and suddenly these stories, these myths, come into relief.[6] It is time to wallow a little less in origins and to come to terms with the appalling consequences that these stories—"our story"—have had. The history this book reclaims cannot be bracketed: it is ongoing. It is not a tale of horrors that have been overcome, where it turns out that we are all better for simply learning about them. If colonial and state violence today are not as brutal as in the twentieth century, our depredation of the earth is yet to peak. The ground we stand on is tarred; dirty water flows over our toes. The human economy has warped the atmosphere and the planet. In this context, the story of how "we" became human eviscerates complexity and responsibility. Myths about human origins offer succor: the less bearable our planet becomes, the more we grab at the idea that our original humanity matured, over untold numbers of generations, out of that earliest infancy. "We" were resilient then, and remain so. "We" survived worse. "We" are natural. "We" mastered nature in the past and we will master it again. Such notions are a dead end, leading not to the past or the future, but to nowhere at all.

Many like to think that to be human today is to participate in a grand adventure that unites everyone since our humble beginnings and that points to a challenging yet prosperous future. But the animals that we call *Homo* had horizons vastly different than we do. They had ecstasies and feelings and terrors we will never comprehend. They flowed in nature and ruptured with it in ways that we cannot imitate. We act as though those beings back then existed for our own pleasure and understanding. Just as it is ludicrous to look fifty or two hundred thousand years into the future and pretend we are the origin of the people hopefully living then, it is folly to glance just as far or further into the past and declare that we are the same as those humans long ago.

What follows is the story of how we came to believe such ideas, and what that belief cost. The humanity we wish for, the world we want, the future we hope to build all depend on a clear understanding of our brutality, our desire, our power, and our eagerness to deceive each other and ourselves.

Part I

SCATTERED SHAPES
OF A FABULOUS PAST

(From the 1750s to the 1870s)

Figure 1.0. *The Customs of the American Indians Compared to Those of Primitive Times.* Frontispiece of Joseph-François Lafitau, *Mœurs des sauvages ameriquains, comparées aux mœurs des premiers temps* (1724).

Chapter 1

THE INFANCY OF HUMANITY

One day in the 1720s, a child left a room without breaking a comb. It happens all the time, it's boring, a non-event, but the comb was found broken, and that child grew up to become the *philosophe* Jean-Jacques Rousseau. No one else was there; he alone could be guilty. To the end of his life, Rousseau protested bitterly. "Do not ask me how the mischief occurred," he wrote forty years later. "I have no idea, and I cannot understand it. But I do most positively know that I was innocent." He was blamed, "lectured, pressed, and threatened," then cruelly punished. That "most grave injustice" by the adults caused a whole "revolution in his ideas, the violent change of his feelings!" When he eventually wrote his *Confessions* in the late 1760s, the episode stirred him still: because of it, "my blood boils at the sight or the talk of any injustice, whoever may be the sufferer and wherever it may have taken place, in just the same way as if I were myself its victim."[1]

By the time that Rousseau wrote his autobiographies, he had withdrawn into his own glass castle. There he could be always pure, "transparent as crystal."[2] It was always the others, shadowy and malevolent, who hurled the stones. His image of a pure childhood,

this memory of an early life prior to artifice, strung together his philosophy. "Man's breath is deadly to his kind"; only the children were
blameless.[3] The first humans—the children of history—even more so.

Rejecting the world of the adults allowed Rousseau to present
the deepest antiquity as an infancy, a place and time of defiance,
strength, and truth. Many before him had tied together nature, "savages," and infancy. But they all concurred: culture involved a long
rise out of that awful, violent pit. Just as an individual grows up, so
too humanity had escaped being dominated by nature in the ways
that Indigenous peoples, the living representatives of humanity's
infancy, still were. Rousseau's intervention scrambled this system.
He expressed deep admiration for the first humans—enough to make
knowledge of the earliest human condition an urgent problem.

OVER THE TWO CENTURIES SINCE EUROPE'S "DISCOVERY" OF THE
Americas, the Indigenous peoples of the New World had been slaughtered, subjugated, tyrannized, enslaved, pushed into alien lands. By
1700, the concepts used to describe them also worked to force them
out of their world and marked them as living in a natural condition
outside of civilization. Gradually, Europeans also used these same
concepts to parachute them back in time. Indigenous peoples would
now inhabit this awkward past before civilization—not the actual
moment in which they shared air, food, and injuries with Europeans.
To encounter them, to map their worlds, was to cross a threshold out
of the lived present and into a strange, almost timeless past.

But if every encounter trapped Indigenous people in a made-up
memory, it also invented surprising places for them. Concepts are not
airtight containers that enclose meaning and let nothing in or out.
By the 1700s, the concept of nature, for example, had established a
peculiar position for "savages" in European thought. If they were in
"nature" and not in "civilization," this had consequences because
"nature" was a fairly plastic concept. For jurists and political think-

ers, "nature" offered the most straightforward language for thinking about a supposed basic condition of humanity. "Natural" could
mean "lacking culture," a wild situation without a state, police, or
civilization. But the adjective "natural" and the concepts "natural
law," "natural religion," and "natural right" also pointed to something shared among all human beings: a humanity in its original
state. When writers wondered about this "Natural Man," Indigenous peoples came straightforwardly to mind. Indigenous peoples
would thus be Natural Man's representatives around the globe, and
in this way they were sometimes invoked so as to criticize current
habits. Were these humans indeed the first Man, writers asked, if
not more ancient than Adam then at least more primal than any
other alive?

In 1701, the traveler and
military commander Baron de
Lahontan's *Nouveaux voyages
dans l'Amérique septentrionale* featured a curly-haired,
dark-skinned (and strikingly
white-faced) Native American on its frontispiece (Figure
1.1), wearing only Adam's fig
leaf and holding up bow and
arrow in triumph.[4] Adario,
Lahontan's fictional protagonist, exudes a comfortable
naturalness as he tramples on
a lawbook and a crown-and-
scepter. He is both a "Natural Man" and an idealized
threat to order.[5] In 1724, the
Jesuit missionary and proto-
ethnologist Joseph-François

Figure 1.1. Frontispiece, featuring the Native
American figure (and hero of the book)
"Adario" standing proud on European
regalia, from *Nouveaux Voyages de Mr.
Le Baron de Lahontan dans l'Amerique
septentrionale* (1703).

Lafitau published a fascinating and tedious four-volume study of *The Customs of the American Indians Compared to Those of Primitive Times*. On its own spectacular frontispiece (Figure 1.0, with which this chapter opened), a seated woman compares antiquities, maps, books, sculptured objects. Two putti are excited to show her idols and scepters from the Americas. Time himself stands before her and enthralls her with a mystical vision replete with Adam and Eve, to draw her back to "our Religion."

Had she strayed? No. She was reconciling the Bible with the idolatrous gentiles who were stubborn enough to exist without being mentioned in it. History was supposed to unfold as Christian Sacred History—in a manner commensurate with the Bible and the Second Coming. But how would Genesis work if it could not account for Native Americans, how they got there, or why they had remained unknown? Some authors like Lafitau adapted the Jesuit idea of "natural religion" to link the humans who populated the New World to those at the start of Sacred History.[6] Their "natural" morality testified to their readiness for conversion—and to their usefulness for criticizing European mores. Lafitau spent some seventy pages debating where to place them—all of them—in the Old Testament. (Before or after the Deluge?) Others thought that the frame of Sacred History was not enough to explain how they had gotten to America.

By 1725, enough scholars were traveling to meet these "first humans" (or sometimes just pretending to do so) for critics to scoff at their pretensions. In *La Scienza nuova*, Giambattista Vico poured scorn on contemporaries who pretended to know "how those first men of the impious races must have thought, or the crude manner in which they formed their thoughts, or the confused way in which they connected them." Travelers were reporting on customs "so extravagant . . . as to excite horror in us." And they were merely describing the "barbaric inhabitants" of Africa and America. By comparison, those "from whom the gentile nations began" had ignored language altogether.[7] So little was known about them—not even where they came from. The horror they might excite was better left unimagined.

Like Lafitau, Vico meant his injunction to reenthrone Sacred History. Only with Providence as a guide could anyone descend the steep gorge into the deep past and find the right place to put Africans and Native Americans. The fog of early humanity was thick, but even for Vico it was alluring: perhaps there *was* a place where these nations arose, and maybe it had laws, the way language had an origin.

By the mid-century, the idea of traveling to meet, study, and understand Natural Man had become a joke, cracked to criticize European pomp. "Everyone who has nothing to do in Europe is running around the world in a kind of philosophical fury," commented Johann Gottfried Herder.[8] Rousseau was still more blunt: "ever since the inhabitants of Europe inundated the other parts of the world and continually published new collections of travels and stories, I am convinced that we know no other men but the Europeans alone." Even the enlightened clutched their ridiculous prejudices tight, he continued: all everyone was doing "under the pompous name of 'the study of man'" was glancing over at their own comrades.[9] Though most of the world is blissfully unaware of winter, Europeans "never fail to show us primitive men inhabiting a barren and harsh world, dying of cold and hunger, desperate for shelter and clothing, with nothing in sight but Europe's ice and snow. . . . To study man, one must extend the range of one's vision."[10]

When Rousseau turned his own gaze to the origins of humanity in 1754, he did it so as to inquire into the gross inequality in his society. If France was marked by extreme inequality, where did inequality emerge and how did it become so dominant?

Several ways of addressing the question and thinking about the past were readily available. He could, for example, turn to the Bible and wonder about its chronology and the peoples not mentioned in it. But Archbishop James Ussher's 1648 chronology, which dated the Creation to 4004 BCE, and which remains an obsession among American evangelicals today, was already discounted when it was first published, and by Rousseau's time it seemed nothing more than an ignorant trick. Some would have turned to the authority of the ancients: Rousseau could have used Hesiod's Greek myth of origins, which declared that in the times of Cronos, the Olympian gods had fashioned a first, golden generation of humans. These first humans had lived in plenty and known no sorrow. In Hesiod's scheme, the paradisiac Golden Age was followed by several others, sometimes tender sometimes vile, until Zeus devised the present generation of humans.[11] But this approach would have tied Rousseau to debates about the authority of ancient texts for which he had no taste. He might also offer a historical treatise that compared different peoples of the past, or a theory steeped in political economy.

Instead, Rousseau chose to think with (and against) the travelogues that he was mocking, and especially with the understanding of nature that he found in them. Enlightenment thinkers were increasingly using natural law as a device to criticize the contorted, monarchic society they lived in, and natural religion to criticize the excesses and irrationality of Catholicism.[12] So Rousseau was hardly alone when, in the *Discourse on the Origin of Inequality*, he identified the "state of nature" with the earliest condition of humanity. Rousseau amped up the device of "nature" to the max, choosing among living "savages" the "Caribs of Venezuela." They, "of all existing peoples, are the people that until now has wandered least from the state of nature." This meant that they were indeed savage in the sense of visceral, strong, unaffected, lacking artifice, animal, but not in the sense of gratuitously violent. They lacked foresight, jealousy, the grotesqueness of culture. But it was "ridiculous to represent savages as continually slaughtering each other."[13]

The "state of nature" concept had a history of its own, with two origin points: the writings of Roman philosophers (especially Cicero and Lucretius) who were rediscovered in the Renaissance, and Jesuit theologians. Both groups had located the state of nature at the beginning of history. For the Romans, state of nature meant basically "the natural state of things."[14] For the Jesuits, it referred to the time after Adam and Eve's expulsion from Paradise, before any civil state had been established.[15] It was a concept that could easily refer to Indigenous peoples.

Surprising as it may sound, most early-modern thinkers who wrote about the state of nature had been at best tentative in identifying it with the deep past. Rather, during the seventeenth and early eighteenth centuries, the state of nature was a dispensation without law and order—a world of anarchy, violence, and cannibalism.[16]

A century before Rousseau, Thomas Hobbes had vividly described the state of nature in 1651 as a place where "man is wolf to man," where life is "solitary, poore, nasty, brutish, and short." For his contemporaries, these descriptions would have conjured up images from the book of Genesis and the newly-founded American colonies.[17] Hobbes did indeed offer some speculation that Native Americans lived in the state of nature (Figure 1.2), writing of "the savage people in many places of America" whose "concord" depended only on "naturall lust," natural men who "have no government at all." But in the same breath, he insisted that a primeval condition was not the example to focus on; his state of nature was best captured by the descent into "civill Warre."[18] The English Civil War had just ended, and so too the Thirty Years' War in Europe. In some regions of the Holy Roman Empire, fully half of the population had died. The jurist Hugo Grotius and the bishop Jean-Pierre Camus also thought of the Thirty Years' War when they imagined the natural state as a life lived in cannibalism. A disordered world was an "amphitheater soaked in blood."[19] Hobbes did not need a history of early civilization: he needed a political technology for switching off the state of nature, because it amounted to permanent war. That technology was the social covenant that established a commonwealth with its sovereign.

Figure 1.2. Detail from the frontispiece of Thomas Hobbes, *De Cive* (1642), featuring Indigenous people living in the state of nature.

In the century between Hobbes and Rousseau, the state of nature was regularly tested against Indigenous "Natural Men." Only occasionally was it linked to early humanity. Thirty-eight years after Hobbes, John Locke treated "the Indian" who knows nothing of property as the human that exemplified the natural state. "Thus in the beginning all the World was America."[20] Locke compared "the kings of the Indians in America" to "the first ages in Asia and Europe," but then pivoted to other matters.[21] Elsewhere, he even likened the state of nature to a Golden Age like Hesiod's.[22] For him too, the state of nature wasn't really about the birth of humanity; it was a device for comparing different polities, and he arguably used it to justify landgrabs in the New World.

Rousseau wrote as though he was singlehandedly saving the state of nature from the powerful Hobbeses and Lockes. To learn about this original human condition, one had to disavow society and begin again *from* the state of nature and the people closest to it (the "Car-

ibs").[23] Rousseau then explicitly dated the state of nature to humanity's start. The fantasy became all-encompassing: "wandering in the
forests, without industry, without speech, without dwelling, without
war, without relationships, with no need for his fellow men, and cor-
respondingly with no desire to do them harm . . . savage man, subject to few passions and self-sufficient, had only the sentiments and
enlightenment appropriate to that state."[24] He installed the inhabitants of that state right at the heart of his argument and avowed
his deep commitment to them. Private property had destroyed the
state of nature, and therefore only Indigenous peoples were still
close to that beginning of humanity. Rousseau's Natural Man, however hypothetical he might be, would now stand judge over human
inequality. Hobbes had not needed for the violent state of nature to
lie in the deep past, but Rousseau did: by existing before human cor-
ruption, the earliest human state might undo it.

Through thought experiments like Rousseau's, it became possible to think of prehistoric humans, to give them some meaning
that was not simply that they had been rude and ugly and we had
plainly improved on them. (Of course people had thought of their
own ancients and human origins before: Judaism, Hinduism, Greek
philosophers—they all had detailed beliefs and ideas. We might even
speculate about how early humans had thought of human origins.
But in the world that came out of Christianity and early modern
science, this much was new: to sidestep the Christian and Classical
approaches for thinking about the earliest past, and to devise a new
meaning for origins.) What was more, the deep past served as a logic
with which to scorn the present. What could moderns even know
about all those centuries of wandering in the forests?

Rousseau assembled his argument by sliding between early
human history and his memories of a childhood destroyed by adults.
Humanity in the state of nature had the same character that allowed
children to resist and withstand the corruptions of adulthood. In the
state of nature, "centuries went by with all the crudeness of the first

ages; the species was already old, and man remained ever a child."[25] Today everyone demanded that the savage cease to be savage, just as everyone dragged children out of childhood.[26]

Rousseau's combination of childhood, the state of nature, and the origin of humanity set him apart. His identification of childhood with early humanity was less original. The creatures of Hesiod's Silver Age—the second of four generations preceding the actual human beings who now roamed the earth—had lived most of their lives as children.[27] Ever since the 1500s, Indigenous peoples had been described as children who would reach maturity upon converting to Christianity. In John Earle's *Microcosmographie* (1628), the child was "the best copy of Adam before he tasted of Eve or the apple . . . his Soul is yet a white paper unscribbled with observations of the world." The child's mind was pure, unacquainted with sin, and happy. Life makes us all into blurred notebooks, Earle continued.[28] Locke would appropriate this idea that at birth, the mind is vacant, and would famously name it the *tabula rasa*, the empty slate.

BY THE EIGHTEENTH CENTURY, THE PATERNALISTIC NOTION OF childhood as lack predominated: children were not mature, not rational, not adult. In the *Encyclopédie*, the great 1750s attempt to reorganize all knowledge that was spearheaded by Denis Diderot and Jean le Rond d'Alembert, the entry on childhood (*enfance*) proposed this was the time that unfolds until "man reaches the use of reason" around age eight.[29] Then came Rousseau's alternative: childhood is not lack, but a robust if endangered purity. It is a crystalline beginning for a human being and for humanity as a whole. Where this childhood of humanity ended, property, inequality, need, power, and true violence all began.

Rousseau's contemporaries elaborated on this problem. The Swiss philosopher Isaak Iselin parroted him and identified the state of nature with childhood, but he could not bear the idea that civilization was at its core violent, and he proposed instead that an intermediary "savage state" had existed between the original childhood of Man and current civilization.[30] Other accounts of the "Ages of Man," notably by Scottish philosophers like Adam Smith, ignored Rousseau's identification of childhood with robustness, fortitude, and strength of character. Childhood to them was pure, yes, which meant it was weak or rude.[31]

Still others simply dropped the comparison to Native Americans. The German philosopher Johann Gottfried Herder agreed that "what is indispensable for every individual human being in his infancy, must surely be no less so for the whole human species in its own infancy as well."[32] But Herder was looking east, to the *Morgenland* (morning land, "the Orient"), and not west to the Americas, to find the beginning. In this vague "Orient" had stood humanity's cradle, he proposed; in Egypt and among the Phoenicians "was formed" the boyhood of mankind.[33] Herder's target was monarchy, and he indulged in an orientalist fantasy in locating its origins. It was the "delicate child sense of the Oriental," he declared, which had brought forth patriarchal despotism.[34] The species had taken millennia to be dragged out of its childhood, barbarism, idolatry, and carnality, and get ready . . . for Christianity.

Depending on their priorities, these and many other authors used the same words and images to celebrate or infantilize Native peoples, to praise or mourn the maturity of civilization. After Rousseau, they were faced with a multiple-choice with several viable solutions: What *did* it mean for the natural state to have been a childhood? Who among all the people encountered around the world could really be the purest? How could one talk about the civilized adults and their relation to these others? Those who followed Rousseau toggled between the different options, but they retained the basic idea: humanity's historical progress resembled the life of the individual from the cradle to the grave.[35]

❈

THE FRENCH REVOLUTION BEGAN IN 1789, AND ROUSSEAU, THOUGH he had died a decade earlier, now offered some of the great mantras for the age. "Man is born free," he had famously written to begin *The Social Contract*, "yet everywhere he is in chains." Each child was born pure and strong: each child was capable of starting history anew. It was possible to transform society because even though society was lost, every child relived the purity of the infancy of the species. By the time the Revolution was over, the political metaphor that celebrated the state of nature had largely died out. But the infant continued to play a role in the theories of human origins. Herder and others gave it a different feeling, more romantic than political: "The childhood of the species will always retain its power over the childhood of every single individual: the last minor will still be born in the first land of the morning."[36] In his *Vestiges of the Natural History of Creation* (1844), Robert Chambers compared childhood and the moral development of humankind: each of them "in its earliest stages" was "sanguinary, aggressive, and deceitful."[37] John Lubbock, who in the 1860s popularized the term "prehistory," kept the comparison going: "the mind of the savage, like that of the child," he wrote.[38] Sigmund Freud loved it, too: the Oedipus complex replayed in each child the history and guilt of humanity since the original murder of the primal father.

Figure 2.0. Depiction of "ancient Germans" in Philipp Clüver, *Germaniae antiquae libri tres* (1616).

Chapter 2

EUROPE'S "INDIGENOUS" NOBLE SAVAGES

lready with the "state of nature" and the "Ages of Man" as its chief concepts, the early inquiry into the human past had developed a geopolitics. Perhaps it is more appropriate to say: prehistory *was* a geopolitics. It was at least as much about projecting power in the present as it was about early human time.

Just as the state of nature relied on the fantasy of cannibals lurking in the New World, so too did other sites of European colonialism influence the perception of the past. Africa, argued the French writer Charles de Brosses, was a world dominated by "fetishism"—the supposed primitive religion of the early humans.[1] The Germans called the Far East the *Morgenland*, land of the morning, to contrast it with the mature *Abendland* of the West. India, then ruled by the British East India Company and celebrated as an advanced civilization, seemed far from Europe's origins. But then Sanskrit was linked to Greek and Latin and came to be seen as important to the origins of language and culture. In other words, the sites of European colonialism were identified with European antiquity. The more easily some peoples were dominated, the further back in Europe's distant past they lived.

The late eighteenth century unleashed a different historical and prehistorical comparison, and with it a different politics. While the French Revolutionaries turned to the Roman Republic for a model of political action, the Germans, both before and especially after 1789, invented a unified identity for themselves as inheritors of Rome's ancient enemies. Modern Germany at the time was divided into myriad states large and small, some Catholic, others Protestant. The dream of modern Germany pined for the "German" tribes that ancient Roman writers (notably Tacitus, Julius Caesar, and Pliny) had described in detail. In the eighteenth century, by a sleight of hand, the peoples that had lived east of the Rhine and north of the Danube in the centuries around 0 CE became identified with the barbarian invasions that destroyed and politically replaced the Western Roman Empire several hundred years later. Crucially, this ancient "German" noble savage was treated as indigenous to Europe, and became the "domestic" deep past.[2]

Roman writers had indeed referred to what is now Germany as Germania, and the peoples living there as Germans. Tacitus insisted that they were *indigenas*, natives, and he seemed awed by their unified and ethnically untainted status. He described their tribes, Tencteri, Chatti, Chamavi, Frisii (Greater and Lesser), Angrivarii, Dulgubinii, Chasuarii, and so on, who had "chosen for themselves" the name Germans.[3] In subsequent centuries, the "migrating races" or "barbarian invaders" known as Franks, Langobards, Vandals, Burgundians, Saxons, Alamanni, Goths, Ostrogoths, Herules, Visigoths, and Alans were also, *at times*, described by observers as Germanic. (Other migrating peoples or invaders were not: Huns, Bulgars, Magyars.) As Patrick Geary, Eric Michaud, and other scholars have shown, all these "Germans" did not share a racial or ethnic identity—not in the modern sense, nor in the way they understood themselves.[4] Rather few were "German," and they could not be connected to the modern inhabitants of these territories.[5]

But in an age where moderns were identified with their imagined heroic forefathers (notably the Greeks and Irish with ancient Greeks and ancient Celts), educated Europeans played at discovering ancestors of their own. Those who wished for a long German pedigree had a whole literary history to back them up. The rediscovery of Tacitus in the fifteenth century had prompted the publication of texts that were hundreds of years old—Jordanes's *The Origins and Deeds of the Goths* (551 CE), Paul the Deacon's *History of the Lombards* (ca. 790 CE), and others. In the 1500s, authors routinely offered detailed discussions of ancient German "barbarism," "purity," and "freedom." By the late 1600s, "Tacitism" was a thing. Ancient Germans were always described as free, pure, ethnically unified, and fearsome. The fantasy of barbaric German tribes started to stampede all over the Republic of Letters. Thanks to it, Europe could be imagined as divided along racial, national, and even class lines. German humanists like Justus Georg Schottelius celebrated their language by describing it as the bearer of bravery, superior to French, the "ugly slime of the Seine" as Herder would later deride it.[6] Even some French intellectuals regarded themselves as (Germanic) Franks—notably the Comte Henri de Boulainvilliers, around 1720. To him, nonaristocrats were non-Germans, mere Gauls.[7] The Baron de Montesquieu wrote at length about the free Germans in *On the Spirit of the Laws*, even pronouncing that "it is impossible to inquire further into our political right if one does not know perfectly the laws and the mores of the German peoples."[8] In taking Tacitus literally, Montesquieu was absolutely unoriginal: even Tacitus's contemporaries had read his book as offering authoritative ethnographic knowledge (which was far from the case). Tacitus's Germans, Montesquieu continued, lived in a "state of nature." Those who followed them adopted laws and built a civil society.[9]

UNLIKE THE GREEKS AND THE CELTS, THE ANCIENT GERMANS HAD supposedly burst out of nature and into history. Ironically, as the "free and pure" ancient Germans left no writing, the moderns could only re-create their "state of nature" lifestyle by relying on the propaganda of their Roman enemies. The "Germanic" enemy of the Roman past became the origin of the self. Gradually this entire way of thinking became politically valuable, especially to the new nationalists of the early 1800s. For some, modern Germans especially, the German movement or conquest was a *Völkerwanderung*, a "migration of peoples." For others, notably the French, it was a series of barbarian invasions (which has a quite different sound to it). The "Germans" from 100 BCE to 500 CE were heroically recast as the destroyers of the declining Roman polity. They possessed great innate strength, were naturally democratic (for they accepted no ruler's yoke), and served as defenders of Europe.

For Edward Gibbon, whose *History of the Decline and Fall of the Roman Empire* was one of the most widely read books of the 1770s and 1780s, the "ancient Germans were wretchedly destitute" of writing, arts, and science. They "passed their lives in a state of ignorance and poverty, which it has pleased some declaimers to dignify with the appellation of virtuous simplicity."[10] They left care of the house, agriculture, and animal husbandry to "the old and the infirm, to women and slaves" in order to waste their time in "the animal gratifications of sleep and food."[11] If ever a noble savage had existed, the ancient German was him. Above all else, he was free: "A warlike nation like the Germans, without either cities, letters, arts, or money, found some compensation for this savage state in the enjoyment of liberty. Their poverty secured their freedom, since our desires and our possessions are the strongest fetters of despotism."[12] With this wink to Rousseau's noble savage, Gibbon distinguished the "warlike Germans, who first resisted, then invaded, and at length overturned the Western monarchy of Rome" from other barbarians and non-Europeans. "The most civilized nations of modern Europe issued from the woods of Germany; and in the rude institutions of those barbarians we may still distinguish the original principles of our present laws and manners."[13]

Gibbon did not lack a sense of humor: "In the days of chivalry, or more properly of romance, all the men were brave, and all the women were chaste." Hard as chastity was to preserve, "it is ascribed, almost without exception, to the wives of the ancient Germans."[14]

Gibbon published his book in 1776—a less-than-auspicious year for the British Empire in whose capital he lived. Johann Gottfried Herder, writing two years earlier in his *Another Philosophy of History* (where he recovered the "infancy" of humanity), was even more melodramatic. Among the Romans, "everything was *exhausted, unnerved, shattered*: abandoned by men, inhabited by unnerved men, going under in excess, vice, disorder, license, and savage martial pride." There was no exit for the sluggish Rome, only "*death!—an* emaciated *corpse* lying in a pool of blood." And then, Providence, which had long favored the Greeks and the Romans, glanced to "the North." There, "*a new man* was born." The usually reserved Herder was moved to enthusiasm for the Germans' vitality and creativity: "*Goths, Vandals, Burgundians, Angles, Huns, Herules, Franks* and *Bulgarians, Slavs* and *Lombards* came, settled down—and the whole new world, from the Mediterranean to the Black Sea, from the Atlantic to the North Sea, is *their work, their race, their constitution!*"[15] Are these the most ecstatic and unsettling italics ever used in historical writing?

As noted earlier, the French revolutionaries looked to the ancient Roman Republic for a prototype of the one they wanted to create. Once Napoleon came to power, with every battle he won he seemed to carry the Revolution on his shoulders, and his enemies relished the fight against a "new Roman Empire." After he declared himself Emperor in 1804 and abolished the Holy Roman Empire in 1806, the parallel to the ancients seemed all too obvious. German philosophers, jurists, and folklorists—from Friedrich Carl von Savigny to the Grimm Brothers—began to look to this ancient "German" history to give their culture a unifying value. Others looked to the Germanic past because sharing blood with distant ancestors generated political authority. Beneath political borders, a nation supposedly pined for its original truth.

Mapmakers too started to celebrate the barbarian invasions as expressions of pure force, as arteries of Europe's renewals. First among them was the Comte de Las Cases, who boldly outlined the "barbarian transmigrations" in an atlas celebrating Napoleon's new regime (see Figure 2.1). Las Cases's map, intended to craft France's "other," was imitated to the point of becoming a "mnemonic code" for the study of European history.[16] With astonishing self-assurance, he and later mapmakers depicted the exact paths of Ostrogoths, Vandals, and Huns as they rampaged through Europe's lands. Versions of this map multiplied rapidly—and survived for at least 170 years, creating for every child at school a visual story of conquering noble savages. Some of them can be found in textbooks even today.

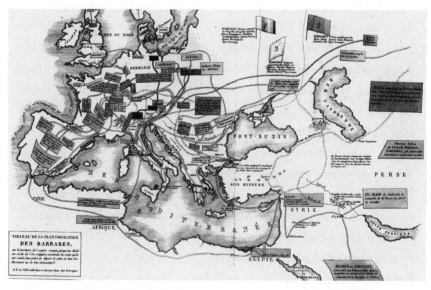

Figure 2.1. "A Map, Exhibiting the Transmigration, Course, Establishment or Destruction of the Barbarians That Invaded the Roman World." Emmanuel de las Cases pioneered the "Barbarian Invasions" map for Napoleon's benefit in 1806; this was his 1827 update.

Pure power became enfleshed—in the ancient Germans. G. W. F. Hegel, a onetime fan of Napoleon, turned them into a force of nature: like a river they "gushed forth over the Roman Empire, something no dam could any longer withstand."[17] What made such poetic liberties useful was the flexibility of key details. Raiders of Rome, the Germanic tribes were also defenders against other "tribes"—notably the Huns. Such ambiguities were easily exploited. Those supposedly "indigenous" to Germany, the good barbarians, embodied the Northern conquest of the decadent Latins, and the resistance against the bad barbarians. For the word "barbarians" could *also* refer, this time in a derogatory manner, to Huns, Mongols, Russians, or other "Asiatic" invaders.

By the 1830s, the moment someone invoked the barbarian invasions, a small army of metaphors and values would be awakened, ready for conceptual battle. They included floods and dams; noble, brute, authentic freedom; war against one's own oppressive Rome; the dream of ultimate victory after long perseverance; and "natural" superiority. The invasions/migrations became a sort of mythic algorithm: each component clicked in place and powered new ones. Authors citing them would get trapped in the logic of the program. They thought they were saying something new but got caught instead in a machine of associations that organized meaning and national feeling.

MOST OBVIOUS WERE THE MOTIF'S USES IN GERMAN NATIONALISM. From the defeat of Napoleon in 1815 through the unification of Germany in 1871 and into the twentieth century, the ancient barbarians became guarantors of German nationhood. The composer Richard Wagner turned the fantasy of a primordial Germany into an art that longed for national regeneration. He also gave it strong antisemitic overtones. Bitterly disheartened that the 1848 revolutions did not succeed in transforming European society, Wagner made clear that he needed something to make up for a modern Germany that "could nowise satisfy my longing." He looked for a "true" German hero, noble and free, by unraveling the old German myths. "As though to get down to its root, I sank myself into the primal element at home." Like a good prehistorian, he claimed to get rid of later historical additions to the myth, until he finally reached down and dredged "the real naked man, in whom I might spy each throbbing of his pulses, each stir within his mighty muscles, in uncramped, freest motion: the type of the true human being."[18] That was the "true," Germanic Siegfried, ready to become the leader of a new myth in Wagner's total work of art. In Wagner's opera of the same name, Siegfried is a force of nature who recognizes himself and his authenticity only when he sees himself reflected in the river and realizes his affinities were with the animals around him, not the (antisemitic stereotype of an) engineer who raised him. The barbarism had not disappeared—Siegfried *is* brutal—but his nobility, freedom, and power carried the promise of a "truer" Germany of the future.

Even non-Germans looked to prehistoric Germany to signal strength. Popular novelists in Victorian England reveled in Gibbon's history, turning it into tales of empire, decline, sex, Christianity, race, and masculine destructiveness.[19] Art historians tried to figure out how the racial origins of artists were reflected in their work, and what this meant for Europe's mosaic of cultures and ideas.[20] In 1840, French Marshal Thomas-Robert Bugeaud, just nominated Governor General of Algeria, proclaimed that the French could not simply stay on the Algerian coast but needed "a grand invasion in Africa that resembles what the Franks did, what the Goths did."[21] Bugeaud put his maxim into extremely brutal practice. But even to those who were politically on the Left, the image of the ancient Germans was useful. The historian and republican Jules Michelet scoffed that France's established order derided "the people" by comparing them to "the barbarians." He turned the accusation around, into a point of pride. "The expression pleases me; I accept it. *Barbarians!* Yes, that is to say, full of new, living, regenerating sap. *Barbarians*, that is, travelers marching towards the Rome of the future."[22] Michelet and Bugeaud could not have been more different. Germanic barbarism just tasted good in the speaker's mouth. As the words came out, they created a hard link to a racial nature, a force dating to a past outside written history.

As we will see, the fantasy would peak with the Nazis. But its geopolitical division of Europe and culture survived World War II, for example in Kenneth Clark's astonishingly priggish 1969 BBC documentary, *Civilisation*, which began with Clark standing before the cathedral of Notre Dame in Paris and portentously worrying that European civilization might vanish thanks to young, leftwing, often non-European hordes. "You know, it *has* happened once! All the life-giving human activities that we lump together under the word civilization have been obliterated once in Western Europe: when the barbarians ran over the Roman Empire. For two centuries, the heart of European civilization almost stopped beating. We got through by the skin of our teeth." "We," for Clark, meant we of a certain class, a certain nationality, a certain art. European Man must not be allowed to fall again.

At the opposite end of Clark, there is the boxer Mike Tyson, who understands himself as a philosopher of war, and who studies Frankish kings for the Europe they made. "Attila the Hun was terrorizing the whole territory . . . and then Aetius, a statesman from Rome, decided he would get all the German tribes together and we're gonna vanquish Attila the Hun. And then he did."[23] Note how Tyson switches from third person singular to first person plural and back: he identifies with their rise from obscurity to triumph. In Tyson's theory, Germanic kings become a source of pure power that vanquished the Huns, then in turn the Arabs. He goes on, echoing a standard modern expression: "Without this victory [at Tours in 732], we would all be speaking Arabic right now."

❖

IT IS HARDLY SURPRISING THAT THE "BARBARIAN INVASIONS" CON-
cept was directed toward chauvinistic ends: other concepts had little
to do with conquest but were deployed similarly. It is even less surprising that eighteenth-century thinkers created selective continuities
between themselves and ancient "conquerors." Writers from London
to Vilnius to Athens were happy to find ancestors and put them on
a pedestal. What is distinctive, however, is how these ancient Germans were braided with European prehistory and became a *name*
for that prehistory.

Today this seems absurd: by the eighteenth century, everyone
already knew that Mesopotamia, Egypt, and Greece had long preceded the entry of the "Germans" into the historical record. But
in the mid-eighteenth century the deep past had become a matter
of states of nature, noble savages, historical stages, the "infancy of
man." Ancient Germans offered a moment where all of those could
be joined together in a single description of early humanity. They
were an archaeological miracle, all the more marvelous in that they
were "domestic" to Northern Europe. With them, history had kicked
off once again. Humans had not emerged from nature because they
were biological beings. Nor had they emerged out of savagery just
once, somewhere, somehow. They emerged several times, and most
effectively as Germans. By rebooting history, the ancient Germans
guaranteed that race provided the scaffolding of history.

Ancient Germany became a synonym for the prehistoric past for
a second reason. In the later eighteenth century, European scholars
encountered Sanskrit and wondered why it related to European languages. The English lawyer and orientalist William Jones, speaking at the Asiatick Society of Bengal in Calcutta in 1786, famously
compared them and postulated that an original language had preceded them all:

> The Sanscrit language, whatever be its antiquity, is of a won-
> derful structure; more perfect than the Greek, more copious

than the Latin, and more exquisitely refined than either, yet bearing to both of them a stronger affinity, both in the roots of verbs and in the forms of grammar, than could possibly have been produced by accident; so strong indeed, that no philologer could examine them all three, without believing them to have sprung from some common source, which, perhaps, no longer exists: there is a similar reason, though not quite so forcible, for supposing that both the Gothick and the Celtick, though blended with a very different idiom, had the same origin with the Sanscrit.[24]

The view that Sanskrit was "more perfect than the Greek, more copious than the Latin, and more exquisitely refined" found a yearning audience in Germany. By the 1800s, Indo-European languages were treated as an ancient family—of structurally similar languages but also of the peoples who spoke them, in areas ranging from Northern India to the Iberian Peninsula. The poet and critic Friedrich von Schlegel, in "On the Language and Wisdom of the Indians," followed up on Jones, whom he praised for revealing "the till-now obscure history of the primitive world."[25] The barbarian invasions produced the model for how these languages had spread—that is, through invasions, except the Indo-European ones had happened a lot earlier. This, conversely, meant that ancient Germans, themselves speaking an Indo-European language, had been part of a primordial movement. Schlegel was interested in what he called the "noble original stock."[26] A small prehistoric people in India had "overflown." Out of necessity, it "naturally" became an invading people and spread beyond India. And it influenced more than language alone: religion and myth, even world history. "All the greatest empires and noblest nations sprang from one stock . . . The colonies planted by Greece and Rome are of small importance when compared with the ancient grandeur of these migrations!"[27] To deconstruct the relations between languages was to reconstruct the history of peoples. German had not sprouted alone but instead in a vast, vague ancient geography that reached from India all the way to the Eastern shore of the Rhine.

Schlegel was dreaming this ancient past at the same time as Europeans were talking about the barbarian invasions and Germans were wondering how to get rid of Napoleon's New Rome. Treatises on linguistics interwove Sanskrit with Tacitus and Julius Caesar's "ethnographies" of the ancient Germans.[28] Jakob Grimm (of Grimm Brothers' fame) urged the study of "Indian" grammar in his monumental *German Grammar* (1818). Sanskrit, "the dialect which history demonstrates to be the oldest and least tainted, must present the sovereign [*tiefste*, most profound] rules for the general exposition of the race."[29] "Race" here translates the German word *Stamm*, which might more exactly refer to "tribe" and, in the literal sense, "root." Were tribe, race, and root interchangeable? Other linguists, including Wilhelm von Humboldt and Franz Bopp, spoke of the "vitality" of languages starting from Sanskrit, which they considered closest to the *Ur*-language of that same Indo-European *Stamm*. Just as Latin had made possible the Romance languages despite becoming corrupted over time, so too had Indo-European formed new linguistic and spiritual developments.[30]

Indo-Europeanism as a system of ideas transposed the barbarian algorithm back in time, turning prehistoric population movements into barbarian invasions *avant la lettre*. It wove language into the tale of noble conquest. And German thinkers styled it not "Indo-European" but *Indogermanisch*, which really was a way of saying that the "original" ancient Germans had in fact existed since much earlier, that they had already conquered the world by spreading their primal language. "Indo-Germanic languages" and "Indo-Germanic peoples" implied that other peoples were but copies of these "ancient Germans." For example, the archaeologist Heinrich Schliemann, who discovered Mycenae and one layer of Troy, imagined his beloved Mycenaeans as proto-German Teutonic warriors. Others fixated on the "Dorian invasion" of pre-classical Greece. Indo-Europeans,

shaped out of the ancient Germans even as they preceded them, operated in deeper time and over a vast geography, where one could seek
the fount of language, the secret to power, and the shape of Europe.

So intense and nationalistically meaningful did this linkage
become that in 1857, in a new preface to his monumental *Comparative Grammar*, the leading comparative linguist Franz Bopp objected
to the use of "Indo-Germanic": "I do not see why one should take
the Germans as representatives for all the related peoples of our continent."[31] His colleagues and students bought the new edition and
promptly ignored him.

All through these different arguments, the same rhapsody predominated, a spirit of free, violent Germans who sacked Rome,
escaping the gloom and crushing norms of Latin classicism, helping

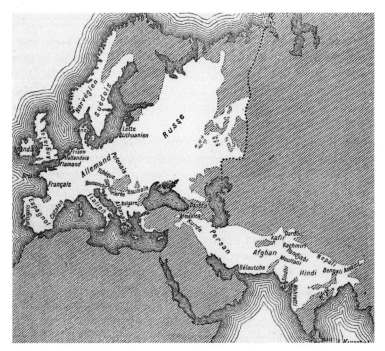

Figure 2.2. Map of the reach of Indo-European languages (and peoples), according to
the prehistorian Gabriel de Mortillet in his 1897 study of the "formation of the French
nation." Europe appears as a space of languages and peoples; areas of Finno-Ugric and
Turkic languages are shaded out, but Semitic languages are completely ignored.

Christian Europe to be born, and inaugurating history proper. By the time that the Germanic tribes had been folded into the Indo-Germanic-speaking tribes of the "noble original stock," this had become one of the most effective and long-lasting theories of the race's birth. "North of the Danube, East of the Rhine" came to signify a Eurasian fatherland for a race that was not bound by history. Where one was born in it equaled what blood one carried, and what kind of conquest one would achieve. The supposed barbarians had created the possibility of modernity itself and offered a model for the spread of language, for an antiquity beyond antiquity that could work as a Wagnerian myth for tomorrow.

Figure 3.0. Pierre Boitard, "Fossil Man" (engraving by Johann Konrad Susemihl) in *Le Magasin Universel* (1838).

Chapter 3

THE CREATURES
DEEP TIME INVENTED

S tates of nature, children, ancient Greeks, Germanic peoples—
prehistory was born in the shadow of these concepts as they
strained and failed to explain human origins. In the early 1800s,
a new science punctured old fantasies about peoples discovered, languages traced to ever-earlier times, natures idealized or terrifying.
Geology became the great science of the abyss of the past. And it
played midwife to the birth, out of this abyss, of a whole swarm of
ostensibly ancient creatures, changing fundamentally how modern
thinkers thought about the present and the past.

From the 1760s through the 1830s, the past distended dramatically. Notions of "deep time," as John McPhee much later and influ-
entially named it, depended on calculations of the earth's age. In the
mid-eighteenth century, *philosophes* knew very well that the earth
could by no means be as young as Sacred History had it, less than
6,000 years, and they also suspected that the history of humanity
did not extend nearly as far back as the history of the earth. Still,
explanations were lacking and tentative. In 1749 Benoît de Maillet
chanced a guess that the earth was several hundred thousand years
old.[1] The naturalist Comte de Buffon hesitated between an answer

of 75,000 and one of unclear hundreds of thousands.[2] As geologists puzzled over particular discoveries and rock beds, they read and tried to outshine each other. In the early 1800s, some of them began to tell a more consistent and riveting story. The earth's age, they claimed, was immense. English geologist Charles Lyell even described it as indefinite time.[3] Where humans fit in this story was unclear.

Lyell built on the work of the Scottish geologist James Hutton, who understood the earth as igneous and covered by geological strata that pressed against each other. Evidence came from sites of pressure between strata, which Hutton called "unconformities." Lyell interpreted these to mean that what to the untrained eye appeared like moments of radical change was simply the effect of slow, uniform processes. Hutton had declared the earth so old that it had "no vestige of a beginning, no prospect of an end."[4] Lyell repeated this: he pushed aside the arguments made by Georges Cuvier and others that the earth was subject to semi-periodic catastrophes like Biblical Floods. These ideas simply reflected their era of political "revolutions," Lyell argued, and they imposed on the earth the ups and downs of human history. With Hutton and Lyell, the earth now had dynamics of its own, and those dynamics were not random: they were immense and uniform.

But if the movements of the earth's crust were uniform, if its laws were eternal, then any calculation of how long certain changes had taken revealed that the earth must be shockingly antique. The numbers were absurdly big to fit into a story. As commentator after commentator has written ever since, it is frustrating enough to wrap one's head around a few hundred years.[5] Tens of thousands is an abstraction. Millions, Lyell wrote, were an "indefinite" history, worse than the purest abstraction, as Darwin later complained: such scales were "so great as to be utterly inappreciable by the human intellect."[6]

The invention of deep time forced into view entire genera and species that had perished, beings that no longer existed, an earth long bygone. Prehuman beings related somehow to the emergence of humanity—no longer culturally, but in their very shape. The "discovery" of the earth's past waved away much of what people had thought of the origins of the world, as if prior theories were pitiful, small-minded jokes. What point was there to speak of a "childhood of Man" in Mesopotamia or India when the timescale showed this infancy to be so recent?

Many concepts were remade by the invention of deep history, and three are of particular interest: dinosaurs, vestiges, and the humans (or ape-men) of the past.

"AMONG ALL THE WONDERFUL THINGS THAT TURNER DID IN HIS day," the art critic John Ruskin declared in 1860, "I think this nearly the most wonderful."[7] He meant the guardian dragon snaking and fire-breathing its way up the mountain in J. M. W. Turner's *The Goddess of Discord Choosing the Apple of Contention in the Garden of the Hesperides* (1806). Ruskin copied the detail (Figure 3.1), and in his draft the dragon stands dark against a white, fiery cloud. To understand the beast, Ruskin explored entire traditions. The Greeks, he began, had dictated the dragon's descent from Medusa and Typhon. Next, he noted to himself, there was Mammon in the Book of Job. But when Ruskin grasped for accuracy, he had to admit that neither of those dragons looked quite right. So he switched lens altogether, and turned to geology: "if I were merely to draw this dragon as white, instead of dark, and take his claws away, his body would become a representation of a great glacier, so nearly perfect, that I know no published engraving of a glacier breaking over a rocky brow so like the truth as this dragon's shoulders would be."[8] The poet Percy Shelley had already compared glaciers to serpents, so Ruskin one-upped him in drama, writing of the "strange unity of vertebrated action" in the Turner dragon, the infinite variations of its vertebrae, the glacial outline. Then he added a beautiful parenthesis: "(and this in the year 1806, when hardly a single fossil saurian skeleton existed within Turner's reach)."[9] Ruskin elsewhere described the dragon as "very nearly an exact counterpart of the model of the *Iguanodon*, now the guardian

Figure 3.1. John Ruskin, "Quivi Trovammo," horizontally inverted copy of J. M. W. Turner's dragon in *The Goddess of Discord Choosing the Apple of Contention in the Garden of the Hesperides* (1806). Ruskin greatly admired Turner's dragon, which he interpreted as a dinosaur.

of the Hesperian Gardens of the Crystal Palace, wings only excepted, which are [in the painting] almost accurately those of a pterodactyle."[10] Ruskin had long obsessed over geology, once complaining that it flattened his faith to gold-leaf. "If only the Geologists would leave me alone, I could do very well, but those dreadful Hammers! I hear the clink of them at the end of every cadence of the Bible verses."[11] In the dragon's strange shape an entire age announced itself. Turner had anticipated the dinosaur discoveries of the intervening years.

Dinosaurs populated a past otherwise uninhabited. Shaped on stone like trilobites, saved in the earth's strata, they told of a natural history before natural history. Hunters and geologists—most famously the paleontologist and fossil hunter Mary Anning—found ever bigger, stranger, more startling fossils. Plesiosaurs were named in 1821, "megalosaurs" in 1824. Gradually, dinosaurs took over from dragons.[12] Something of the Romantic beast was lost in the translation, but its grandeur was not. Words like "stratigraphy," with its staccato sound, may wield Ruskin's hammers, but geological research was a patient and generally quiet business. Dinosaurs were anything but quiet: they gave beautiful, terrifying life to stones, noisy history to bones. And with their fossils came butchery. The fossilized litter of dinosaurs was found to contain bones of other dinosaurs, and suddenly scientists fantasized about a carnivorous scene. The English geologist William Buckland wrote: "Coprolites form records of warfare, waged by successive generations of inhabitants of our planet on one another . . . the general law of Nature which bids all to eat and be eaten in their turn, is shown to have been co-extensive with animal existence upon our globe."[13] The first images of the age of lizards depicted a version of this carnage. Inspired by Buckland, Henry De la Beche's illustration *Duria Antiquior* (1830, Figure 3.2) depicted massive bulging eyes; teeth, so many teeth, always razor sharp; a neck-less, big-headed, long-mouthed ichthyosaur that crushes the long neck of a plesiosaur between its jaws; pterosaurs that face off in an aerial battle; fish that swallow other fish; ammonites and coprolites that litter the ocean; and a plesiosaur that gleefully raises its neck out of the water to snatch a pterosaur in flight.

Beche's watercolor is known primarily as a gift to Mary Anning—its sales were meant to celebrate and finance her fossil-hunting research. Yet it is so much more: a violent landscape of the dinosaur state of nature, a model for Darwin's struggle for survival. The battlefield is also replete with *tropical* flora. Beche would have encountered the palm trees in Jamaica, where, while still a child, he had inherited a plantation and 207 enslaved people. As an adult, he traveled to his plantation repeatedly and worried about the declining income he received from it. In an 1825 apologia, Beche pronounced that slavery was not in itself evil, except for particular practices like branding and whipping, because slaves benefited from work, Christianity, and moral improvement.[14] Just a year after the illustration, the rebellion known as the Baptist War would break out in Jamaica. In Beche's mind, ancient Dorset *was* colonial Jamaica, only with

Figure 3.2. Henri De la Beche's *Duria Antiquior* (1830) was intended as a gift to the fossil-hunter Mary Anning, whose research had a major impact on geology and the understanding of the Jurassic period. Yet his watercolor was shaped by British colonial anxieties and by a Malthusian understanding of resource competition. It became the prototype for the depictions of dinosaur life (see Figure 3.3).

warring dinos instead of slave rebellions. Civilization stood apart and stared in, exactly like the illustrator peering from London into the deepest past. The sea serpents that had once peppered maps with *"here be monsters"* now found a new form, the ichthyosaur/plesiosaur battle, and a new home, the scientific illustration.[15] The Jurassic era, only just dragged out of the earth, acquired an exotic, tropical, colonial feel. The Oldest world made sense of the violence of the New.

Within a couple of years, professors like Buckland were handing out copies of Beche's *Duria Antiquior* to students, and paleoartists around Europe were designing nearly identical versions of Beche's image. Its ecosystem of brutality became the prototype of dinosaur life (see Figure 3.3). In their books and their lectures, geology students encountered ichthyosaurs beheading their "dire enemies" the plesiosaurs, and plesiosaurs grabbing screeching pterodactyls out

Figure 3.3. "Lost animals" from Félix-Edouard Guérin-Méneville, *Dictionnaire pittoresque d'histoire naturelle et des phénomènes de la nature*, vol.1 (Paris: au Bureau de souscription, 1833–1834). This is one of several artworks from the 1830s and 1840s that clearly derive from Beche's watercolor (Figure 3.2).

of the air.[16] Seen from the fortress of civilization, the earth before humans became the dinosaur state of nature.

Was this depiction of nature as furious abundance an effort to cram every idea about violent dinosaurs into one drawing? Yes, and more than that also: it played out Thomas Malthus's principle of population, his famous idea from 1798 that populations expand in a geometric ratio that forces them to compete over food. Survival is constantly threatened, as Darwin later echoed: "More individuals are born than can possibly survive. A grain in the balance will determine which individual shall live and which shall die,—which variety or species shall increase in number, and which shall decrease, or finally become extinct."[17] Darwin was famously indifferent to dinosaurs, but he knew well Buckland's Malthusian quip about coprolites being records of nature's warfare. Could his "struggle for existence" have been better illustrated? We forget, Darwin claimed, that birds, "these songsters, or their eggs, or their nestlings, are destroyed by birds and beasts of prey."[18] "What a struggle between the several kinds of trees must here have gone on during long centuries," he went on, "what war between insect and insect—between insects, snails, and other animals with birds and beasts of prey—all striving to increase, and all feeding on each other or on the trees or their seeds and seedlings!"[19] Even if Darwin did not much care for dinosaurs, his readers knew to read dinosaurs into his work, and his work into dinosaurs.

DINOSAURS WERE NOT THE ONLY INVENTION OF DEEP GEOLOGICAL time. Ever since the middle of the eighteenth century, Europeans had been obsessed with ruins, and now they gradually became passionate about fossils and biological vestiges. For a whole century, architects and artists, art historians and philosophers and poets drew and wrote about cities and buildings and cultures lying in ruin. Collapsed temples and buildings—of which there were many—confronted them with worrying signs of what could befall the living.[20] Decay, they often explained, had been the signature of a civilization's old age. What if culture were to be destroyed again, like the ancients had been?

In an era fixated on barbarians destroying ancient Rome, this was no rhetorical question, no idle nightmare from which you just wake up. What was more, they noted, natural processes could be themselves catastrophic, perhaps even more so. Vesuvius erupted with some regularity throughout the eighteenth century, and all the while excavations were ongoing in Pompeii's ruins just beneath it. A massive earthquake leveled Lisbon in 1755, leading *philosophes*, most famously Voltaire, to mock trust in God's judgment and an a priori just world. Ruins were both fascinating and terrifying. Giovanni Battista Piranesi gave life in his etchings to the same Roman ruins that Edward Gibbon fetishized in *Decline and Fall of the Roman Empire*. "The earlier ages of mankind's childhood have passed," Herder wrote, "but plenty of remnants and monuments remain."[21] Constantin-François Chassebœuf, comte de Volney, writing about

Palmyra's ruins in 1791, saw in them the "genius" of liberty, the punishment of oppressors, the love of human beings. Entire cultures, he went on, and the labor of untold numbers of people, were crumbling as "the palaces of kings become the den of wild beasts . . . filthy reptiles inhabit the sanctuary of the gods. . . . Thus perish the works of men! thus do nations and empires vanish away!" Who could assure Volney that the same would not befall France's Revolution?[22] He meant the question to inspire the revolutionaries onward, but others were shocked by the idea that today's life was but tomorrow's debris. Who could forget Shelley's "Ozymandias" once they'd read it? In it, an ancient Egyptian sculpture, wrecked in the endless desert, threatened the reader with its inscription: "My name is Ozymandias, King of Kings! Look on my Works, ye Mighty, and despair! Nothing beside remains."

With the discovery of the earth's indefinite past, ruins became interesting to scientists—geologists and biologists especially. Were not fossils nature's ruins? Dinosaur bones were remains of a lost world. They too described bodies wrecked and left behind to rot.

By 1830, scientists were looking at fossils as signs of destruction caused by time's passage. Charles Lyell, the geologist, set as the frontispiece of his *Principles of Geology* (1830) a drawing of the Macellum of Pozzuoli in Naples (Figure 3.4).[23] Its columns were marked by holes left behind by mollusks. Lyell was eager to show that the mollusk marks were not signs of a massive catastrophe, like an explosion of nearby Vesuvius, but rather evidence of slow, steady tectonic movement. Over two thousand years, the building had slowly been submerged in the sea, where the mollusks did their work, only later to rise again.

Lyell turned the classic image of the destroyed temple on its head: destruction and ruins are merely a part of the earth's life. Like a romantic poet, he begged his reader to imagine a future where the Mediterranean's sea level would rise so that its "tidal current should encroach on the shores of Campania, as it now advances upon the

Figure 3.4. Mollusk-damaged
ruins of the Temple of Serapis
(today called the Macellum
of Pozzuoli). Frontispiece of
Charles Lyell's *Principles of
Geology* (1830).

eastern coast of England." Towns would be inhumed, skeletons
entombed in lava, streets would be built upon old streets.[24] Whole
worlds had already been made and broken thanks to the slow and
sure hum of the earth.

Over the 1840s and 1850s, it was common to look at nature and
see not a pristine landscape but ruination. Fossils were the "monuments and inscriptions constructed by nature."[25] For archaeologists,
geological deposits became "archives" or "ruins of the old world"
where one could search for flint tools and crania that "proved the
existence of Man as surely as the whole Louvre does."[26]

The same logic that sucked ruins into nature also imported them
into the body. The human body, and sometimes the "body of the
race," became a visceral, lived, biological carrier of past change and
destruction. The body too became full of "vestiges"—creatures cre-

ated by deep time, devices or monsters that lived "within" oneself. If ruins were visible in a landscape, vestiges would be the debris carried in the body. No longer would ruins be about buildings, exactly, but about remnants of a past that survived in its descendants.

Robert Chambers's influential and controversial *Vestiges of the Natural History of Creation* proposed that non-white peoples set the stage for true civilization; some occasional elements of their life and culture, he declared, survived and became ordered in European society. Europe had absorbed earlier cultures, and what had once been their innovations were now mere worthy leftovers.[27] In *On the Origin of Species*, Darwin, who was inspired by Chambers, spoke often of the "monuments" of the "lapse of time," and discussed "rudimentary" organs.[28] These were incomplete organs to be found in the body in different stages of development or decline. They testified to natural selection: sometimes they showed a species in transformation; sometimes they were evidence of an organ abandoned, like the human appendix and the tailbone. In his *The Descent of Man* (1871), Darwin abandoned the idea that rudiments anticipated a more fully fledged organ, an idea he had been fond of. Human rudimentary organs were now ruins—they "are useless or nearly useless, and consequently are no longer subjected to natural selection. They often become wholly suppressed."[29] Disuse had caused them to become collateral. Rudiments carried a past within, but except for rare occasions of "reversion," they offered no advance.

Once ruins had entered the body as vestiges or rudiments, Darwin and his contemporaries found them everywhere. In language, for example: Darwin and Lyell compared rudimentary organs to redundant letters (like the "silent h") that had once been useful but now were without function.[30] In emotions: Darwin was interested in the survival of powerful emotional expressions and religious symbols in modernity.[31] In ornamentation: anthropologist Edward Burnett (E. B.) Tylor wondered aloud why women "mutilate" their ears by piercing them and he declared earrings to be "relics" of lower civilizations.[32] He introduced the term *survivals*, defined as "processes, customs, opinions" that, despite great social evolution, "remain as proofs and examples of an older condition of culture."[33] John Lubbock, the Englishman most associated at the time with the term "prehistory" for his authorship of *Prehistoric Times* (1865), actually gave the book the subtitle *As Illustrated by Ancient Remains and the Manners and Customs of Modern Savages.* Prehistory illustrates such remains and survivals, he claimed: "customs which have evidently no relation to present circumstances, and even some ideas which are rooted in our minds, as fossils are imbedded in the soil."[34] Flints, which became accepted as objects dating to the Paleolithic in the early 1860s, were now ruins of prehistoric culture. Ernst Haeckel, Darwin's grand proponent in Germany, understood "rudiments" as *remnants* and pronounced them the most significant demonstration of human descent from apes.[35] In animal embryos, he went on, rudiments from earlier species would show the potential of development, only to be stunted and left aside. More obsessively than anyone else, Haeckel argued that humans carried ruins within, cultures carried them as well, entire countries lived with them. Contemporary geographers wrote of *Ruinenländer*, ruined countries, countries of ruin.[36]

It's not simply that people talked about ruins. Rather, what had even recently been a cultural anxiety ("Look on my Works, ye Mighty, and despair!") became, thanks to geology and its deep time, the basis for experiencing national and human prehistory as a past reality.

No longer was the past simply to be seen (or not seen) in destroyed buildings or imagined connections with ancient barbarians. Think about that nineteenth-century moment like this: there are ruins all around you, and inside you, and in your language, and the past is here but somehow degraded and your life is itself this strange assemblage of things living and past that are dead but are nevertheless everywhere. The ravaged past was physically *inside*, as easy to touch as a tailbone. Its original, unbroken forms needed to be studied, discovered, unearthed.

What was more, Europeans decided that the ruins of ancient European cultures mirrored the contemporary reality of Indigenous peoples elsewhere. Megalithic menhirs and cromlechs made by "races of the mysterious past" in England and France "have been kept up as matters of modern construction and recognized purpose among the ruder indigenous tribes of India." Ancient Swiss settlements "have their surviving representatives among the rude tribes of the East Indies, Africa, and South America."[37] Living Indigenous peoples were vestiges of ancients, and their cultures degenerated just as ancient Western ones had done.

BY THE 1850S, A NEW VARIETY OF RELICS HAD COME TO STAND IN for this past: human bones. Literate Europeans in the nineteenth century "knew" themselves to be historical beings, descendants, bearers of cultures past. As a political feeling, this proved powerfully effective, as with the nationalist celebration of the ancient Germans. But as the growing craze for archaeology and ruins created a market for antiquities, flints, and bones, it was not merely a political matter. Consider for a second what it would mean for those living in the later nineteenth century to take in the lessons of geology and biology, of vestiges and indefinite time—to recognize as never before that each human being carries untold numbers of generations, stockrooms of biological ruins past, in body and culture and spirit. To look down at their body only to realize that they were machines of history, products of deep time that carried too many vestiges to count. It must have been a heart-stopping realization, impossible to pass on to someone else until the *eureka* moment hit them too. We know well how threatening the linking of Man to Ape was (and Darwin was hardly the only author to propose it). The idea that across evolution, some intermediary figures had prevailed, for a while, just as we prevail, for a while, was not an idea that offered comfort.

Naturalists had long reported on the discoveries of extinct mammal remains in Europe and, well before *On the Origin of Species*, on the discovery of human-like bones. Mammoths and hyenas were interesting, but hominids were disturbing. Were these the remnants of some precedent, and, if so, whose? Fossil skulls did for ideas about humanity what dinosaur bones had done for ideas of nature. They pushed time back, but confusingly. As Alfred Russel Wallace put it: "you must go to an enormous distance of time to bridge over the difference between the crania of the lower animals and of man. I said, perhaps a million, or even ten millions, of years."[38] Such was the new antiquity. But could geologists really be right? Could limestone and other deposits be trusted to protect skulls accurately and hence reveal deep time? Some skulls had a flat back, others a dramatic eyebrow ridge. Did they look different because of some pathology—or was each variation representative of an extinct "race" or "species"?

Already in the early 1820s, Buckland had found bones in Wales that he interpreted as belonging to a Roman-era female ("the Red Lady of Paviland"). Wrong on the gender, and still unable at the time to push the date beyond the historical record, he ended up misdating them by some 30,000 years. In 1828, Paul Tournal found hominid bones in a cave in Bize-Minervois in France, alongside animal bones. A year later, further discoveries were made in France (this time near Montpellier) while in Belgium the Dutch naturalist Philippe-Charles Schmerling discovered two skulls alongside hyena and rhinoceros bones in the Engis (now Schmerling) Caves. The smaller of the two skulls was a morphological mystery and became known as the Engis skull. In 1837, Edouard Lartet described the jawbone of an ancient gibbon discovered in France and later classified as *Pliopithecus antiquus*, a Miocene species. In 1848, a large skull fragment with much of the crown missing but with a protruding eyebrow ledge and other differences from "modern" humans was lifted out of a quarry in

Gibraltar. Four years later, railroad tunneling in Gloucestershire in England led to the discovery of another skull, whose antiquity was immediately debated. The first skull described as a Neanderthal was famously discovered in 1856 in the Neander Valley, though in the ensuing debate, other skulls were recategorized as Neanderthal, including the Gibraltar skull. Whether Engis was a Neanderthal, another species, or an anatomically modern human was unclear. In 1858, human and extinct-hyena bones were found in Brixham cave (now Windmill Hill Cavern) in South-West England, scattered alongside stone tools. In 1868, French geologist and paleontologist Louis Lartet (son of Edouard) found four skulls in Les Eyzies, in France, and classified them as "Cro-Magnon Man." Other skulls were also confusing matters: in 1891 a skull in Brunn (Brno) seemed atypical, but scholars noticed that it resembled another one that had been found twenty years earlier in Brüx (also in the Habsburg Empire) and that also seemed odd. Whether these were Neanderthal, Cro-Magnon, or something else would be debated for decades (one factor was how their original Bohemian location was understood). Heated disputes followed findings everywhere from Ohio and Mississippi to Sweden, from Wales to the Nile. Most of them would prove to be "anatomically modern," as we now (strangely) call Homo sapiens, but what they meant was rarely self-evident.

These (and many other) fossils were named after the spots where they were found. In English, the location of discovery was attached to "Man"—Cro-Magnon Man, Neanderthal Man, Galley Hill Man, later Heidelberg Man, and so on. In French, the term race was liberally applied to past peoples and identified with particular locales—la race de Canstadt, la race de Neanderthal—partly because the term was even more elastic than in English.[39] Not only geographical sites but ancestors and eras, races, and even species could be personified in skulls.

The debates became all the more pressing after 1859, when flint implements were accepted as evidence of prehistoric tool use, and hence of early human existence, and when Darwin's *Origin of Species* was published. The filiation of species had been clear well before Darwin, and other naturalists like Thomas Henry Huxley had started insisting on gibbons as the ancestors of humans.

The skulls also became scientific relics for the modern age—means of pushing humanity way back in time. They were not perfect evidence for the passage from ape to human because they were rare, confusing, hard to date. Darwin complained in *Origin* of the "poorness of our paleontological record."[40] Casts were made to be reexamined in London, Paris, and Berlin, lifted up like Yorick's skull by those Hamlets curious about human space-time and the workings of deep time. Even when they were contested—which they all were—they provided intensely physical and visual connections to worlds bygone.

Suddenly, new books by geologists and prehistorians took on a strange structure: they all discussed first the neolithic cultures of Europe, then they moved back to the earliest times and looked at extinct megafauna (mammoths), and then forward again, till they finally reached "fossil men" in Europe and in non-European Indigenous cultures. Charles Lyell pioneered the form in his *Geological Evidences for the Antiquity of Man* (1863); John Lubbock perfected it in his *Prehistoric Times* (1865). Prehistory meant you began with ancient cultures, then plunged back in time all the way to animality, and then pivoted to the present day, the skulls, and the living "primitives." This even became a literary form. The French botanist Pierre Boitard, in the rather odd novel *Paris Before Man* (*Paris avant les hommes*, 1861), had his protagonist travel with a little devil back in time, only to find, to his immense surprise, that the ultimate antiquity was to be found in the Paris area. Europeans were proud of their digging and discoveries. There—right at the center of Europe—the primordial origin was breaking free.

Still, colonial analogies prevailed. The skulls were constantly compared to those of living Indigenous peoples; by extension, living peoples could also be included among "fossil men."[41] The French pre-historian and racial theorist Jean Louis Armand de Quatrefages de Bréau did this in the bluntest of ways: in his *Crania Ethnica* (1882), he identified skulls discovered in Europe and around the world with specific "races." In his *Fossil Men and Savage Men* (1884), nine of his eleven studies came from Indigenous peoples, including the recently annihilated Tasmanians.[42] The Englishman William King, who first described the Neanderthal skull in English and named it *Homo neanderthalensis*, immediately compared it to the people of the Andaman Islands. The Andamanese were British colonial sub-jects at the time—and they, too, were being wiped out.[43]

How the deep past was connected to Native peoples in the pres-ent was a question that now seemed easy to resolve.

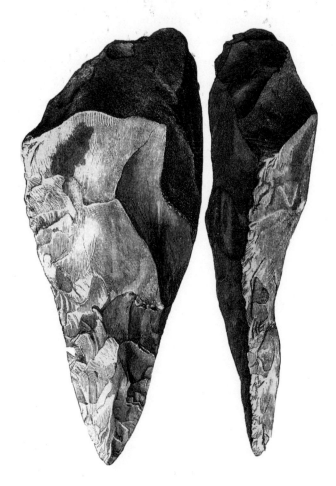

Figure 4.0. Flints from Abbeville (where Jacques Boucher de Perthes and his workers were digging for "antediluvian" implements).

Chapter 4

HUMANITY, DIVIDED BY THREE

After his defeat at Waterloo in 1815, Napoleon was exiled to St. Helena. For more than two decades, France had promised that the future was at hand, that the Revolution was the vanguard of humanity, that it embodied progress, equality, social renewal. Now, the country was humbled and exhausted. As in the rest of Europe, the pressure for order was intense. No political force could be neatly identified with progress anymore. Europeans faced a strange new era: not the forward march, but the aftermath of total war. The empty rituals of the old regime were resurrected as the guarantee of a stable future.

Several ideas about history also faded. The concept of a state of nature seemed very much passé. No less outdated was a division of history into four stages, which had been common in the eighteenth century. The Scottish economist Adam Smith had logged "1st, the Age of Hunters; 2dly, the Age of Shepherds; 3dly, the Age of Agriculture; and 4thly, the Age of Commerce."[1] Edmund Burke, the English political thinker, had also used four. The past could be seen all around us, he proposed: "the Great Map of Mankind is unrolld at once. . . . the very different Civility of Europe and of

China; The barbarism of Persia and Abyssinia. The erratick manners of Tartary, and of arabia. The Savage State of North America, and of New Zealand."[2]

Around the outbreak of the Revolution in 1789, naturalists and philosophers generally abandoned the four-stage schema for much more complex ones. During the worst period, the Terror of 1793–94, the mathematician Nicolas de Condorcet outlined a historical schema with no fewer than ten stages. Condorcet was hiding from the Jacobins; he would soon be caught and executed. But he believed that humanity was close to the mountaintop of history, having walked the long path up the slopes of time. If one looked back to the very beginning, he concluded, the earliest sight would have been of families uniting into tribes, giving birth to society. From there on, each stage improved on the last, adding science and social progress. The last stage impended: it would flower with the true conclusion of the revolution.

The idea of a long series of stages captured the urgency of change. In the *Mémoire sur la science de l'homme* (1813), Henri de Saint-Simon thought Condorcet's decimal schema wasn't enough, so he added two more stages.[3] But the theories were unwieldy. In a list of ten or twelve stages, where exactly did one stand in the movement of history? Who could remember whether some people or other should be placed in the sixth or the eighth stage, or the fifth? What is more, none of these theories could account for Waterloo, nor even for Napoleon's ascension to Emperor after the Revolution. Was history really still advancing? After 1815, the idea of progress came to be more associated with the sciences, industrialization, and technology than with politics. A new schema for progress emerged, and it used a more convenient number of stages: three. Every major nineteenth-century stadial theory used the three stages as a way of dividing human time. To use the three stages became a way to talk about progress and technology and to divide the past, but also to create a hierarchy of peoples—from Indigenous peoples who could be lodged in deep time, to Europe's industrial superpowers which, commentators were sure, represented the future.

One triad was already in wide circulation when Napoleon was exiled: *savage, barbarian,* and *civilized.*[4] It was less a three-stage theory than a group of two duets. The first duet contrasted *the civilized* to *the rest. Civilized* Europeans had stepped up from the rest of the world, and so perhaps had some Asian societies. The second duet, *savage* versus *barbarian,* allowed interpreters to compare the rest. Those cultures deemed more advanced than abject savages would be categorized as barbaric. This approach seethed with racism: the French biologist Georges Cuvier insisted, all the way to his death in 1832, on identifying *savage* with "Negro or Ethiopian," *barbarian* ("stationary") with "yellow or Mongolian," and *civilized* ("progressive") with "white or Caucasian."[5]

A different triad emerged in the 1830s in Nordic prehistoric archaeology. Its inventor, the Danish archaeologist Christian Jürgensen Thomsen, worked as a curator at the Copenhagen Museum. In the 1810s, he began to organize ancient tools, pottery, and weapons by their basic material: stone, bronze, and iron. Once he published his approach, it became common to divide early human history between the *Stone Age,* the *Bronze Age,* and the *Iron Age.*

At a time when colonial powers battled enemies with weaker technologies, Thomsen's approach helped travelers and armies to compare the material cultures of entire economies and peoples. The Stone Age told of worlds prior to written history, and it seemed to apply also to some peoples living in the modern era. Once archaeologists such as Jacques Boucher de Crèvecœur de Perthes began scavenging flints from the ground, the Stone Age was extended indefinitely backward, and it was divided into three parts of its own, Paleolithic, Mesolithic, Neolithic. The Paleolithic was eventually broken into three more: Lower (that is, earlier), Middle, and Upper Paleolithic. Paradoxically, the finer subdivisions only strengthened the comparison to Indigenous peoples.

A THIRD NOTABLE TRIAD EMERGED IN THE 1820S OUT OF HENRI DE Saint-Simon's ideology of industrialization. The idea was that civilization moved from a *theological* stage through a *metaphysical* to finally a *positive* (meaning scientific) stage. The old worldview had been mere conjecture. Astronomy, then physiology, then even psychology had slowly grounded knowledge in observation, and the theology of the past had collapsed. This triad worked well with the other two: the three schemes of three told the same story of a progress from a *theological, savage* antiquity of *stones* through *metaphysical, bronze,* or *barbarian* times and toward a *civilized, positive* age of *iron*.

Saint-Simon died in 1825, leaving behind a motley crew of acolytes who hailed mostly from the wealthy, growing industrial class, and who studied in Paris's new engineering polytechnic. The Saint-Simonians followed their founder's worship of industrialization and technoscientific governance.[6] They are treated, mostly, as an embarrassment in the history of socialism: in 1828 they founded a commune and lived rowdy until 1831, when several were imprisoned for their sexual politics, then a rump group followed Prosper Enfantin, the "Supreme Father" of the "New Christianity," on a quest to Egypt in search of a female Messiah whom Enfantin would marry.

Still, Saint-Simonians demonstrate how different ideas about space, time, and stages were fitted together. In the 1820s, they jettisoned Saint-Simon's twelve-step theory and pushed instead for his *theological/metaphysical/positive* triad as the basis of history. The positive stage had not bloomed yet, not fully.[7] When it did, it would enable an "administration of things" rather than a "government of men"; everything would be based on true knowledge.

One outsider to the movement was even more important than Saint-Simon himself. This was Auguste Comte, who had been Saint-Simon's secretary but fled and quickly became the Saint-Simonians' favorite enemy. As though to outwit his former comrades, Comte called his philosophy "positivism." In lectures starting in 1824, he hardened his argument into a tripartite division of history and knowledge. Knowledge, like history, had *theological, metaphysical*, and *positive* stages. Each new science was built on the back of more precise sciences, becoming ever more positive over time. And each new stage of history established superior morals and politics. Comte also divided each part of the triad into (surprise!) new triads. The theological stage, for example, had fetishistic, polytheistic, and monotheistic substages. "Fetishism," a term used at the time to describe religion in Africa, supposedly described the earliest religion and the earliest life of humanity.[8]

Just after Saint-Simon's death, and just as the official Saint-Simonians were adopting the terms "theological," "metaphysical," and "positive," Comte published his theories in their journal *Le producteur*.[9] On the basics, he agreed with them: you could only speak of humanism, industrialization, and technoscience if you believed that the world was moving from fetishism, which had once been universal, toward a European "Religion of Science."[10] But by 1828, the "official" Saint-Simonians, particularly their leaders Enfantin and Saint-Amand Bazard, were huffing that Comte was a usurper and a plagiarist. He couldn't be allowed to treat the three-stage theory as his signature innovation, and they tried to it claw back.[11]

As the movement fractured, its tenets gained in influence. By the late 1830s just about all the Saint-Simonians were leading perfectly boring bourgeois lives, often in service of the French state and especially its empire. Their ideas were no longer mere theories: the three stages of history began creeping up in policy proposals. Saint-Simonians were fanatics of industrialization; supported the rapid expansion of French railways; and believed in colonialism and France's "civilizing mission" in Algeria. Enfantin urged the French government to construct a Suez Canal decades before it was actually built, claiming it would enshrine Man's domination over the same nature that, back in "fetishistic" times, had dominated Man.

IN 1836, THE FRENCH GOVERNMENT SENT THE LEADING SAINT-
Simonian economist Michel Chevalier to visit the United States and
Mexico. Upon his return, Chevalier promptly published on the value
of materials, minerals, and mining for industry, railways, and canals.
The book helped him get elected to one of France's most hallowed
academic institutions, the Collège de France, where he lectured on
industrial politics and political economy. He invoked the triads regularly, and now something interesting happened in his language. In
a famous 1842 course, Chevalier outlined a movement from stone
to copper to iron. Native Americans were not aware of the met-
als, he claimed, until they came to know European iron. Chevalier
then turned to Homer and Virgil to compare these "primitives" with
Europe's deep past.[12] The "progress of civilization follows technical
progress," which meant that the history of the West was, in reality, a
history of iron.[13] Iron, "in universal use today," was the foundation
of freedom, and the measure of progress.[14] So important was iron
that Chevalier even asked his audience to imagine a new barbarian invasion, headed by a Genghis Khan, whose ultimate savagery
would not be mass murder, but a simple ban on iron. Civilization
would collapse: steam engines kaput, sciences trampled, transportation halted, printing stopped, liberté shredded. In just two pages, not
even five minutes of lecturing, Chevalier had spooked himself enough
to move from building modernity out of iron to prophesying that his
age would collapse into cannibalism.

This was because to Chevalier, as to many of his contemporaries, "Iron Age" meant two things at once. It meant Thomsen's
Iron Age, which he discussed with reference to Native Americans
who ostensibly hadn't reached it, and it also meant the modern Industrial Age. So Chevalier folded the three triads on one another: the old
savage/barbarian/civilized distinction was for him the same as the
Saint-Simonians' theological/metaphysical/positive distinction and

Thomsen's *Stone/Bronze/Iron Age* distinction. The triads were three versions of the same movement of history.

After Louis-Napoléon Bonaparte, the nephew of Napoleon, staged his 1851 coup, founded the Second Empire, and declared himself Napoléon III, Chevalier rose through the ranks of government to a position of great influence as an economist. He negotiated a key 1860 free trade agreement with Britain and ended his career a Senator. Enfantin, the hippie guru of earlier years, became involved in business and railway construction. He dedicated a book to Napoleon III to "reintroduce" him to Saint-Simon, and he even proposed that in Napoleon III's positive age, "railways, steamboats, electric communications constitute . . . a *new nervous system*" for the state.[15] Napoleon III did not need the reminder: if anything, he was as much a Saint-Simonian as an Emperor.

Comte, for his part, was a marginal figure in France, but he broke into the English intellectual scene.[16] Harriet Martineau translated his massive book of lectures into English. John Stuart Mill corresponded with him and published a book on him. Herbert Spencer blended evolutionary thought with Comte's progress theory.[17] In a sense, Comte won the marketing contest: lawyers, anthropologists, even geologists identified him, not the Saint-Simonians, with the *theological/metaphysical/positive* triad. And they used it constantly: even when he was criticized, it was on the terms he himself had set. The anthropologist E. B. Tylor, in his foundational book *Primitive Culture* (1871), chastised Comte for not getting the stages exactly right. He replaced Comte's "fetishism" with "animism" and built a three-stages argument of his own: *animism/religion/science*.[18]

Like Chevalier had done, English writers equated the triads and "translated" the terms of each into those of others. In the beginning was the Stone Age, a "savage" era ridden with vague fetishistic religious beliefs. Charles Lyell, for example, blended the Stone with what he called the Savage Age. In 1863, technology seemed to him an excellent lens for comparing ancient "Acheulian culture" with contemporary Australian aborigines. "To a European who looks down from a great eminence on the products of the humble arts of the aborigines of all times and countries," he wrote, "the knives and arrows of the Red Indian of North America, the hatchets of the native Australian, the tools found in the ancient Swiss lake-dwellings, or those of the Danish kitchen-middens and of St. Acheul, seem nearly all alike in rudeness, and very uniform." Savages continued to exist despite the "mighty empires that . . . flourished for three thousand years in their neighbourhood."[19] Savage life showed such slow progress that bronze tools simply took the shape of earlier stone ones. Let's spell out their concept: barbarians and savages populate the world. Savages refuse to give up their stone tools. Mighty barbarian empires have better tools, but nothing like Europeans. Industrialized Europeans have science and modern iron. They are dynamic and in control of time, not captives to it.

The three-stage theory became standard in anthropology. James Frazer transformed it into *magic/religion/science* in his bestseller *The Golden Bough*. Across the Atlantic, Lewis Henry Morgan included it in *Ancient Society* to distinguish among different Native American nations. In German-speaking lands, influential theorists—psychologists, anthropologists, art historians even—modified the contents and kept the structure going.[20] Even Sigmund Freud, in *Totem and Taboo* (1913), gave over a chapter to animism and magic, linking them to religion, and contrasting all three with science.

In England as in France, the theory had real policy consequences. James Lorimer, the Regius Professor of Public Law at Edinburgh, had a penchant for editing his book *The Institutes of Law*. In 1872 he contrasted "the civilized and savage races of mankind."[21] In 1880, he edited the book to expound on his opinions on the racial superiority of Aryans. In 1883 he switched out the civilized/savage binary for an explicit three-stage history. It made his racism respectable: humans, he declared, could be divided into three spheres: civilized, barbarous, and savage. In international law, each type deserved a different kind of recognition: "plenary political recognition, partial political recognition, and natural or mere human recognition."[22] Aryans deserved the first—naturally.

French prehistorians applied the triad to prehistory. Édouard Lartet had identified a key correlation between the principal animal found at a certain depth and early humans' use of tools. He thus bequeathed to us subdivisions of the Stone Age that still survive: the Cave Bear Age, the Mammoth Age, the Reindeer Age. His student Gabriel de Mortillet added his own way for studying the Stone Age. He proceeded by linking tools, and the beds in which they were found, with particular local cultures. He, like many contemporaries, called these toolmaking cultures "industries" (see Figure 4.1). In turn, these cultures or industries became periods of the Stone Age: Acheulean, Mousterian, Solutrean, Magdalenian. The Solutrean period was named after the rock formation in Le Solutré in Eastern France, the Magdalenian after the rock shelter formation La Madeleine in Southwest France. Each site was a place where

Figure 4.1. Emile Bayard, "The First Atelier of Human Industry," in Louis Figuier, *L'homme primitif* (1870).

a culture's characteristics were clearest or most explicit. The exact timing in thousands of years was less significant than the types of stone tools that defined the culture. All of human time and culture, in other words, could be divided on the basis of a Eurocentric account of the Stone Age.[23]

BETWEEN AROUND 1750 AND AROUND 1850, INDIGENOUS PEOPLES were primitivized and turned into living representatives of the first humans. "Primitiveness" was their identity. Before triads, scholars had not quite worked this out. Yes, Rousseau had compared "the Carib" with the first humans, and Smith and Burke had spoken of a savage state, but they did not talk of a single, clearly demarcated scheme that "explained" the cultures of colonized peoples, the structure of progress, the emergence of humanity out of prehistory, and the future transformation of the world. Now, as a result of the elastic triads of stadial history, Indigenous peoples could be systematically identified with the first humans. Their technology was presented as though it had not really moved past the Stone Age, their religions were treated as animistic, their customs as savage. They fit as "early" in all the triadic orders, just like the original humans did: thus the "animistic savages of the Stone Age" became literal primitives who could not escape the past. Civilization was moving forward, modernity was geographically bounded, the deep past could be seen in the technologically backward, religiously inchoate people outside Europe.

The triads might all seem faintly ridiculous (theories proposing ten or twelve stages all the more so). But let us not forget that until very recently, many in the West still spoke of the First, Second, and Third World. Or that after World War II, grand theorists of development like W. W. Rostow and innumerable ministers, technocrats, journalists, and readers used these same terms, these same triads, to locate in time these supposed primitives (even when they did not use the word) and to propose gilded paths to modernity.

Figure 5.0. Jacques Boucher de Perthes, photograph to document the location of a Paleolithic tool for the English geologist Joseph Prestwich. "This is the first axe that was authenticated . . . by Mr. Prestwiche [sic] and Pinsard in 1855 . . . The worker points with his finger to the axe engaged in the mass of pebbles. Amiens, St. Acheul."

Chapter 5

THE CONFLICT OF THE SCIENCES

Few books are as frequently identified with their year of publication as Charles Darwin's *On the Origin of Species* is with 1859. Long anticipated among his colleagues and friends, explosive in argument, the book was immediately celebrated and despised. It remains so to this day. Its model for evolution famously but implicitly proposed that humans emerged out of apes. The idea was not new, and some scientists had suspected monsters and other beings (real or imagined) might be either offshoots or intermediary species (see Figure 5.1). Novelists too: in one of his earliest writings, "Quidquid Volueris" (1837), the French writer Gustave Flaubert featured a scientist who coerces an enslaved woman to mate with an orangutang, their offspring being a mute, confused, brutish ape-man. Flaubert sympathized with this figure, treating him as far less savage than his colonial Frankenstein-style creator.[1] Emmanuel Frémiet exhibited the first of his several sculptures on the theme of a gorilla kidnapping a woman in 1859. He would continue to make them through the first decade of the twentieth century. Still, Darwin's masterwork claimed to dispassionately overcome such dra-

Figure 5.1. "Anthropomorphs" by C. E. Hoppius, in Carl von Linné et al., *Amoenitates academicae* (1756–90).

matic speculation, and to demonstrate evolution once and for all; its consequences were, as is well known, profound.

At first, Darwin dodged the question of specifically human origins. He only promised that through his work "light will be thrown on the origin of man and his history."[2] To Charles Lyell he sent a letter waving away polygenist theories of human nature, which proposed separate creations for each race. Nonetheless, "The Races of Man offer great difficulty." While he had "some rays of light on the subject," he claimed that the Indian Mutiny of 1857 had held up his research.[3] One wonders what that meant. At any rate, humans in *Origin of Species* played the role of observers of nature and enforcers of artificial selection on domesticated animals.

Still, Darwin allowed the reader to draw analogies from animals to humans. This oft-quoted passage, for example, was meant to be about animals: "In the preservation of favoured individuals and races, during the constantly-recurrent Struggle for Existence, we see the most powerful and ever-acting means of selection. The struggle for existence inevitably follows from the high geometrical ratio of increase which is common to all organic beings."[4] The controversies that followed only grew more intense once Darwin elaborated on human origins in 1871 in his *The Descent of Man and Selection in Relation to Sex.*

In the twelve years between Darwin's two books, "prehistory" became a systematic field. The debate on human origins did not quite hinge on Darwin. So much else was going on—in British, European, and global politics but also in competing sciences. In fact, every scientist who could make use of the deep past did so, claiming to show how it confirmed his theory about the place and meaning of humanity. New institutions and societies for the "sciences of man" created new spaces for discussion, new rules for research, new fields of inquiry. What was the human—what is it *now*? That was the central question as prehistory became the leading laboratory for conceptions of humanity.

WHEN DARWIN PUBLISHED *ORIGIN OF SPECIES*, A RETHINKING OF human origins was well underway. Thomas Henry Huxley had recently demonstrated structural homologies between human and ape skeletons: this could be no coincidence. If Huxley had started it all, at least in public, his argument only clicked once Darwin had made his own move by tying "variation" in and between species to "natural selection." Darwin was much more prudent an intellectual, using masses of tedious evidence to establish a position others would find hard to assail. Huxley, for his part, tried a much more confrontational approach. The 1860 Oxford debate with bishop Samuel Wilberforce is remembered for Huxley's wit. Wilberforce is reported to have asked whether Huxley preferred to have his grandfather or grandmother for an ape. Asserting the independence of science from religion, Huxley is supposed to have retorted that he would be ashamed instead to be associated with someone who used his talents to distort scientific truth. His attitude expressed the naturalists' claim that true science lay with them. And as the evolutionists pushed competitors out of the way, Huxley's anatomy mattered even more than Darwin's theoretical structure. In 1863, Huxley published an image that staged four ape skeletons side-by-side with a single human one, and told a story. It became famous as it begged readers to look from left to right (see Figure 5.2). Compare the heads, it said, the chests, the forearms, the hands, and you see evolution in action. Human beings derived their nature from their biology. They were a single species, however stretched out they were by a plethora of different natural "characters." They drew on their ape history. Natural selection was a law, and biological evolution downplayed human and social history in favor of species history.

Darwinians were hardly the only ones to claim humanity's great story for their camp. Also in 1859, English geologists verified that tools made of flint and found in geological strata were as old as those strata. Stones were more durable than bones, and easier to recover and study. The French antiquarian Jacques Boucher de Perthes had collected such tools since the 1810s and more regularly from the later 1830s on (see Figure 5.3). By the 1840s he was distinguishing

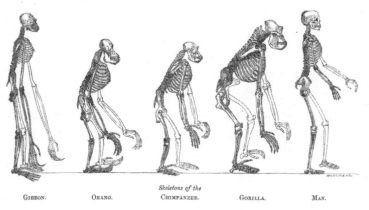

Skeletons of the
GIBBON.   ORANG.   CHIMPANZEE.   GORILLA.   MAN.

Figure 5.2. A drawing aimed at showing structural analogies between ape and human skeletons. Frontispiece of Thomas Henry Huxley's *Evidence as to Man's Place in Nature* (1863).

between the Celtic and the "antediluvian" (that is, prior to the Biblical Flood) on the basis of their location in gravel pits. For Boucher
de Perthes as for his supporters, stone tools showed humans living in
France alongside large extinct mammals already in the Pleistocene
era (more than 12,000 years ago). French scholars found Boucher de
Perthes uncouth and nonsensical, not least because he paid bounties
to local miners for any stones that they brought him. (The bounties
also elicited forgeries.) But the systematic quality of his work was
undeniable: Boucher de Perthes catalogued and presented alongside
one another stone tools of often considerable difference, arguing for
their potential date and value.[5]

   This was a very different game than Darwin's or Huxley's.
Boucher de Perthes did his best to show himself to be a good school-
boy: he stuck with old terminology like "antediluvian" and he followed the methods of geologists with diligence. In his lectures he
avoided the controversy that evolutionists basked in—he was sim-

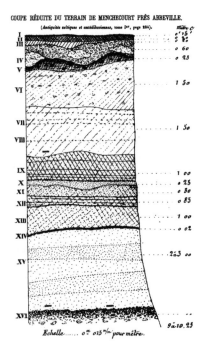

Figure 5.3. Vertical cross-section of
the ground at the Abbeville quarry
(1860). Jacques Boucher de Perthes
was anxious to show the great depth
at which stone implements were
discovered. Tools are marked with
black horizontal dashes, toward
the bottom and in the middle of
the drawing.

ply offering geological proof. Ancient humans had flourished in pre-
cataclysmic times, breaking rocks to make tools, and using them
to feast on large herbivores. Sure, they had been biological beings,
but above all, Boucher de Perthes's ancients had been dexterous and
industrial ones.

What Boucher de Perthes lacked all his life was recognition. So
once Lyell's students turned him into a success story in 1859, he
jumped at the chance to take credit. He alone had discovered "primitive industry." He alone had proven the "great antiquity of Man,"
the existence of humans on "a virgin earth, stranger to civilization."[6]

The English geologists who authenticated Boucher de Perthes's
discoveries were not very interested in him: they were using him to
angle on behalf of their science. Lyell, in his *Geological Evidences
for the Antiquity of Man* (1863), presented Boucher de Perthes as
helping the dashing young English geologists destroy the formerly
dominant but outmoded French establishment. Lyell's title says it
all: the real evidence came from geology. Huxley's skeletons, Darwin's natural selection, and Boucher de Perthes's implements were
but cherries on his own cake, or small domains of the empire he had
established since the 1820s. They offered good grounds for thinking
about human antiquity but even so, the real forces could be found
in the study of the earth. (In a sneaky move, Lyell pushed natural
selection down to his subtitle: *With Remarks on the Theories of Origin of Species by Variation.* Darwin was unimpressed—"fearfully
disappointed"—with Lyell's *Antiquity of Man,* and he recognized
that Lyell had dodged the criticisms levied at evolutionary theory.[7])

QUESTIONS ABOUT HUMAN NATURE BECAME ALL THE MORE THRILL-
ing in the years after 1859. The West was ablaze. Italy was mostly
unified by 1861, though the Papal States held out for a few years. Britain was fresh from major military victories—in the Crimea against
the Russians (1856), in China during the Second Opium War (1860),

in India after the Indian Mutiny of 1857. Now it moved to consolidate the Raj and held another International Exhibition in 1862, to repeat the feat of the Universal Exhibition of 1851 that had showcased the empire's industrial supremacy and global reach. From 1864 to 1870, Prussia fought and won three wars—against Denmark, Austria, and France—and then it strong-armed the other German states and unified Germany. Imperial tentacles slithered ever deeper into Africa. Industrialization accelerated. States expanded their involvement and control over every-day urban life. Capitalism, strengthened after the failure of the Revolutions of 1848 across Europe, pressed on with the oppressive proletarianization of large sections of the working class. The Americans fought their Civil War (1861–65) and continued their wars and expropriation policies against Indigenous North Americans.

In such a triumphal context, it was little surprise that the geopolitics of evolution centered on civilization's development. The earlier theories of human origins that had treated races as distinct species recalled slavery—which Darwin abhorred. If neither slavery nor polygenism were acceptable, what could explain the difference between "cultures," between "races"? Simple: that some cultures had evolved further than others. This idea allowed Darwin to reject racial hierarchy while allowing for major differences between cultures. His solution placed "all living societies . . . on a stream of Time—some upstream, others downstream" and thus recognized an image of British and European force as "civilization."[8] Other peoples would be dragged up to its level.

Theories of language, like theories of life, the earth, and technology, also underwent a transformation. Philosophers and linguists had long argued that language was foundational for humanity, there since the beginning. With Darwin and the biologists now countering that what mattered was the passage from animal to human, the importance of language was suddenly under threat. However, opportunity for linguists also beckoned.

A century and a half earlier, philosophers like Gottfried Wilhelm Leibniz had asked: What language was spoken in Paradise?
Could it *really* have been Hebrew? Scythia, Leibniz decided, had witnessed the birth of language.[9] Jean-Jacques Rousseau wrote a famous
Essay on the Origin of Languages that offered an even blunter starting point: "Speech distinguishes man among the animals."[10] Once
humans learned to use figurative language, they broke the intimacy
of the human community living in the state of nature.

As we've seen, in the nineteenth century these ideas had been
left behind by the comparative study of Indo-European languages
that asked about the roots of words to find a path to the roots of
humankind. The leading intellectual in this endeavor was August
Schleicher, a German linguist and, as of 1862, author of *A Compendium of the Comparative Grammar of the Indo-European Languages*.[11] The *Compendium* was a textbook—in other words an ideal
propaganda project. Schleicher distinguished the Indo-European
from Semitic and other language families and then tracked from living languages back to an original Indo-European (*Indogermanisch*)
one (Figure 5.4). By his time, Indo-Europeanism had become a
conduit for nationalist feeling—my language is more primordial

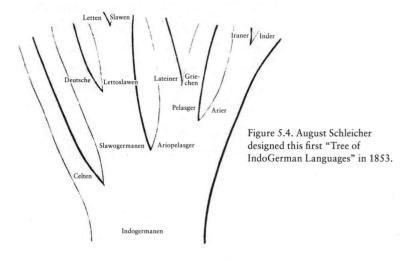

Figure 5.4. August Schleicher
designed this first "Tree of
IndoGerman Languages" in 1853.

than yours. (It's not that different today: Lithuania still celebrates its language as the oldest Indo-European language.) For some of Schleicher's contemporaries, language was all about "our" race: "fecund," "grand," and "ancient."[12] The influential Geneva-born scholar Adolphe Pictet published his magnum opus *Indo-European Origins, or the Primitive Aryas* in 1859. He declared his "linguistic paleontology" would bring back to life the "cradle of the world's most powerful race, the very race to which we belong."[13] After he died, a young linguist by the name of Ferdinand de Saussure recognized the perniciousness of these ideas: in a highly ironic tone, Saussure wrote that research on these "Aryas" was almost consciously reinventing them as "an ideal humanity."[14] Saussure correctly accused his teachers and contemporaries of fantasizing about prehistory-as-Arcadia and inventing an imagined linguistic motherland. Out of that motherland, waves of Indo-European invaders supposedly spread—half noble-savages, half proto-Germans.

Naturalists never really thought much of the linguist interlopers: language, Darwin mused, was only a more complex form of animal speech. Yet Schleicher, Pictet, and contemporaries like Friedrich Max Müller, another German-born Oxford professor of comparative linguistics and Indo-European mythology, were riveted in the 1860s by questions like: Are nations and languages identical? Are languages expressions of racial difference? The antiquity of Indo-European language groups suddenly made a major difference. So too did the promise to reconstruct the "inferred Indo-European original language," and the idea that languages and races had traveled via violent conquest.[15] A hierarchy of Indo-European languages meant a hierarchy of peoples (see Figure 5.5).

Lecturing at the Royal Institution in 1861, Max Müller declared that language "establishes a frontier between man and the brute, which can never be removed."[16] Max Müller admired Darwin, but he advanced a strong criticism: there was no way to jump from animal noises to complex human language.[17] Animals have no language; but all humans speak. In which case, language is the human minimum. Who cares about apes if you can't use them to explain what distinguishes animals from humans?[18] Humanity descended from apes, yes, but its true origin, and with it the similarities and differences between cultures, could only be found by studying linguistics.[19]

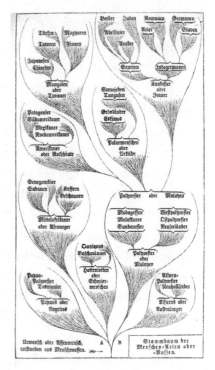

And what came first, race or language? Were racial differences real, or were they the collateral effect of linguistic differentiation? Max Müller thought language came first, and even some Darwinians agreed with him.[20] Language, he noted, was the foundation

Figure 5.5. As Ernst Haeckel's "Genealogical Tree of Human Species or Races" (1868) shows, the distinction between languages and races was weak. Linguists and biologists were comfortable depicting ethnic differences in terms of linguistic origins. Note the celebration of IndoGerman on the upper right.

for religion, for thought, for culture, for myth. The study of myth (the field was often called "mythology" at the time) merely addressed "a disease of language."[21] Other linguists insisted that thanks to their mind-numbingly dry comparative analyses of phonemes they could explain all these bigger issues, from religion to thought to power. Human origins had little to do with the physical body, and everything to do with vowels and phonemes. Ethnic and racial distinctions simply echoed linguistic differences (see Figure 5.6).

The linguists convinced the anthropologists, most importantly

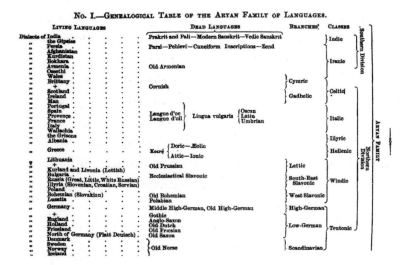

Figure 5.6. Friedrich Max Müller, "Genealogical Tree of the Aryan Family of Languages" (1862).

E. B. Tylor.[22] In the 1860s, Tylor was doing his best to figure out what evolutionary theory meant for the masses of information available on distant societies. He followed Max Müller on the importance of language and even agreed with him on the ancient Indo-European myths and "the great Aryan race," whose birthplace his contemporaries were seeking out.[23] In his first major work, *Researches into the Early History of Mankind* (1865), Tylor opened by declaring that he was going to study the foundations of "knowledge and art, religion and mythology, law and custom." Then he put all that aside to spend his first 150 pages on language.

AS A FIELD, PREHISTORY (VORGESCHICHTE, *PRÉHISTOIRE*) WAS BORN amid the political and intellectual chaos of the 1860s. Geology, biology, anthropology, linguistics . . . what took precedence? Geologists were scouring caves and gravel pits looking for more tools. In 1867, Boucher de Perthes exhibited his stone tools at the Universal Exhibition in Paris; other exhibits included bone tools and perhaps most notably, small engravings. Prehistorians had begun to discuss portable art, usually figures of animals engraved in bone.[24] Skulls that appeared ancient fascinated the educated public. On top of which, they often reflected nationalist arguments: when Cro-Magnon skulls were discovered in France, they were taken to indicate greater refinement compared to the Neanderthals of the German states (bone-headed Prussia especially). Images of early human life began to circulate more broadly: crude-looking cavemen and ape-men spread the news and poisoned the air with stereotypes that would persist for well over a century. And some forty years after the death of Sarah Baartman, the "Hottentot Venus" who had been brought to Europe and put on exhibit as a specimen of savagery, the practice of carrying Native people off to Europe was still ongoing.

With the establishment of the Royal Anthropological Institute in 1871, ethnology and anthropology moved to capture the title of the "Science of Man."[25] Theories of culture (and race) would henceforth rely on both recorded history and evolution. Peoples would be distinguished on the basis of where they stood in the development of humanity. This was convenient for European empires that imagined their purpose as "civilizing" the Natives—the "white man's burden." Tylor was the most famous proponent of this approach, and in time he would be crowned "the father of anthropology." In *Primitive Culture* (1871) he recast society as having laws of its own. These, like biology's laws, were unambiguous and enduring. But they were not based on some racial or biological essence: Tylor looked down on those who would compare societies using just biology, geological tools, or language. Rather, "mankind is homogeneous in nature, though placed in different grades of civilization."[26] He blended together the concepts of historical change that were available to him: Darwin's natural selection, Comte's three-stage theory of history, and notions of remnants and vestiges. Above all, he decided that distance in space and distance in time were the same: today's "savage races" were "the nearest modern representatives of primaeval culture." They showed the way toward the earliest "condition of man . . . a primitive condition, whatever yet earlier state may in reality have lain behind it."[27] Tylor patted himself on the back that in his book, for the first time, "the European may find among the Greenlanders or Maoris many a trait for reconstructing the picture of his own primitive ancestors."[28]

Of course, Tylor was not the first to compare living Natives with early humanity. But for him, to be an anthropologist was to practice a "science of culture" that looked to that past and compared it to the present: this meant deciphering primitive codes. Such anthropologists, Tylor went on, needed a way to organize "scientifically" all the information circulating about distant societies, their myths, practices, body modifications, art, law, and language. Sometimes statistics were available, sometimes ethnographic reports, sometimes archaeological remains. The anthropologist could generalize, could *own* the science of social organization. This would be a new science, one that, despite its opposition to racism, also served as a convenient one for Europe's hegemony. In other words: Tylor merged all the approaches available at the time to present himself as the inventor of a sort of super-science that would speak with ease about the deepest past.

Tylor's compatriot John Lubbock used much the same approach to found prehistoric archaeology. A baron who was a banker by day and an archaeologist by night, Lubbock had proceeded in his own 1865 book *Prehistoric Times* to guide readers from the Stone Age toward Indigenous peoples in the present.[29] Lubbock too utilized the entire available arsenal of concepts: Tacitus's Germans; the technological *Stone/Bronze/Iron* triad; bronze-bearing, Indo-European conquerors who had destroyed Europe's Stone-Age natives; Boucher de Perthes's stone tools; evolution.[30] Prehistoric archaeology could tie together these strands. No longer would Europeans have to turn to the East: "suddenly a new light has arisen in the midst of us" and the oldest relics were not to be found in Egypt or Assyria but "in the pleasant valleys of England and France."[31] His book became a bestseller of enduring influence. What motivated him, even more so in *The Origin of Civilisation and the Primitive Condition of Man* (1870), was the sense that in the ground, in the deepest past, lay an explanation of society, culture, and humanity itself. And what was the first reason to study prehistory? Its value "in an empire such as ours."[32] To know humanity was to master and own it.

WITH THE DESCENT OF MAN (1871), DARWIN FINALLY REFUSED TO BE talked about and talked over, and he offered his own understanding of human origins. His position in the science versus religion argument was well known. So were the criticisms of his notion of natural selection.

Tylor and Lubbock had used his work as one of the foundations of their own. Little did they understand how indifferent Darwin the stately naturalist was to thinking historically or culturally. Yes, he confirmed, "the high antiquity of man has recently been demonstrated by the labours of a host of eminent men, beginning with M. Boucher de Perthes."[33] He never bothered about that eminent man again. But he did shift and shove Lubbock, Lyell, and Tylor as necessary for his own points to work. His concerns were human descent from a preexisting form, the problem of race, and the continuity of mental abilities from animals to humans. He now proposed a secondary mechanism, sexual selection, to avoid reducing humanity to a plaything of natural selection. This new mechanism did not really involve random innovations, seemingly out of nowhere, some of which survived; it involved a series of *restrictions* on what was possible.

Since his earliest notes, Darwin had wondered about human continuity. In his *The Voyage of the Beagle* (1839) he did express shock at the mistreatment of Natives, but also pride in Britain's power. While traveling on the *Beagle*, Darwin had been surprised to meet, in Tierra del Fuego, a "barbarian—man in his lowest and most savage state." He went on: "One's mind hurries back over past centuries, and then asks, could our progenitors have been men like these?—men, whose very signs and expressions are less intelligible to us than those of domesticated animals; men who do not possess the instinct of those animals, nor yet appear to boast of human reason, or at least of arts consequent upon that reason."[34] Savages, he proposed, are to the civilized as wild beasts are to tame, domesticated animals.

Given his monogenistic convictions, Darwin "knew" that the man he encountered was akin to his ancestor; but he could not yet bring himself to assert as much. Modern humans were no longer the creature in the "savage state" that he'd encountered in Tierra del Fuego and whose humanity he could hardly believe. In *The Descent of Man*, Darwin both avoided and endorsed domestication as an explanation. Humans differed "widely from any strictly domesticated animal."[35] Domestication meant the selection of individuals with particular characteristics in order to guide reproduction and accentuate these characteristics. It was a favorite subject of Darwin's—he had published another enormous book on it in 1868. Now, unlike animals, humans had never been subjugated, he argued, to the point of being selected for breeding. Still, he referred to "civilized men" as "highly domesticated," and he distinguished between races as between different "domesticated animals."[36] While it was not *conscious* and fully intentional, the movement of history was less an ascent of civilization than a guiding of cultural and racial differentiation into refined groupings.

For *Descent*, Darwin needed a mechanism to explain this guiding of selection. After all, natural selection lacks all intention. It simply means that, in a world where death reigns, a world where all organisms die, the environment kills everyone off with absolute certainty but with unequal efficiency. Some organisms have offspring;

some characteristics of the organisms survive across generations more easily than others. The specter of chance hung over the theory. Was natural selection too random?

Sexual selection was the solution. Females of a species, Darwin argued, select among the males: whenever the males fight or display themselves, they offer a female certain (visual) cues about their potency, which is useful for the species' survival. Female choice—or submission to the winner—enabled *particular perceivable* variations to survive. Such choice also allowed for gradual racial differentiation and the evolution of mental faculties. In other words, sexual selection protected natural selection, while explaining that individuals do make choices that help lead down particular paths. It was a variety of domestication and it offered room for intention. Darwin insisted that as a training and refinement mechanism, sexual selection had been the biological basis of the development of human mental faculties. There is no separate creation, he insisted, for mental faculties. If "Man" ruled over the planet, he owed his "immense superiority to his intellectual faculties, his social habits, which lead him to aid and defend his fellows, and to his corporeal structure."[37] Natural and sexual selection—domestication with an eye to intelligence, attractiveness, and social superiority—had made this possible.

Domestication is why Darwin found the "lowest savages" interesting. Though the difference between them and the highest ape was "immense," they represented humanity in its least domesticated and earliest state.[38] To move up from these lows of savagery was to move forward in time. It was also to move *inward*, from the jungle toward the home. It was to switch from the brute force (and also the weakness) of humans living in the wild to complex structures of economic and political power. If Boucher de Perthes's human was a tool-user, and if Tylor's human charged across the history of culture, Darwin's human was a biological self in which the individual vanished. Individuals made only small choices, evermore constricted by the slow training of the species. The species had become properly human by obeying natural selection, and then, through sexual selection, it had become differentiated into particular groups, each with their own cultures and power.

Did humanity enjoy a single origin? Could racial or linguistic origins stand in for it? Should authors look to cultures on other continents, or deep in the ground? What mattered more: biological nature, diversity of cultures, or sheer power? Many answers were proposed during the critical decade between *Origin* and *Descent*. Darwin, for his part, believed in a universal humanity, and that civilization was the path toward it. Take the following words from *Descent* for an example. They can be read as a celebration of British power, and yet they also have a more utopian cast:

> As man advances in civilization, and small tribes are united into larger communities, the simplest reason would tell each individual that he ought to extend his social instincts and sympathies to all the members of the same nation, though personally unknown to him. This point being once reached, there is only an artificial barrier to prevent his sympathies extending to the men of all nations and races.[39]

We will return to these words more than once in the chapters to come.

Part II

‗‗‗‗‗

THE CONCEPTS THAT TIED IT ALL TOGETHER

(FROM THE 1830S TO WORLD WAR I)

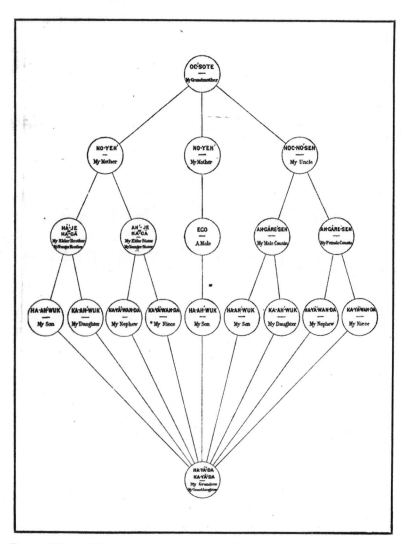

Figure 6.0. Seneca/Haudenosaunee Consanguinity Diagram by Lewis H. Morgan (1871), Plate VII.

Chapter 6

MOTHER LOVE:
PRIMITIVE COMMUNISM

By the early 1880s, Friedrich Engels had been writing books, pamphlets, and essays for forty years. Most famous among them—the books that secured his place in the communist pantheon—were *The Condition of the Working Class in England* (1844) and the *Communist Manifesto* (1848) that he and Karl Marx had coauthored. But in 1883 Marx died, and Engels took over his manuscripts and notebooks. He committed to them with the same passion that he had for the man himself. He edited volumes 2 and 3 of Marx's *Capital*. He stitched together *The German Ideology*, a strange, dysfunctional, influential book.[1] He added two new prefaces to the *Communist Manifesto*. Even one of the *Manifesto*'s most famous sentences—"The history of all hitherto existing society is the history of class struggles"—was not safe from his editorial obsession. Having promised not to change the text, Engels instead appended a long footnote.

The famous sentence applied, he declared in that footnote, only to "all written history." Back when they had been writing the *Manifesto* in 1847, "the pre-history of society, the social organisation existing previous to recorded history, was all but unknown." Since then, though, researchers had clarified the importance of "primeval" common ownership from India to Ireland, from Russia to the United States. Much credit was owed, he went on, to the American ethnologist Lewis Henry Morgan, who had reimagined family and property. Engels concluded: "With the dissolution of the primeval communities, society begins to be differentiated into separate and finally antagonistic classes. I have attempted to retrace this dissolution in *The Origin of the Family, Private Property, and the State*."[2]

Why did Engels think such an obscure and self-serving paragraph would clarify the famous line in the original? Largely because anthropologists and political economists had forced a shift from the 1848 image of precapitalist society as agrarian, feudal, preindustrial. So Engels had reason to believe that his *The Origin of the Family, Private Property, and the State*, which he published within a year after Marx's death, was more accurate, more strategically useful than the *Manifesto*'s generalities. *The Origin of the Family* declared it possible to imagine family, gender, and property in a fundamentally different way than capitalism had. It promised that the state would "wither away" after the revolution: like the family, the state was the product and champion of bourgeois domination. Once again, prehistory offered a model for the present—now for true socialist kinship.

OVER A CENTURY EARLIER, ROUSSEAU HAD LAMENTED PROPERTY AS the "greatest lie in the history of civilization," the exile from the state of nature, the fall that had begun with "the first man who, having enclosed a piece of land, thought to say 'this is mine' and found others simple enough to believe him."[3] Adam Smith, for his part, had located the establishment of property between the "First" and "Second" Ages, the transition from hunter-gatherer to agricultural societies.

The question of the origins of property persisted. In the 1870s, those interested in "primitive" property had a standard reference to turn to: the work of British jurist and historian Henry Sumner

Maine. In *Ancient Property* (1878), Maine linked questions of property to the development of law.[4] Contract dominates modern society,
he argued, but in the ancient world everything was determined by status. Blood ties had preceded property, so property had hinged on the
particular status of individuals within a family hierarchy. Of course,
Maine's theorizing wasn't only about distinguishing moderns from
ancients. The British were very anxious to decide if the "backwards"
peoples they had colonized could sustain modern ownership—if they
were capable of a free market—and Maine effectively answered no.[5]
In an 1873 lecture on the property of married women, Maine also
turned to the ancients, first to Roman and "Hindoo" law, and then
all the way back "to the earliest institutions of so much of the human
race as has proved capable of civilization."[6] In "the vast antiquity
now claimed for the human race," the establishment of the family
was "the one condition of [women's] progress to civilization." The
patriarchal family had instituted law and private property, he added,
and this had strengthened the position of women. Before, they had
had no real status, no rights, no property.[7] The "end to the seclu-
sion and degradation of an entire sex" would occur thanks to legal
reform: to him, progress, for women too, was a fact of civilization.[8]

Marx mocked Maine as a donkey who believed his neighs spoke
for all civilization. Other socialists were less dismissive, because
Maine's view that precapitalist property had been communal made
it possible to lament its destruction.[9] By the mid-1870s, well before
Marx and Engels, "primitive communism" was circulating as a concept. Socialists mostly derided capitalism as being itself a "primitive" way of life.[10] But "primitive communism" took on added value,
because it allowed advocates to slide easily from nostalgia for preindustrial communities to imagining a pure prehistoric world. In
The Atlantic in 1878, for example, Arthur G. Sedgwick discussed
the "arrival" of socialist ideas in the US. He was the editor of the
American Law Review, a decorated veteran of the 20th Massachusetts Infantry Regiment in the Civil War, and later became a Har-

vard Law professor. He declared primitive communism a historical fact that offered a path for property reform. And he, like others, treated it more as a matter of property and inequality than of gender. Although it could be combined with racial origin stories—for example in dreams of the powerful, barbaric ancient Germans living in common—it generally undermined racial exceptionalism because it depended not on nature but on how property organized family and society.

Parallel to this work on property emerged a "primitive marriage" literature, most famously in books like John F. McLennan's *Primitive Marriage: An Inquiry into the Origin of the Form of Capture in Marriage Ceremonies* (1865), Darwin's *The Descent of Man* (1871), and Alexis Giraud-Teulon's *The Origins of the Family* (1874). Ancient marriage became a matter of debate, partly as Europeans sought reasons why Indigenous peoples around the world disdained monogamy.

McLennan, the most original—that is, wildly speculative—of these authors, argued that "primitive marriage" in exogamous societies (societies whose members did not marry one another but looked outside the group) had originally required the kidnapping of a bride. Kidnapping from another tribe or clan was neither occasional nor a mere practicality: it lay at the fount of all non-Western (and perhaps even originally Western) marriage.[11] He was not shy to call it a savagery—the result of the supposedly warlike nature of *all* Indigenous life. So ubiquitous was it, he concluded, that wife-capture was standard in the state of nature.[12] It was the symbolic foundation of elopement and rape.

The drama of McLennan's theory gave it enduring popularity, despite its lack of evidence and its overall ridiculousness, which drew Maine and Morgan's ire. Today, still, McLennan alarms the reader: he decries violence against women by projecting it onto awful rapist savages. More interesting is *why* he did so: for McLennan the entire "primitive" economy was based on bodies. Bodies meant status;

marriage founded power. Society, law, and money all depended on a melodrama of some bodies capturing others so as to usurp power.

Darwin was cruder still: two-thirds through *The Descent of Man*, he famously postulated that just before human society proper, clans had taken the form of hordes. "Primeval man aboriginally lived in small communities, each with as many wives as he could support and obtain, whom he would have jealously guarded against all other men." Darwin then quoted someone called, quite apropos, Dr. Savage, that only "one adult male is seen in a band; when the young male grows up, a contest takes place for mastery, and the strongest, by killing and driving out the others, establishes himself as the head of the community." Darwin continued: "The younger males, being thus expelled and wandering about, would, when at last successful in finding a partner, prevent too close interbreeding within the limits of the same family."[13] Sexual selection involved violent contests for mastery, as male contestants battled one another and the victor came to own the females at stake. Polygamy ("most savages being polygamists") meant a half-formed, still brutal society; monogamous marriage meant civilization.

Shortly before Darwin and McLennan released their books, Johann Jacob Bachofen, a philologist living in Basel, Switzerland, had published *Das Mutterrecht* (1861), translated as *Mother-Right*.[14] His theory that societies were originally matriarchal made him into a strange intellectual star. Bachofen identified an original state in which sex was free and a polyamorous society celebrated a Mother Earth–type figure. This "hetaerism" was followed by a properly matriarchal "lunar" stage, characteristic of early agriculture, where maternal authority persisted. In turn, a "Dionysian" stage marked the advent of the patriarchy, a warrior ethic, the masculinization of power. Bachofen concluded that civilization proper was born and the patriarchal social order fully established with the next, "Apollonian," stage. The last shift was visible, Bachofen declared, in ancient

Greek tragedy, a point Friedrich Nietzsche would draw on in *The Birth of Tragedy*.[15]

LEWIS HENRY MORGAN, WHO BEGAN HIS CAREER AS A RAILROAD lawyer and later became a state senator in New York, proved far more inspiring and influential on the kinship/property question. He had published a book on "the Greek genius" early on, then dove into quasi-ethnographic pursuits in the 1840s and 1850s, beginning with the Seneca people on whose behalf he had advocated as a lawyer in the 1830s and 1840s. (The Seneca had adopted him in thanks for his legal and political activism—though today we would see Morgan's role as much more problematic.) In the golden year 1871, when Tylor published *Primitive Culture*, Lubbock *The Origin of Civilization*, and Darwin *The Descent of Man*, Morgan chimed in with his first major work, the *Systems of Consanguinity and Affinity in the Human Family*, following it up with *Ancient Society* in 1877. His was a very different approach to the family.

Rather than imagine the past, Morgan wondered how the structure of a society reflected that of the family. His best-known point is that *some* peoples had organized themselves around a matrilineal system—but this was not new. The issue was *how*. In *Systems of Consanguinity and Affinity*, Morgan constructed detailed kinship tables on terms like father, mother, daughter, husband, brother. And he compared the way that these terms were deployed in actual lived relations between human beings.[16]

Morgan observed that terms used to designate family relations sometimes differed from current ways of life. He had expected to study the family *as it actually exists*, but particular peoples used terms that did not exactly apply. An individual would call people brothers or sisters even when they technically were not actual blood relations as brothers or sisters—and whom she did not treat as such (see Figure 6.0). She might refer to the children of such "brothers" or "sisters" as her own children—without, crucially, caring for them in the same way as for her own children. Something was off; there was a gap between the *language* of kinship and the *actual* family Morgan was observing (what we call "blood relations"). He thought of this gap as an interval, a time lag. Language does not change at the speed of life: it has a time all its own. And thus two levels were in play. One level was about actual kinship, the expression and feelings of familial attachment; the second level was linguistic. The language, Morgan concluded, indicated the "ruins" of an earlier era, residues of a time in which one's siblings' children had indeed been the same as one's actual children. People had lived differently, he theorized, in group marriages or non-paired relationships where polyandry (many husbands to one wife) or polygyny (many wives to one husband) predominated. Morgan's gambit was to breathe life and imagination into a past structure that could be imputed but not observed. Between the remnants of those earlier times that endure and what is observed now lay the gap for thinking about the original form of the family. If it was agreed that monogamy was anything but ubiquitous, *Systems of Consanguinity and Affinity* achieved a *linguistic* explanation of this problem: a set of social relationships that had since collapsed.

Like his contemporaries, Morgan patched together existing concepts and images. He too used three-stage divisions, from "savages to barbarians and barbarians to civilized men."[17] He was a convinced monogenist, declaring that "mankind were one in origin, their career has been essentially one, running in different but uniform channels upon all continents."[18] He repurposed the rhetoric of ruins. He too was interested in property and, like Maine, tried to figure out if Native Americans "fit" with "civilized" expectations about it. He identified living Indigenous peoples with European ancestors: "the history and experience of the American Indian tribes represent, more or less nearly, the history and experience of our own remote ancestors."[19]

Nevertheless, he organized all these concepts very differently from anyone else. Where Darwin explained the original family through a horde of man-apes fighting for mastery, and McLennan promoted wife-capture theories, Morgan offered his readers no horde, no disorganized violence, only a power struggle inside the family. He threaded together his ethnography of Native Americans with his study of ancient Greeks, Romans, Germans, Celts, and Sanskrit-speakers. He made a supposedly "intuitive" leap from Native American "barbarians" to ancient European "Aryans." In *Ancient Society*, he tracked the buildup of the Iroquois (or Haudenosaunee) gens, phratry, tribe, and confederacy, one chapter for each, and then repeated the entire movement with the Greeks and then the Romans. In a very real sense, he was pushing the very same people he talked to back in time so that they lived in his own deep past. And to follow them back in time, he again chose specific navigators into the past: the linguistic and cultural leftovers that retained something of societies before colonization. Ruined temples could be seen all over Greece; old Iroquois buildings may not have survived, but other old structures did, in the odd language of family. Iroquois social organization—*from* the family *outward*—reflected the deep European past prior to the Greek city-states.

The "original" kind of marriage, Morgan went on, had been "Hawaiian" group marriage, a world of promiscuous intercourse akin to Bachofen's haeterism model. Consanguineous intermarriage between brothers and sisters followed; then, gradually, the separation of brothers and sisters and the construction of a "gens" (a set of families sharing a common ancestor); much later, a paired family; finally, a patriarchal family. Morgan divided the gentes into either *matrilineal* or *patrilineal*. Whereas the gentes had become disorganized in modern societies, Morgan argued, to Native Americans as to early Greeks and Romans, it had once been the basis for government, "a social organization . . . which had prevailed from an antiquity so remote that its origin was lost in the obscurity of far distant ages."[20]

In their original form, the gens—and society—had been matrilineal. This meant that descent, family organization, and property went down the female line: a mother and her children, in turn only the children of her daughters, and so on. The various fathers were excluded, as were the children of her sons, and those of her daughters' sons.[21] Property and power stayed within the gens, which left males (who couldn't marry within the gens) bereft of authority.

By Morgan's time, the original dispensation was gone: matrilineal descent lay in a prehistoric past that could not be accessed. All that survived from this past were some myths and these surprising linguistic kinks, where members of a society continued to refer to each other in the old ways, ways that no longer reflected their feelings.

But if only residues of the matrilineal age survived, whether among Hawaiians, Native Americans, or Europeans, what had happened to bring about its end? Halfway through the "barbarian" stage, Morgan decided, a shift had taken place from descent through the female line to descent through the male line. Now, "descent is in the male line—*into which it was changed after the appearance of property in masses.*"[22] When did this happen, though? How did a venerated gens up and change its line of descent? Marx and Engels would find Morgan's simple answer—property—ideal. Agriculture and animal-rearing had allowed men to control subsistence and to own their home and land—permanently. But in a matrilineal society, their children (and their children's male children in turn) couldn't inherit. Thus began a "contest for a new rule of inheritance, shared in by fathers and their children." The accumulation of property by men dramatically altered the status of women in the family. Rather than abandon the gens as a form of social organization, societies kept the gens but "overthrew" the female line.[23] The male hero and the property owner became associated with authority and sovereignty, the new basis for gens. The use of an animal name to define each gens was laid aside as outdated. Animality lost its connection to humanity, the meaning of totems faded, whatever was left of savagery was over. Property forced a renaming of the gens after some ancestral hero—a conqueror or someone who had "made it." Fathers now had

all the power. And once the former hero or property owner receded
into the mists of the past, the gens found a new hero to identify with
and declare the originator of the family.[24]

THE PERIOD WHEN ENGELS WAS WRITING *THE ORIGIN OF THE FAMILY*
witnessed considerable agitation and change in the familial, legal,
and economic status of women. In Britain, where Queen Victoria
was widely perceived as a national matriarch, Parliament passed the
Married Women's Property Act (1882), which reestablished married
women's legal identity and allowed them to own and control property separately from their husbands. In the United States, some states
had been granting women the right to own and sometimes control
property since the late 1860s. In France, trade unions, secondary
schools, and universities opened to women, and divorce at the wife's
request was legalized in 1884. That same year, Germany granted
women legal majority. Women activists were becoming more visible
across Western Europe. Socialists had an added imperative: outdo
and outshine the feminist reformers.

Primitive communism became attractive in this context of family reform. In his last years, Marx was turning increasingly toward
anthropological matters, and in 1880–81, he started a detailed notebook. He listed virtually all the relevant and known works on early
society, copied and commented on almost a hundred pages of passages from Morgan's *Ancient Society*, and criticized other works,
including Maine's and Lubbock's. How could one best confront capitalism with anthropology? Marx was recognized by his supporters as
the thinker who unveiled the law of human history, so the expanding
literature on the ancients and Indigenous peoples was hardly a matter
of indifference. His health was precarious, however, his energy for
writing faltered, and he dedicated long periods to travel for medical
purposes. He returned to the subject in 1882. But he was preoccupied

with his own family as much as with the family as an institution: his wife died in 1881, his eldest daughter in January 1883. In the weeks before Marx's own death, on March 14, 1883, Engels, who was regularly on Marx's side, was penning long letters to young socialist leaders Karl Kautsky and Eduard Bernstein to refute Kautsky's recent articles for offering an unimpressive theory of capitalism's effect on the family.[25]

Engels shared Marx's passion for the sciences, and he had long tried to formulate a theory of precapitalist societies, at least since his essay "Precapitalist Economic Formulations" of 1857. In 1876, he also wrote on "The Part Played by Labor in the Transition from Ape to Man." After Marx's death, Engels began taking cues from the notebook that Marx had left behind. He latched on to Morgan's argument that property had deformed the original family. Other authors offered help—though not McLennan, the "pedantic Scot" as Engels put it, who had placed a premium on primitive violence. Engels relied on the same tropes as Morgan and Marx: a three-stage theory of history (savagery/barbarism/civilization), an infancy of the species, a fundamentally peaceful past.[26]

In the eulogy to Marx with which he began *The Origin of the Family*, Engels praised Morgan for discovering, independently of Marx, the *materialist* basis of society.[27] What is materialism? Engels redefined it as the "production and reproduction of immediate life."[28] If materialism is about our means for survival and our place in the economy, Engels offered, it also concerned the reproduction of humanity. Sexual reproduction and the ordering of society in relation to children became, in Engels's hands, an unexpected center of the communist theory of society.

Engels retold the story Morgan had proposed: pure consanguinity (marriage of brothers and sisters) *must have* come first.[29] Hawaiian group (or "punaluan") marriage had followed, with wives sharing husbands within a group that excluded siblings.[30] Then the

gentes emerged, first through a matrilineal gens and then, eventually, the couple: the "pairing" family. Such monogamy, Engels commented, "did not by any means dissolve the communistic household
inherited from earlier times."[31]

One might expect Engels to have begun the book with the "primitive communistic household," but no. He only introduced it at this
point, when it was *not yet* being dissolved, as though he was anxious to avoid telling us when or how it began, or whether it had been
the same as the punaluan group marriage. Engels immediately gendered it: primitive communism meant "the supremacy of women in
the house, . . . the exclusive recognition of a natural mother . . . high
esteem for the women."[32] Yet, while the communistic household was
not destroyed by the "pairing family," it was not long for this world.
Agriculture and husbandry, technology and wealth enabled private
property. With man's greater access to technology, and his need for
physical power, his claim to wealth had increased. Thus began the
oppression of women: technology and property "raised the man's
status in the family, and created an impulse to exploit the strength
and position in order to overthrow, in favor of his children, the traditional order of [matrilineal] inheritance . . . Mother right had to
be overthrown, and overthrown it was."[33] Notice that last twist:
"Mother right." Engels had followed Morgan to the letter—even using
the word "overthrown" (*umgestossen*). But what was overthrown
was not just matrilineal inheritance, but something more powerful.
The men's "revolution" had overthrown Bachofen's matriarchy. In
Engels's mind, the two had become one: a matrilineal matriarchy.[34]

How had this shift taken place? Through private property.
Where? *Everywhere*, as evidence from "a whole number of American
Indian tribes" showed.[35] Both ancient Europe and its mirror, Indig-
enous America, had been transformed by proto-capitalism.

Morgan had been blunt on the corrosive effects of civilization.
"The outgrowth of property has been so immense, its forms so diver-
sified, its uses so expanding and its management so intelligent in the

interests of its owners, that it has become, on the part of the people, an unmanageable power. The human mind stands bewildered in the presence of its own creation." Nevertheless, he had expressed some hope: democracy, fraternity, equality in rights, and education would help humanity reach "a new and higher plane of society."[36]

Morgan was not at all a socialist sympathizer, but to Engels, certain affinities with Marx confirmed that the two had reached the same conclusions—and that Marx was correct. A society attached to capitalism, "with its control centred in the state," had enforced a situation where "the family system is entirely dominated by the property system."[37] Marriage and monogamy became the "first class oppression," and so began "the content of all written history"— exactly what he would add in that first footnote to the *Manifesto*.[38]

Worse still: "The overthrow of mother right was *the world historical defeat of the female sex*. The man took command in the home also; the woman was degraded and reduced to servitude; she became the slave of his lust and a mere instrument for the production of children." In the modern era, the degradation of women might not be as visible, but "in no sense has it been abolished."[39] The argument was more than a means to win more women to socialism; Engels, and after him many others, felt it deeply. If the state was the machine that ensured class oppression—an apparatus for plunder and coercion—it had really begun with the division of labor in the home. A new revolution would completely reverse the one that had ended matrilineal matriarchy, the one out of which capitalism had arisen.

IN THE DECADE AFTER ENGELS'S BOOK, INFLUENTIAL MARXISTS COPied his argument and turned primitive communism into common currency and even into a catchword.[40] *The Origin of the Family* quickly became, as Vladimir Lenin later wrote, Engels's most popular book. August Bebel's *Woman and Socialism* (1879) was already a classic. Little did this matter: Bebel rewrote it so as to follow Engels's lead.

Younger socialists used the concept of primitive communism to imagine a moment outside of capitalism, a family outside of the patriarchy, a world where gender and property relations had been very different. As importantly, they could argue that communism was not simply a politics, but rather the way of nature itself. Revolutionary leaders, including Rosa Luxemburg, followed suit. In her *The Accumulation of Capital*, Luxemburg celebrated Arab/Kabyle families in Algeria as matriarchal units where much property was shared. She argued that Ottoman rule and especially French colonization had sought to destroy this "natural economy."[41] For his part, hiding from Russia's Provisional Government in August 1917, Lenin composed his infamous *The State and Revolution*. He used *The Origin of the Family* to explain how the post-revolutionary state would depart so radically from its existing forms as to no longer really merit the name of a state. Two months later, he led the October Revolution to the first triumph of communism in any nation. Alexandra Kollontai, who became People's Commissar for People's Welfare in Lenin's first revolutionary government and also served as a member of the Central Committee of the Communist Party, found inspiration in both Engels and Bebel.[42] In 1920, she included Engels's matriarchal family in her essay "Communism and the Family," which focused on the way the Soviet Union was ending patriarchal oppression. Family arrangements change, and the bourgeois one was over. "There was a time when the kinship family was considered the norm: the mother headed a family consisting of her children, grandchildren and great-grandchildren, who lived and worked together."[43] (Kollontai nonetheless thought that Engels's understanding of female sexuality was crude, and that what mattered most was the problem of reconciling sexuality and labor in the post-revolutionary era.) And in 1936, Joseph Stalin, at the height of his power as General Secretary of the Communist Party and leader of the Soviet Union, penned an essay, "Dialectical and Historical Materialism," that declared primitive communism as the first of five stages in human history, "succeeded precisely by the slave system, the slave system by the feudal system, and the feudal system by the bourgeois system."[44] He thus made primitive communism central to Soviet dogma: the book in which the essay figured, *The Short History of the Communist Party of the Soviet Union* (1938), was printed in tens of millions of copies, and schoolchildren all over the USSR studied it as the truth. Discoveries of "Venus" figurines during this period, both in the Soviet Union and abroad, came to be interpreted as "Earth Mother" figures (see Figure 6.1).

Primitive communism mattered, in part, because conservatives increasingly presented socialism as simply "antithetical to human nature": it was capitalism that was "natural," communism the aberrant form. So whenever it was reported that some Indigenous society shared property and the catch of the hunt, debates would erupt as to whether this meant that communism represented humanity's universal past.[45] In 1922, for example, the influential English psychiatrist and anthropologist W. H. R. Rivers ran on the Labour Party ticket for Parliament. As part of his campaign, he gave a series of lectures,

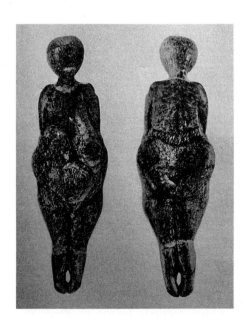

Figure 6.1. P. P. Efimenko excavated this and other "Kostienski Venuses" in 1936. They were among the early Soviet considerations of women in the Paleolithic era and were sometimes interpreted as evidence of a primitive matriarchy, an interpretation important in the USSR but also in the West where "Venus" figurines were treated as earthen Mother goddesses.

including on primitive communism, to defend against the claim that the Labour Party was opposed to human nature.[46] The debate continued, on and off, all the way to the 1980s.[47]

Feminism, already in Kollontai's and Luxemburg's time, was riveted by such questions, especially regarding the sufficiency of Engels's ideas. For Kollontai, Engels's reference to a matriarchal period was welcome, but unsatisfactory as far as the current struggles of the women's movement were concerned. Feminist thinkers grappled with this problem. In *The Second Sex* in 1949, Simone de Beauvoir followed the pattern of both accepting Engels and criticizing him as insufficient. She dutifully began by writing out Engels's primitive communism story. "Private property appears: master of slaves and of the earth, man becomes the proprietor also of woman. This was 'the great historical defeat of the feminine sex.'"[48] But then she turned on him. Engels's fantasy of a prehistoric family with powerful women was crude. Worse, for Beauvoir, was that oppression did not simply end with a vague proclamation of some new equality and an analysis of prehistory. One is not simply born a woman, Beauvoir famously argued, nor does one simply become a woman because one is situated in a socioeconomic milieu. Every woman has to transcend her condition, to make and remake herself.[49] But, always, "He is the Subject; he is the Absolute. She is the Other." Every situation posits her as inessential; Beauvoir went on: "How, in the feminine condition, can a human being accomplish herself? What paths are open to her? Which ones lead to dead ends? How can she find independence within dependence?"[50] And so, where Kollontai and Luxemburg had negotiated feminism and socialism, Beauvoir presumed socialism but then, for the sake of feminism, shifted terrain altogether. Engels and other communists had counted on an alternative, an "outside" to this history—but they were using matrilineality, matriarchy, and primitive communism as feminist props. They promised true equality after the communist revolution, but as Beauvoir and many others saw it, this was an empty promise. Feminism had to begin *after* Engels and not simply on his premises.

WE MIGHT CLOSE, HOWEVER, WITH ANOTHER INTELLECTUAL, ONE who invented a matriarchal paradise to reject hierarchy and war. This was the English archaeologist Arthur Evans, who in his youth traveling through the Balkans—then in the midst of revolts against the Ottoman Empire—before returning to England to become Keeper of the Ashmolean Museum. Evans was fascinated with Heinrich Schliemann, who had discovered Troy and Mycenae. In 1899 Evans moved to Crete, which had also risen up against Ottoman rule in 1897–98. Massacres had followed, and a Cretan state was briefly established under the Great Powers' control before it united with Greece. In 1900 Evans began unearthing ruins of a palace, which he identified with Bronze-Age Knossos and named after the fabled King Minos. Over the next twenty years, he and his collaborators

Figure 6.2. "It is the Divine Child adoring the Mother Goddess," wrote Arthur Evans, who discovered the ruined palace in Knossos and regarded ancient Minoan society as matriarchal. Evans commissioned this design linking the two statuettes, at least one of which (the one on the right) is regarded as a forgery.

"rebuilt" that palace using their knowledge, imagination, and a lot of reinforced concrete.

Commentators like the historian Cathy Gere have repeatedly pointed out that the "restored" Minoan palace came to resemble modernist works—Lenin's Mausoleum or Le Corbusier's Villa Savoy.[51] The novelist Evelyn Waugh noticed at the time that Evans and his collaborators blended contemporary styles into antique ones: they had "tempered their zeal for accurate reconstruction with a somewhat inappropriate predilection for covers of Vogue."[52] Over-interpretations and forged antiquities rounded out the invented world.[53] And this world, Evans insisted, was led by women (see Figure 6.2). The massacres of the Balkans and Crete in the 1890s, then the astonishing cruelties of the Balkan Wars of 1912–13 and World War I, led him to look at Knossos as a matriarchal paradise straight out of Bachofen and Morgan. Morgan had proposed that matrilineality had survived along with the Lycians in Crete—which worked just fine for Evans.[54] Rather than imagine his ancient city as the home of the Minotaur, Evans privileged Ariadne. In the process, he influenced artists from Pablo Picasso to Hilda Doolittle (H.D.) and replayed a fantasy of motherly serenity, without needing communism to complete the image. On the island now most distant from the European theaters of war and revolution, Mother-Right was dug up out of the earth and shared as a promise for the future. If primitive communism offered a past that could be re-created for a utopian future, artists could deposit into their dreamt-up Knossos what seemed most distant in modern life: peace, love, nonhierarchy.

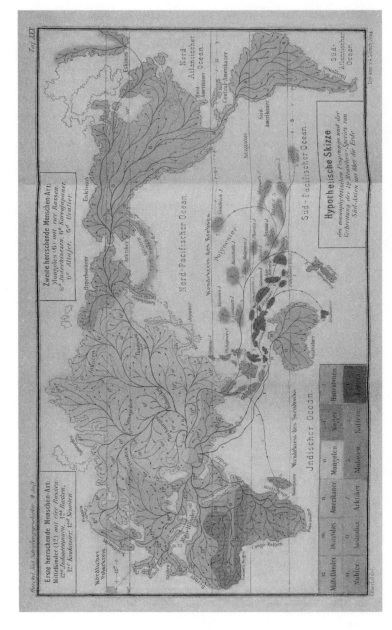

Figure 7.0. Where It All Began (according to Ernst Haeckel): "Hypothetical Sketch of the Monogenist Origin and Spread of the 12 Species of Man from South Asia to All over the Earth" (1889).

Chapter 7

THE DISAPPEARING NATIVE

In 1838, and not for the first time, the Aborigines' Protection Society in London reported with utter horror that Indigenous populations around the globe were declining dramatically. It noted the population collapse of the Khoikhoi ("Hottentots") in the Cape Colony from 200,000 to 32,000. It directly blamed the Colony for unrelenting violence "until the Hottentot nation were cut down and the small remnant left were reduced to abject bondage."[1] In its plea for a fundamental change in the British treatment of Indigenous peoples, the Society argued that the "perishing races of uncivilized man" posed a moral problem akin to slavery, and perhaps still worse. "Whilst we hesitate to plead their cause, they cease to exist, and we shall inquire after them in vain."[2]

European and American scientists, lawyers, politicians, all knew that Indigenous communities were under frequent, even constant, assault and expropriation, and that many had entered death spirals from which they could not escape. Not a decade had passed since the mass removal of Native American "Civilized Tribes" from the Southeast United States across the Trail of Tears, and expulsions were ongoing. No less notorious, already at the time, was the genocidal Black War in Tasmania (then called Van Diemen's Land), where in the course of a decade, white settlers choked off the food sources of the Aboriginal population, kidnapped women, and responded to native violence with escalating attacks of their own, culminating in the "Black Line" where, by decree of the then-governor of Van Diemen's Land, three lines of white settlers attempted to push all remaining Aboriginal peoples to the Tasmanian Peninsula. Fewer than a hundred of the hunted Aboriginal Tasmanians survived the 1830s.

These were the very same decades that slavery was being denounced by some as the height of immorality and was, however gradually, being abolished. By comparison, the dispossession and destruction of Indigenous populations—which the Society, for one, consistently linked to slavery—was largely ignored by the public. For the most part, Britain and the US treated Indigenous peoples as belonging to the past, as active threats to modernity.[3] Above all, they described Natives as "perishing," as "disappearing."

FOR HOW LONG HAVE PEOPLES BEEN "DISAPPEARING"? IN RECORDED history, at least since the Assyrians, who exiled many of those they conquered in order to break their social cohesion.[4] Many more surely since 1492, when Native peoples in the New World were "discovered" and war, disease, enslavement, and settler colonialism began. Over the following centuries, these practices destroyed, displaced, and disempowered communities around the world. The language of "the disappearance of the native" became a convenient euphemism. Natives don't die of diseases introduced by settlers, they're not murdered in asymmetrical warfare; really, they disappear. It's a "natural process"—not our fault. Disappearance connected "the romance and the racism," in terms at once rhetorical and emotional.[5] You could plead for the "disappearing" and yet rationalize that "disappearance." What was more, devices like "disappearance" repeated the damage wielded against nonstate peoples by producing distance at the intellectual level. To this day, we do not speak of peoples who are voluntarily isolated. We call them "undiscovered" tribes that are "endangered" and "vanishing."[6]

Because nineteenth-century intellectuals spoke about indigeneity as synonymous with human antiquity, they imagined disappearance as continuous with "missing links" from ape to human. The language of disappearance naturalized extinction and expropriation: it produced a continuum from the hominid skulls discovered throughout the nineteenth century to the active destruction playing out simultaneously. Naturalists and biologists—the Darwinians first among them—carried out an intellectual operation that shocked already in its own day. For the struggle for survival explained away colonial devastation as biologically inevitable. Some of their positions, including Darwin's, constituted a watershed in the rationalization of extermination.[7]

In *The Voyage of the Beagle*, Darwin reported from Sydney: "there appears to be some more mysterious agency generally at work. Wherever the European has trod, death seems to pursue the aboriginal." He had the Americas, Polynesia, Southern Africa, and Australia in mind. To Darwin, this "more mysterious agency" bringing death to Indigenous peoples was not exactly about Europeans: "The varieties of man seem to act on each other; in the same way as different species of animals—the stronger always extirpating the weaker."[8] Darwin attributed this reality mostly to disease and the changes in diet obliged by European control. Reporting from Van Diemen's Land, he bemoaned the forced removal of the Indigenous population, then blamed both Tasmania's Aboriginal peoples and the "infamous conduct of some of our countrymen." He did not recognize it as wholesale annihilation, but merely allowed that such cruelty was the way of the world.[9]

In his primary works, Darwin continued in the same vein. In *On the Origin of Species*, he compared "savage races" with "the various beings . . . left stranded" by rising tides—as if Indigenous peoples resembled the differentiated animal species he found on the Galapagos, and were left behind like Robinson Crusoes, isolated by oceans from the continents of civilization.[10] Central to Darwin's work and its reception was the concept of "struggle for existence," hardened in the mid-1860s by Herbert Spencer as "survival of the fittest." In the *Descent of Man*, Darwin argued that both mind and body emerged thanks to evolution. It followed that intellectual and moral faculties played a role in the advance of certain groups over others, their "fitness to survive," their control of their environment. Offhandedly Darwin included phrases like "numerous races have existed, and still exist"—extinction seemed normal to the life of humanity. He debated with himself the reasons for that "extinction." Extinction, he insisted, follows from racial competition and natural checks on the population: "periodical famines, the wandering of the parents and the consequent deaths of infants, prolonged suckling, the stealing of women, wars, accidents, sickness, licentiousness, especially infanticide, and, perhaps, lessened fertility from less nutritious food, and many hardships." Unbothered, he went on to say that any unequal competition would be "soon settled by war, slaughter, cannibalism, slavery, and absorption." And in a passage that rephrased Hobbes's dictum on life in the state of nature being "solitary, poor, nasty, brutish, and short," Darwin added: "When civilised nations come into contact with barbarians the struggle is short, except where a deadly climate gives its aid to the native race."[11] To present the "struggle for existence," Darwin relied on native extinction.

Three years later, in a second edition of the book, he added several pages to further illustrate this argument. He returned to the destruction of the Tasmanians. By then they had been "literally 'civilized off the face of the earth,'" as the anthropologist George W. Stocking, Jr. has put it. Because Europeans did not count mixed-race people as Tasmanian, they noted that only a handful of women had survived the Black War and removal, and they were well past childbearing age.[12] Darwin also noted the dramatic population reduction

of the Māori of New Zealand and even more so that of Hawaii (then
the Sandwich Islands) from a presumed 300,000 in 1778 when Captain James Cook landed to 132,000 in 1832, when a first census was
taken, and then to 51,500 in 1872. He distributed blame among colo-
nists and colonized (taking time to reproach "profligate women," a
classic if nonsensical target). He eventually decided that "the wilder
races of man are apt to suffer much in health when subjected to
changed conditions or habits of life" and they are "in this respect
almost as susceptible as his nearest allies, the anthropoid apes."[13]
Read alongside some strongly eugenicist passages in the *Descent*, the
argument was clear: "savage races" were undomesticated leftovers,
fossils who had failed to transform and therefore survive and master the pressures of nature. Their fate might be heartbreaking but it
was altogether unsurprising. They were passé, the last "living representatives of the early Stone Age."[14] Alfred Russel Wallace, usually
credited with discovering natural selection independently of Darwin,
went further still. In "the future of the human race," he concluded,
natural selection meant that "the higher—the more intellectual and
moral—*must* displace the lower and more degraded races."[15] When
pressed on the point in front of the Anthropological Society of London, he continued: "the mere fact of one race supplanting another
proves their superiority . . . that two races came into contact, and
that one drives out the other. This is a proof that the one race is better fitted to live upon the world than the other."[16] The French biologist Armand de Quatrefages, in his book considering *Fossil Men
and Savage Men*, spent two chapters and over a hundred pages on
Tasmania and the Black War.[17] Tasmanians, he pretended, finally
"disappeared" because of tuberculosis. They were vestiges for him,
human ruins: the intermediary between the ancient species whose
fossils were being recovered and other peoples now "disappearing."
What had happened to the extinct ancient humans with the strange
skulls was "taking place in our days, under our eyes," among Pygmy
peoples of the Congo basin, "a relatively feeble race" now in "one

of the last scenes of a drama whose earlier acts go far back into the past."[18] In both cases, their milieu killed them; they failed to be the "fittest." The same fate, he predicted, would soon befall the people of the Andaman Islands, Tahiti, New Zealand, and others.[19] When Thomas Henry Huxley looked to Haiti, he outlined extinction as an effect of enslavement.[20] However sad, it confirmed that the strong destroyed the weak.

Did Darwin and his notable followers, most of whom became committed to eugenics, simply know no better? No: first of all they knew well from Thomas Malthus that "war and extermination" were crucial checks on population growth. Darwin, like Wallace, relied on Malthus to understand competition between populations. But Malthus had been troubled by the moral implications of colonization.[21] Darwin & Co. wrote a half-century after Malthus, by which time they regarded colonization and its effects to be a simple reality.

Why then was slavery abhorrent and extinction normal? Slavery was a social imposition, therefore immoral. But if war and disease had resulted in extinction throughout human history, then fault lay with nature. Humanity's first duty was indeed to survive and improve. This included peoples in Europe who had been rendered extinct by the ancient "Germans" or, earlier still, by Indo-European conquerors. Cruel as contemporary extinction was, it entailed no surprises: the Indigenous peoples living today were quite exactly living vestiges of the Stone Age. Even the term "fossil men," which was quite common in French, carried within it a paradox: though alive, these Indigenous humans were fossils. Societies had to be made responsible for their own future, and this was a matter of eugenics, not just an issue to be resolved through politics and economics.

FOR AGENTS OF EUROPEAN STATES—WHETHER AT HOME OR STA-tioned in the colonies—extinction figured quite differently. Recall that a century earlier, the "state of nature" concept had helped to justify the idea that "savages"/"natural men" know little about property—which meant expropriation was fair play. In the nineteenth century, the language of disappearance offered a rhetoric for usurping and managing Native lands. If the inhabitants were "disappearing," then the territory was as good as empty. What was more, the British in West Africa, the French in Indochina, the Americans on their continent consistently treated Native peoples as decidedly cut off from modernity—as committed to corporal violence, human sacrifice, polygamy, fetishisms, and so on. To assimilate, to modernize, to bring the rule of law was to ruin these societies and to redeem their people—to *make* them disappear for being backward, brutal.

It is worth noting that the destruction of Indigenous peoples from the Americas to the Caucasus to Tasmania also shocked some commentators. Missionaries and anthropologists, in particular, wrote powerful, outraged texts decrying forced disappearance. To some, defending nonstate peoples became the new abolitionism. Generally, though, what marked them apart was how few of them there were. International law was supposed to carry the moral force of humanity, to offer something superior to imperialism and ugly state power. The Swiss jurist Joseph Hornung maintained *both* that those who had fallen behind in the advance of progress were degenerating vestiges *and also* that protections must be extended to them. If populations lacking sovereign statehood were unprotected, the whole dream of law and justice was lost. In a series of articles in 1885, he began with the example of Russia's mass killing and expulsion of the Circassians. He then listed atrocity after atrocity perpetrated by empires. The inability of "civilization" to stop itself from committing massive violence against those it did not recognize was a deep embarrassment to the law.[22] The French jurist Charles Solomon made the point more fervently: colonization "begins with violence, injustice and shedding of blood: the result is everywhere the same; the disappearance of the native races coming into contact with civilized races."[23] The expression is as clear a rejection of Darwin's "the struggle is short" as any.[24]

Still, it was mere theory and critique: colonial wars and "civilizing missions" continued unabated.

For anthropologists, disappearance became a real obsession, even a justification for their science. Their stories were all about what it means to consign peoples to the past, what it meant for "primitives" to be alive today, what it meant to imagine policies to manage them. The tears they shed on behalf of Native peoples sometimes contributed to state policies and sometimes pioneered ways to fight them. Lewis Henry Morgan, for example, understood his anthropological work as extending his legal and political activism on behalf of Native Americans. He had begun his career lobbying against the plunder of Seneca land by the Ogden Land Company, and he continued his advocacy right through the Indian Wars in the American West and the Civil War. He traveled to Kansas and Nebraska in 1859–60, Fort Garry (now Winnipeg) in 1861, and Missouri in 1862 for ethnographic work. When he published his magnum opus *Ancient Society* in 1877, he was acutely aware that the Great Sioux War was ongoing.

Right from its opening pages, Morgan declared that, as a result of dispossession, his project had become urgent. "When discovered," Morgan informed the reader, Native Americans had existed in, well, three distinct stages: savagery, barbarism, civilization.[25] His science was new, and only "feebly prosecuted among us at the present time, the workmen have been unequal to the work." Native Americans "are perishing daily, and have been perishing for upwards of three centuries. The ethnic life of the Indian tribes is declining under the influence of American civilization, their arts and languages are disappearing, and their institutions are dissolving."[26] Note the passivity with which Morgan presents Native Americans: they were "perishing . . . perishing . . . declining . . . dissolving." Peoples were "disappearing" in much the same way that matrilineal society had vanished into the mists of the past. He claimed that ending the disappearance was the raison d'être of his writing.

Morgan was hardly alone: German ethnologists like Adolf Bastian and Georg Gerland used the same language to describe their aims. To the next generation of anthropologists, working around the turn of the century, Native peoples' "disappearance" became their field, and often a rallying cry. Many shared Morgan's ostensible priorities: activism and advocacy on behalf of those who were "disappearing." For the most part, they did not recognize that when they declared peoples to be disappearing, states actually benefited— that states could now be free to pursue expropriative policies, free to claim that they were "preserving" cultures in museums, free to treat land as empty. Instead, to collect images of "a world whose culture is dying out" was, as the German-Jewish collector and art historian Aby Warburg put it, a way to reject capitalism's destruction of the intimacy of the cosmos.[27] We must defend mythical and symbolic thinking, Warburg urged, or else all chance for reflection and care would disappear with them.[28] Human existence would be worthless, the French art theorist and ethnographer Victor Segalen agreed. The destruction of Indigenous peoples led grotesquely to the erasure of all human difference, to "the Kingdom of the Lukewarm; that moment of viscous mush without inequalities, falls, or reboundings."[29]

Warburg's and Segalen's worries were existential: civilization was crushing something fundamentally, universally human. Others were more prosaic. Activists routinely repeated Morgan's words about civilization's destructiveness so as to better advocate on behalf of particular Indigenous peoples. Museums tried to "salvage" and house what was left. For them, as for many anthropologists, "disappearance" became their rationale or their excuse. Against some forms of dispossession and violence, they signaled their humanitarianism and they promoted an engagement that often amounted to a different form of dispossession.

As I noted earlier, concepts tend to escape their human designers and the institutions meant to house them. Such was the case with "disappearance." Writers could use it to refer to the misery of a people or else the purity of others. Some writers lacked any interest whatsoever in Indigenous peoples; others signaled how appalled they were by state practices and how much these peoples needed to be protected from civilization itself.

Among the most forceful activists was the German-Jewish anthropologist Franz Boas, the first anthropologist with a university chair in the US (at Columbia University), who twisted the motif of disappearance around in his *The Mind of Primitive Man* (1911). Boas found "disappearance" useful as a means to target theories of blood purity and racial domination. He refused the biologists' "ill-fitting analogies with the animal and plant world," and insisted that intermixture between "races" was ever-present. Racial purity and white supremacy theories—for which he blamed Darwinians as well—were nonsense, and so was disappearance.[30]

The British psychologist and ethnographer W. H. R. Rivers went further. Rivers had been a member of the Cambridge Torres Strait Expedition of 1899, an important early attempt at ethnographic fieldwork. In his *History of Melanesian Society* (1914) he remarked on a recent, devastating volcanic eruption. It paled, he emphasized, in comparison to the damage wrought by European "civilization" on Melanesian Natives.[31] In 1922, when Rivers ran with the Labour Party for Parliament, he equated the British Empire with "depopulation." "The decline is taking place so rapidly that at no distant date the islands will wholly lose their native inhabitants unless something is done to stay its progress."[32]

BUT WHAT COULD BE DONE? STATES WERE NOT ABOUT TO ENDORSE Indigenous claims. By proposing to protect what was left, anthropologists imagined saving the past, the world of the extinct. They usually meant artifacts, bodies, skeletons, even living people, all of which museums housed, often with zero regard for their real significance or meaning. Today, "salvage anthropology" rightly has a dirty reputation—for the pilfering it enabled so that state institutions could store (and own) the disappearing past, but also for its humanitarian pretense. Take one example, hardly the worst: the most famous of the individual people studied in the US was Ishi, meaning "man," the "last" of one of the Yahi tribes that had been massacred by white settler militias during the California gold rush. Brought by anthropologists to live at the Museum of Anthropology in Parnassus, he performed for them (and for the public) traditional practices—building fires, making bows—before succumbing to tuberculosis and being dissected, his brain ending up at the Smithsonian for several decades. How much does it matter that Alfred Kroeber (Boas's former student), who studied and worked with Ishi to preserve something of the cultural traditions of the Yahi, also became his friend, his host and advocate? Is this a story of the awful effect of good intentions?

In the first decades of the century the trope of disappearance did more than any other to convince the public to repeat that Indians suffered because they were paleolithic men from a bygone age. The aim of studying "primitives" was, in part, the light they shone on humanity's origin.[33] And yet, colonialism and industrial capitalism were destroying these "fossil men." Had European and North American audiences become so greedy about knowing, owning, and managing the world that they were ready to consume even guilt over its destruction?

"DISAPPEARANCE," BESIDES "INNOCUOUSLY" RECORDING A REALity, began to figure into state affairs. It allowed anthropologists to believe that they were real scientists and not drivers of colonial violence. Regardless of how they imagined themselves, the scholars who deployed it ended up strange bedfellows with others who saw it as an unalloyed, Darwinian good.

In 1914, the Polish-born, Vienna-trained, London-based anthropologist Bronisław Malinowski arrived in Australia from Britain. When World War I broke out, he promptly found that his Austrian citizenship made him an enemy alien. He opted not to return and fight nor to be interned, and spent the war years first in New Guinea and then in the Trobriand Islands, where he studied the economics of the kula ring. He finally published his Trobriands research eight years later in his magnum opus *Argonauts of the Western Pacific*. Malinowski's book would become as famous as Boas's and Morgan's. For sixty years, it was taught as *the* blueprint of anthropological fieldwork. And in it, Malinowski urged his reader to make a profound attempt to "understand the native's point of view," not just to study a people but to feel, intimately, "by what these people live" and to understand "the substance of their happiness." Yet, even for Malinowski, "tragedy" lay not in the destruction of Native peoples but in the predicament of ethnology, for which "the material of its study melts away with hopeless rapidity. Just now, when the methods and aims of scientific field ethnology have taken shape, when men fully trained for the work have begun to travel into savage countries and study their inhabitants these die away under our very eyes."[34]

In another text he also published in 1922, Malinowski made clear the consequences: anthropology was all "for the future scientific guidance of human affairs." He listed the worst recent and ongoing crimes of the colonial powers and the euphemisms they used: "slavery, 'dispersing of natives,' indiscriminate 'punitive expeditions,' and the grossest forms of reckless injustice to natives . . . the opium war on China, the Australian 'blackbirding' in the South Seas, the even more recent 'transplanting' of the Hereros in German South-West Africa and the Belgian and French atrocities in Africa." In their place he advocated anthropology, that same tragic discipline, which offered "a *scientific* management of natives or native affairs."[35] He italicized scientific, for the "management of natives" was ongoing; the point was to turn scientists into state agents. Anthropology appeared the ideal science of control: Hire anthropologists to best manage your natives! It was at once resigned to the reality of biological and cultural destruction and morally appalled by it.

Very real villains spoke the same language.[36] In a notorious 1932 speech before an audience of two thousand people, "Climate and Man in Africa," Jan Smuts, then-former Prime Minister of South Africa (and perhaps the most important figure in the development of apartheid) presented the San people as a vestige, "a mere human fossil, verging to[ward] extinction." Smuts fashioned himself a scientist, and he followed a by-now longstanding tradition that saw the San as almost animalistic nomadic marauders.[37] Unlike Europeans who had dominated nature, "bushmen" were incapable of controlling their desert milieu.[38] They should be protected, he continued, for the purposes of science. To Smuts, they should form part of South Africa's great zoo.

By 1930, eugenic policies were explicit in their ambition to get rid of the vestiges of "Paleolithic Man." Some anthropologists spent their time measuring skulls and celebrating racial norms. Just down the hall from them worked biologists, few of whom saw something wrong in nature "taking its course." Other anthropologists objected, often loudly, to eugenics and to the biologists' indifference. They sought to "protect" the most downtrodden. Well-meant and noble as this aim seemed, it also led them to a rather gruesome competition. Who could find and study the most wretched savages on earth? The most primitive of primitives, the most uncultured and uncouth and desolate, those who could offer the best lessons about human nature, those who would die out soonest?

Some writers, even politically progressive ones, continued to talk about the "disappeared" and "disappearing" Native peoples in the same past tense as "ancient races" or "species."[39] Other accounts of disappearance recognized the stark reality: Indigenous peoples were being actively destroyed, often haphazardly, sometimes systematically. But these anthropologists' writings created a pattern: shock at the desolate condition of people pure of civilization, outrage and sometimes acquiescence toward their genocide, melancholy clarion calls about the failures of humanitarianism, appeals for better anthropological "care" or "management" (which often entailed just-as-awful policies). This continued throughout the century. *The Last of the Mohicans* (James Fenimore Cooper's novel) was made into a film or a TV series no fewer than twelve times before Daniel Day-Lewis acted him out yet again in 1992. Disappearance remained an obsession and a metaphor for primitiveness.

Let's pause in 1959, when Claude Lévi-Strauss updated disappearance for the Space Age. Lévi-Strauss had long struggled with the fact of extinction: in 1936 he had written that Brazil's Kaduvéo people were in "complete decomposition"—"ten years from now, they will have completely disappeared."[40] In 1952 he proposed that anthropology be renamed "entropology," a name appropriate to its object of study: the disintegration of peoples. And he wondered about the anthropologist's compulsion to spend energy, money, time, and even health in order to make himself "acceptable to a score or two of miserable creatures doomed to early extinction."[41] Mimicking Victor Segalen, he was arguing that without profound differences in ways of being, humanity was no humanity at all. Now, in 1959, he asked his readers to imagine an unknown planet moving through space and approaching the earth. It would hover nearby for a while, then disappear. How much treasure would go, he speculated, into studying it? The study of Native peoples was more urgent. "Native cultures are disintegrating faster than radioactive bodies." In a world disappointed by politics and any attempts at utopia, anthropology carried the only humanism possible, but that humanism depended on the celebration of other peoples, cultures, lives "which will be lost and gone forever."[42] His sense of purpose retained nonetheless one thing from Malinowski: disappearance was the last act in a tragedy, one impossible to stop, however hard it was to accept.

Figure 8.0. "Neanderthal Man." Detail of an abbreviated version of Rudolph Zallinger's "The March of Progress" (1964).

Chapter 8

Neanderthals,
"Our Doubles"

Neanderthal stands third from the right in "The Road to Homo Sapiens," Rudolf Zallinger's design for the centerfold of the 1965 *Early Man* volume of the *Life Nature Library* (Figure 8.1). Both *Homo erectus* and Cro-Magnon Man steal an upward-sideways glance at us, as if to remind us that they are our direct ancestors. But the Neanderthal male looks mutely on. He stands shorter than the "Early Sapiens" who precedes him. He hunches. He holds neither weapon nor tool. There is no promise to him, only a finality to how he puts down his foot. Further to his right, Modern Man marches on.

By this point, "Neanderthals" had already been a subject of discussion for a century. Since 1857, the scattered, mute ruins that Neanderthals left behind have been taken up by scientists and the public, who have spoken for them, classified them, given them meaning and, time and again, a thoroughly modern image and voice. Pliable, unsettling, endlessly recyclable, Neanderthals offer the most startling of mirrors—an image of what "we" might have been and yet are not. Even today, as we learn more about their way of life, as our scientific methods improve, what they mean is still decided in

Figure 8.1. Rudolph Zallinger's enormously influential "The March of Progress" (1964), first published in F. Clark Howell and the Editors of Time-Life Books, *Early Man* (1965).

advance by the concepts and biases they are made to embody. One can make them do anything: a recent article in *Scientific American* even blamed "trysts" between Neanderthals and *sapiens* for depression, obesity, and nicotine addiction.[1]

The Neanderthal has two faces. On one is written a racial and colonial fantasy that accentuates their distance from humanity. The idea of a Neanderthal fall, to be contrasted with modern European grace, began in the 1860s and peaked between the 1910s and 1960s. It is still alive. Tortured rationales have drawn meaning out of imagined Neanderthal inferiority and from their extinction. The second face came into view through spurts of sympathy. It features Neanderthals as postcolonial subalterns whose voice and humanity wait to be freed. Its advocates stress Neanderthals' intelligence, their kindliness, their affinity with nature, and their resemblance to "us." Over the last two decades, the complexity of Neanderthal life has come into sharp relief. For now, the anticolonial Neanderthal has won the day—though he is also becoming a darling of the extreme Right.

THE NEANDERTHAL WAS BORN DURING AN 1857 DEBATE THAT resulted from a chance discovery in a quarry near Düsseldorf. A set of bones and the crown of a cranium had been found and shared.

HOMO ERECTUS   EARLY HOMO SAPIENS   SOLO MAN   RHODESIAN MAN   NEANDERTHAL MAN   CRO-MAGNON MAN   MODERN MAN

Ironies abounded. The Neander Valley, where the fossils were found, and after which they were christened, was named for the Germanized Greek term for "New Man." The bones weren't even the first to be found: other skulls had been dug up, notably in 1848 in Gibraltar, but also in Minsk in Russia, in Mecklenburg in Germany, in Denmark, in Engis in Belgium. But their similarities had gone unnoticed. The last of these, discovered in 1833, would only be confirmed as a Neanderthal child after a whole century, in 1936.

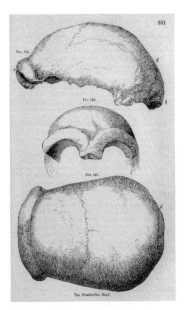

Figure 8.2. John Lubbock's representation of the Neanderthal skull (1865). Lubbock's book was among the first in which the Neanderthal dome appeared; this drawing became ubiquitous in publications about prehistory.

The first Neanderthal debate, in Germany, pitted Hermann Schaaffhausen against Rudolph Virchow on the question of whether the cranium from the Neander Valley was something new or merely a diseased specimen. Virchow was as famous for his major contributions to cell theory as for his temper and hatreds. When challenged by Chancellor Bismarck to a duel, he supposedly countered that they spar by eating two sausages, one of them poisoned with trichinella. Virchow vigorously disparaged evolutionary theory. And as head of the Berlin Society for Anthropology, Ethnology and Prehistory, he pronounced the new cranium a mere pathological aberration: its owner "suffered from two separate head injuries, one of which led to decay of the bone. . . . An atrophied brain . . . led to the growth of additional bone matter around the forehead and fusion of the sutures, which further deformed the skull."[2] Rickets and severe arthritis completed the diagnostic picture: Virchow remained unconvinced that this skeleton was meaningful. Schaaffhausen thought otherwise. He identified, based on the skull fragment, an entire aboriginal people who had inhabited Europe long before "the *Germani*." He compared the dome favorably to Seminole, Peruvians, and "other races," then consulted with the classics, including Tacitus and Julius Caesar, to ask how the skull recast knowledge of the ancient Germans.[3] This was a new being, not merely a pathological case (see Figure 8.3).

Over in Britain—which was then in the throes of debate over Darwin's *On the Origin of Species*—the key question concerned race. Was this a distinct species or an extinct, "disappeared," race? Geologist Charles Lyell included the cranium in his book on *The*

Figure 8.3. Early representation of a Neanderthal, in Hermann Schaafhausen, *Der Neandertaler Fund* (1888).

Geological Evidences for the Antiquity of Man (1863), and admired how "it departs so widely from the normal standard of humanity."[4] He and his colleague Thomas Henry Huxley—and seemingly half of London with them—wondered whether the skeleton was to be treated as "a distinct race of mankind."[5] The Neanderthal skull fascinated them, as did the Engis skulls (one sapiens, one Neanderthal). The Neanderthal one was "the most brutal of all known human skulls, resembling those of the apes." Yet, despite its "pithecoid bony walls," it was in size "very nearly halfway" between a "European" skull and a "Hindoo" one, and "far above" the gorilla maximum.[6] Elsewhere, Lyell decided it was barely inferior to "the Australian" skull (see Figure 8.4). Eventually they linked the skull to the one from Denmark and wondered if the Neanderthal had been "a race allied to the Borreby people" of the Danish Stone Age.[7] Meanwhile, Armand de Quatrefages in Paris generalized Neanderthal into a broader "race of Canstadt" whose characteristics he identified among Australian Aboriginal peoples, Khoikhoi, and some Europeans.[8]

A year later, the Gibraltar skull was recategorized as Nean-

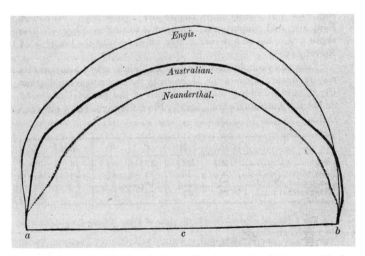

Figure 8.4. To estimate the Neanderthal skull's relationship to living races, Charles Lyell compared the Neanderthal dome to the "Engis skull" (now recognized as *Homo sapiens*) and to that of an Australian Aboriginal man (1863).

derthal, and William King, an English/Irish geologist, proposed to name the species *Homo neanderthalensis*. "Considering that the Neanderthal skull is eminently simian," he felt "constrained to believe that the thoughts and desires, which once dwelt within it, never soared beyond those of the brute." An elevated cranial dome was essential for true humanity, so King compared the Neanderthal unfavorably to the Andamanese people, his choice for "most backward" people. (British colonial subjects at the time, the Andamanese were being wiped out.) Their "brute benightedness" and "dimmest conceptions of the existence of the Creator" placed them, for King, alongside the Neanderthals: "very little above animals of marked sagacity."[9] Schaaffhausen, Lyell, and Huxley also positioned the cranium on a continuum that spanned from the supposed high dome of "modern Europeans" to their choice among the "least evolved races."

Figure 8.5. "The Neanderthal Man" as desperate caveman, in *Harper's Weekly* (1873). Right above this drawing, *Harper's* reproduced the representation of the skull from Lubbock's *Pre-historic Times* (Figure 8.2).

It matters that the British, French, and Germans were not alto-gether clear on the border between "race" and species. Could a being as ambiguous as the Neanderthal be still within the domain of the human, as yet another extinct race? On what grounds was it a truly *other* being? Thanks largely to Huxley, the British decided that the newcomer could be glimpsed just beyond the colonized and "least evolved." If there had really existed other hominids and miss-ing links, this was most likely one of them, but it could also be just another vanished race.

IN THE 1870S EMERGED THE FIRST POPULAR DEPICTIONS OF NEAN-derthal man, and this figure did not differ much from the then-modish club-wielding "troglodyte" or caveman. In America, *Harper's Weekly* featured him in situ, in profile, barrel-chested, with rippling muscles, and disturbingly similar to representations of the colonized Anda-manese (see Figure 8.5).[10] He seemed sad, looking uncomfortably upward toward the light outside his cave, surrounded by his dogs, stone implements, victims' bones, and a reclining half-naked Cave-woman with her face obscured. Domestication had become a matter of debate, partly thanks to Darwin's *Variation under Domestication* (1868), and this pitiful Neanderthal was depicted as having already managed it. Still, the family romance here is grim: in the absence of children, the puppies stand in for progeny and signal that the couple has no future. Ernst Haeckel had proposed for Neanderthals the name *Homo stupidus*. The term didn't stick but the meaning—speechless, dumb, incapable of coping—did.[11]

The pace of skeletal discovery picked up after 1880. In 1908, in a cave near La-Chapelle-aux-Saints in France, one finding finally outshone the original cranium: the relatively complete skeleton of a Neanderthal man was discovered in clay and loam. The new skeleton was soon celebrated in a publication by Marcellin Boule, the leading

French geologist and anthropologist of his time.[12] Judged at
first to be fifty to fifty-five years
old, this Neanderthal was baptized the "Old Man." The name
might strike us as more appropriate than "New Man," but the
choice rested on the assumption
that hunter-gatherers and early
hominids usually died young—
that grandparents and old age
are recent developments.[13] Boule
had more than enough to show
that this skeleton was definitively
different from *Homo sapiens*,
and not a racial or mere pathological aberration. But he also
played down signs of arthritis and other pathologies—
including fused vertebrae—in

Figure 8.6. Paul Jamin, *Flight from the Mammoth* (1885). Jamin painted several "scenes" from prehistory (see Figures 9.2 and 16.5); this painting casts the cavemen/ Neanderthals as incapable of coping with the stomping power of the mammoth.

his specimen so that when he described the Neanderthal's normal
posture, it seemed badly hunched.[14] Boule thus helped establish the
famous look of Neanderthals that would persist for a century.[15] Just
as significantly, he declared that the skeleton had been intentionally
buried, and hence that Neanderthals had rituals, a religious sensibility, and clearly human intelligence.[16]

Neanderthals now became sensational. Decoupled from the
caveman image, they were regularly compared to *sapiens*. These
were the Piltdown Man years: the discovery/forgery of the Piltdown
fake in 1912 only helped harden the Neanderthal-*sapiens* divide. As
Piltdown had been declared a human ancestor, Neanderthals now
seemed very far from the precursors of "modern" humans. Boule,
despite doubting the Piltdown mandible, pointed to "inferior traits"

and even "degeneration" in the Neanderthal, and described a separate family of hominids rather than a direct line of descent.[17] Neanderthals had "primitive characteristics" in common with Australian
aborigines, he insisted, which had generally disappeared "*chez
nous*."[18] (Others might not sign on to these descriptions but rode their
own hobbyhorses. Arthur Keith in Britain thought the "Old Man"
was closer to forty years old but detected racial distinctions among
Neanderthals.[19]) Boule also helped introduce the idea that Neanderthals had been exterminated by *sapiens*, rhetorically asking "was
there a mere displacement, a migration, or on-the-spot extinction?"[20]

Alongside the "Old Man" appeared a slew of new visual depictions. In Brussels between 1909 and 1914, Louis Mascré sculpted
away for Belgium's Royal Academy of Science, creating a plaster
cranium, a mother and child, the upper body of an elder. At a time
when Belgium was brutalizing the Congo, the country's Royal Academy compared Africans to apes and also depicted Neanderthals as
simian rather than caveman-like (see Figure 8.7).[21] In Paris, Boule
spoke with the well-known primitivist (and
later cubist) painter, František Kupka, who
drew a massively-built, hunched, super-
latively hairy, club-wielding aggressor
(see Figure 8.8). The drawing was published in *L'Illustration* in Paris and
in the *Illustrated London News* in
1909, somewhat to Boule's embarrassment. Its racist implications
were hard to miss.[22] In response
to this crudeness, Arthur
Keith and *Illustrated London News* commissioned a
1911 work by Amédée For-
estier, in which a nature's-

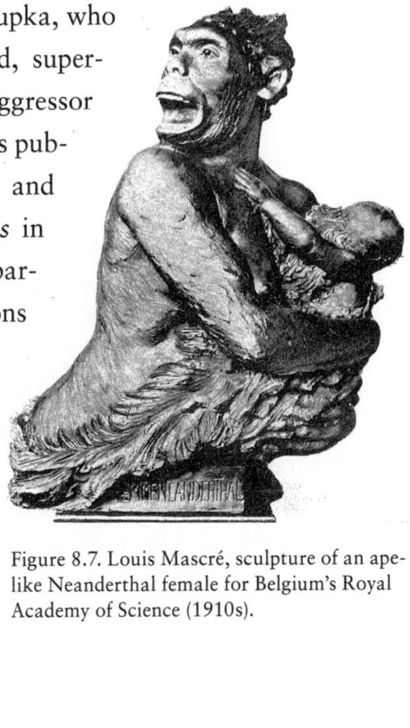

Figure 8.7. Louis Mascré, sculpture of an ape-
like Neanderthal female for Belgium's Royal
Academy of Science (1910s).

Figure 8.8. For a long time, this was perhaps the most influential representation of a Neanderthal. František Kupka, "An Ancestor: The Man of Twenty Thousand Years Ago. The Man of La Chapelle-Aux-Saints." *Illustrated London News* (1909).

Figure 8.9. Amédée Forestier's "gentle" Neanderthal tool-maker in 1911, was supposed to debunk Kupka's "Ancestor" (Figure 8.8).

child Neanderthal became the tool-maker of "500,000 years ago" (see Figure 8.9).

Natural history museums debuted Neanderthals along these lines. The Paris Muséum National d'Histoire Naturelle opted for a wise-looking druid-like bust with a well-tended beard, while the Italian criminologist and race theorist Cesare Lombroso commissioned a ferocious bust for his museum in Turin (see Figure 8.10). Neanderthals also entered the history books. Another angry profile glowered across the page in H. G. Wells's *The Outline of History* (1919). Wells's *Outline* was an attempt to tell a comprehensive history of humanity, to show grandeur and violence, to advocate for socialism and world-government as humanity's great goal. The book sold over two million copies and was widely translated, which probably means that its angry Neanderthal was the most widely seen image of the period.

These depictions were talking to one another. Forestier's objected to Kupka's. Kupka's clearly inspired the one in Lombroso's museum and the one in Wells's book (right down to the jutting teeth and the tiny nasty eyes inside the big sockets, see Figure

Figure 8.10 (*left*). Norberto Montecucco's bust of a Neanderthal (ca.1909), designed at the request of the Italian criminologist Cesare Lombroso. Figure 8.11 (*right*). "Neanderthal Man" in H. G. Wells, *The Outline of History* (1920).

8.11). Although the violent "Abominable Snowman"–style figures blatantly play out the "savage beneath the thin veneer of civilization" metaphor prevalent at the time, the ones at the kindlier, thoughtful, nature's-frightened-child end of the spectrum were also rooted in the human-animal distinction. Somewhere in between the two options we find the works of Charles R. Knight, an American paleoartist who worked often for *National Geographic* and for the American Museum of Natural History in New York, where he painted several great murals. He began depicting Neanderthal life around 1920. His paintings retain some of the caveman aesthetic: his Neanderthals are strategically half-naked, invariably set against forbidding rock overlays, abandoned to the elements. They wear bundled leopard rags and hold only basic tools and weapons, particularly spears, stones, and bones. They—including a few cowering females—are generally found in a startled defensive posture, always responding to aggression, their hair often whooshed back

Figure 8.12. Charles R. Knight, *Neanderthal Flintworkers, Le Moustier Cavern, Dordogne, France* (1920). Painted for the Hall of the Age of Man at the American Museum of Natural History.

by a strong wind. Knight depicted Neanderthals almost entirely in three-quarters or full profile—perhaps in an effort to avoid granting a more complex humanity to the nose and long face. Profile was more effective for conveying the effect of living alone, brutishly, endlessly surprised by an empty, hostile world—terms consistently replayed in then-contemporary writings.

Images began to define the Neanderthal and not merely decorate it. A 1928 article by Grafton Elliot Smith in *Scientific American*, arguing against the Bohemian-born American anthropologist Aleš Hrdlička's claim that Neanderthal was "our ancestor," used Knight's painting to indicate Neanderthal life.[23] Smith denied any connection between Australian aboriginal peoples and Neanderthals and stressed that the difference was of species and not race; he also

Figure 8.13. Field Museum of Natural History (Chicago): Restoration of a Neanderthal boy by Frederick Blaschke (1927). The Field Museum's diorama featured two males, a half-naked female carrying a child, this boy, and another female skinning a deer. The diorama heightened the distance between the *sapiens* viewers and the disconnected "Neanderthals" on view.

fed right back into Knight's and Kupka's imagery with diagrams of
the Neanderthal neck, proclaiming a strong inclination of the sixth
vertebra.[24] With each work, the Neanderthal slid further down the
slippery slope to subhumanity: in 1929, the Chicago Field Museum
built a diorama of an especially dim, static, heavy, bone-sucking
Neanderthal family (including the boy in Figure 8.13). It stayed on
view for decades.

The Neanderthal/*sapiens* estrangement had everything to do with
modern European views and very little with the actual bones and skulls
that had been found. Discussions of "disappearing natives," natural
selection, racial purity, and the "human minimum" created specific
constraints for interpreting the Neanderthal.[25] Imputed Neanderthal
traits could be combined in a variety of ways: select among the options
of intelligence-vs.-stupidity and then combine them with inferiority-
vs.-proximity, with disappearance-slaughter-interbreeding, with colonized or aboriginal analogies. The result was an echo chamber where
multiple stereotypes were available, where they supported other ideas,
including the Darwinian struggle for survival. In the highly racialized discourse of the late imperial period, Neanderthals served as met-
onyms for colonial subjects, for Europeans of a past that had been
overcome, for proof of white success in a "struggle for existence."

So, for example, in *The Outline of History*, Wells discussed
whether *sapiens* and the "Neanderthal race of pre-men" interbred,
citing the influential botanist, adventurer, and colonial official Harry
Johnston.[26] Johnston pronounced colonialism the natural state of
human affairs: "man has been the greatest of colonizers for something like half a million years." Privileging Kupka's brute over Forestier's "thinker," Johnston presented their fate as the consequence
of colonization, sex, and war-to-the-death:

> Neanderthaloids developed large brains, but they must have
> been hideous-looking savages, with a bowed, shambling gait,
> short-necked, pulled-back heads, eyes glowing under project-

ing, bristling brows, long arms, and no doubt hairy bodies. Probably it was war to the death between the two species, though it is not impossible that the women of *Primigenius* were sometimes captured and taken to wife by *Sapiens*, and that mixed races arose through the mingling of the two species.[27]

A Darwinian struggle for existence, replete with colonial wife-capture rhetoric like John F. McLennan's, led Johnston to harbor "no doubt" that interbreeding produced "persistent types which have been great colonisers."[28] For his part, Wells demurred on the subject, and declared Neanderthals "on a different line of development." Then, perhaps to tamp down the racial language of Johnston and others, he pointed to "neanderthaloid" features in living populations, a solution that did not escape racial connotations.

In subsequent decades, notions of interbreeding between *sapiens* and Neanderthals sounded the bells of racial sex panic and fed into ideas about colonial and geopolitical hierarchy. They also played into eugenic dreams about the past and future of humanity, partly thanks to the intervention of Julian Huxley, who collaborated with Wells and Wells's son, G. P. Wells, on a massive, multivolume project entitled *The Science of Life* (1929–31). Huxley also coauthored *We Europeans: A Survey of "Racial" Problems* (1936) with Alfred Cort Haddon, a Cambridge anthropologist of race. These works were at once critical of race theory (Huxley called race a myth and preferred the term "ethnic group") and committed to eugenic ideas about the vague improvement of humanity.[29] In both works, interbreeding between sapiens and Neanderthals was presented as unlikely, "pure guesswork," or lacking evidence.[30]

In the US, by contrast, anthropologists like Earnest Hooton, Aleš Hrdlička, and Carleton Coon argued that intermixture had taken place, and suggested that the Neanderthal was a direct human ancestor.[31] Coon, who later came to represent the old and racist school in anthropology, used the Neanderthal in 1939 to claim that

Figure 8.14. Carleton Coon,
"Neanderthal Man in Modern
Dress" (1939).

hybrids had contributed to European racial formation. Could anyone really tell apart a shaved Neanderthal in suit and checkered tie?
he asked.[32] The illustration that he appended (Figure 8.14) hinted
no. But was Coon really so concerned about the survival of Neanderthal features, or about seeing African Americans in suits and
checkered ties?

AFTER WORLD WAR II, ALTHOUGH THE US-VERSUS-THEM LOGIC PER-
sisted, the border between *sapiens* and Neanderthal became haz-
ier, and the imperial angle less evocative.[33] The French philosopher
Georges Bataille credited Neanderthals with all sorts of achievements, including toolmaking, fire, stone-carving. Only art set *sapiens* apart, he mused.[34] So, he wondered aloud: did Neanderthal's
appearance "cause the same horror in these men that the sight of a
monkey induces in us?"[35] His contemporary, the French prehistorian and anthropologist André Leroi-Gourhan, complained that we

unjustly "disown the Neanderthal man with the flattened cranium and big jawbones just as the Lord of high lineage pretended not to know that the grandfather of the first Baron tilled the soil in the time of the Merovingians."[36] Neanderthal/*sapiens* hybrids lost their contentiousness. A 1958 state-of-the-field article in *Scientific American* declared that unlike *Homo sapiens*—who, it was now being recognized, appeared first in Africa—Neanderthals were from Asia, were genetically influenced by *Gigantopithecus* apes, became habituated to a very cold North, and "matured" far earlier. When the glaciers receded, they died out, whereas *sapiens* now

> showed an aggressiveness of spirit and a burst of cultural advance which had been foreign to his nature in the preceding hundreds of thousands of years . . . Sapiens began to live regularly in caves, to bury his dead with grave offerings, to hunt big game, to use fire, to make specialized tools, to clothe himself and to produce religious art. Probably all or most of these new accomplishments were heritages from Neanderthal man.[37]

The idea of "heritages from Neanderthal Man" raised, once again, the specter of race. Were Neanderthals really a separate species? Or did interbreeding signal that *sapiens* and Neanderthals were not really divided? Scholars like Wilfrid Le Gros Clark at Oxford could classify the Neanderthal at the "extreme limits of variation" between race and species.[38] Le Gros Clark used a pathologizing idiom to compare Neanderthals to normal *sapiens*: "*massive* supraorbital torus . . . *marked flattening* of the cranial vault . . . *massive* development of the nasomaxillary region . . . with an *inflated* appearance . . . a *heavy* mandible."[39] He was far from clear on what these differences meant.

The colonial reference frame was giving way, and two popular Neanderthal depictions from the 1950s defined a new range of pos-

Figure 8.15. Zdeněk Burian, *Spears* (1952). This painting had many of the key details Burian used for Neanderthals: a hairy centerpiece male, dark in color, with hunched shoulders, who is at once capable of complex handiwork (singeing spears) but retains a closed-off, vacant, rather unreadable expression.

sibilities. These were the paintings by Czechoslovak artist Zdeněk Burian, and the novel *The Inheritors* by the British author William Golding. For all their differences, the two helped establish Neanderthals as characters on a new political and anthropological stage.

Burian illustrated publication after publication by his compatriot, paleontologist and popularizer Josef Augusta. Later, his paintings also adorned Western publications. Burian used a specific look (clear in Figures 8.15 and 8.16): he staged Neanderthals in a diorama-like space, standing inertly in a landscape, seen mostly in profile, in temperate (rather than glacial) climates. The oversized shoulders of the centerpiece male jut forward to show how his spine is hunched. Strung-up animal hides cover his crotch. His skin is much darker and hairier than in Knight's depictions. His posture indicates a limited openness to the world. His face is labored and often withdrawn,

Figure 8.16. Zdeněk Burian, *A Group of Neanderthalers outside the Kulna Cave* (1960).

Figure 8.17. Zdeněk Burian, *The Cannibals of Krapina Cave* (1950s).

disengaged, the gaze closed off, even slightly dead. We are invited to seek out some interiority, but we discover we lack all access to it. Typically, another male crouches, usually at the bottom left, his back to us, bent over and working. In one image (Figure 8.16), Burian features a cave, a family-sized group, and a little social differentiation. The male at the center seems not to notice the two rhinoceros in the background, nor his friend calling on him and pointing to them. Even more happens *around* the bodies: to complement his characters, Burian crammed the paintings with artifacts and easily accessible references. His Cro-Magnons from the time were lively and resemble his Native Americans, but his lethargic Neanderthals receded toward African stereotypes and the animal kingdom.

Another ideological implication lurks in Burian's dead gazes. It is as though Neanderthals lost out because they were overawed by the world. They are "close to nature," but not in a way in which they would master it. Romanticized landscapes saturated with tension often entrap, confront, even condition them. In one of Burian's works, Neanderthals that look more ape than human behold a forest fire with neither fear nor particular engagement—in a kind of dumb absorption. Nor are they Marxism's primitive communists, even if Burian often shows them hard at work, preparing arms, conserving hides. In his grisliest work, *The Cannibals of Krapina* (Figure 8.17), Burian staged a scene based on an 1899 discovery in a cave in what is now Croatia. Dragutin Gorjanović-Kramberger had found fossils at Krapina that included signs of possible cannibalism.[40] (That debate would continue all the way until the 1980s.) Scholars like Boule who believed in Neanderthal burial did not touch on cannibalism; others argued that as savage brutes, Neanderthals obviously ate one another.[41] Burial and cannibalism *had to be* mutually exclusive, as though cannibalism indicated animality and burial aligned with religion and humanity. Elsewhere, Burian did depict a burial; but in this specific painting, he featured a corpse all cut up and parceled among the other Neanderthals. The alpha male in the center holds a lopped-off head that closely resembles his own. His companions feast on flesh; one hammers into a bone for its marrow. Some cowed, huddled figures pick at a skeleton. The middle area behind the leader resembles a sacrificial ground: is this a killer ape's totemic meal? And there is a touch of self-reference to this painting: an object in the leader's extended right hand resembles a drawing tool.

The stylistic distance from Burian's Neanderthals to Rudolph Zallinger's in "The Road to Homo Sapiens" (the image with which this chapter opened) is minimal—and Burian's paintings illustrated part of the same 1965 *Time-Life Nature Library* volume. While Zallinger's Neanderthal lacks the protruding shoulder blades and barrel chest, he otherwise has the same bearing, the same lack of a meaningful gaze.[42]

THE NEANDERTHALS IN WILLIAM GOLDING'S *THE INHERITORS* WERE as far from Burian's as Golding could take them. Published in 1955, on the heels of the success of *Lord of the Flies*, *The Inheritors* basically repeated the prior book's formula: it transported *Lord*'s squabbling children back into prehistory, and with them the motif of humans as *homo homini lupus*, wolves to other humans.

A small band of Neanderthals travel from their winter refuge to their spring domain in search of food. We follow the story from their point of view as they run into a group of "new people" (*sapiens*) who, fearful of them, start to pick them off one by one; the new people eventually snatch the Neanderthals' only living baby to raise as their own. Golding opened with an epigraph, taken from Wells' *The Outline of History*—where Wells, citing Johnson, described Neanderthals as repulsive, cunning, "gorilla-like monsters." In his novel, Golding mocked Wells's depiction of Neanderthals as inferior, brutish, and socially inept. His Neanderthals are capable of emotions altogether lacking in the "new people." They communicate their feelings telepathically to one another as pictures—pictures that they remember from the past or that they project as they somehow feel the future. While capable of language, they find it deceitful and avoid it. His characters' faces are highly expressive, and on one occasion, the narrative clearly plays out one of Knight's paintings of a fight against a hyena.[43] The "new people" are technologically advanced— they possess bows, arrows, even a boat. They evince more deceit than kindness and they lack any telepathically driven community. Their accurate, complex language becomes more cunning. What each group lacks, the other is exquisitely good at. Eventually the Neanderthal community falls apart, and the "new people" inherit the earth, but their fear and guile remains; it seems that only the Neanderthal baby they kidnap can turn them into a warm community.

Golding even fashioned the Neanderthal characters after his daughter, his wife, and himself. This trick was not entirely unprecedented: in 1950, painter Maurice Wilson had playfully used paleontologist Kenneth Oakley and his squabbling children as models for a Neanderthal family romance.[44] Like Burian, who saw Neanderthals as almost closed-off, Golding saw them as alien to *sapiens* sensibility. Golding, however, asked his reader to identify with Neanderthals, and then he used them to express pastoral nostalgia in late-industrial Britain. His virtuous hunter-gatherers lingered under the constant threat of a dehumanizing, emotionally deadening technological progress and extinction.

The Inheritors brought into popular consciousness the Neanderthal as a subaltern bearer of lost natural values and confronted him with humans who employed deceit and wholesale murder. This idea stood at the opposite end of a spectrum of possibilities from Burian's. Where Burian described Neanderthals as hairy, slow man-apes perhaps intent on cannibalism, Golding blamed humans and modernity for the disappearance of Neanderthals. In retrospect, he appears as the precursor of a *decolonize-the-Neanderthals!* view that is more prevalent today.

In the late 1970s and 1980s, the spectrum was much the same. Albert Barillé's children's TV animation series *Il était une fois . . . l'homme* (*Once Upon a Time . . . Man!*) dedicated its second episode to Neanderthals, treating them as complex tool-makers living in differentiated societies similar to those of the Cro-Magnon that followed them (see Figure 8.18). *Once Upon a Time. . . . Man!* was a 1978 French-Belgian-Canadian-Italian-Swiss-Dutch-Norwegian-Swedish-Spanish-Japanese coproduction. It was released on TV in at least thirty-four countries, including the United Kingdom (where it was dubbed into English) and Greece, where it was standard after-school fare throughout the 1980s and where my sister and I watched it as children. *Once Upon a Time . . . Man!* granted its female Neanderthals inventiveness and agency and resolved that Neanderthals were eventually unable to withstand the severe cold of the ice age. In their final sequence, the Neanderthals bury the old bearded "Maestro" who serves as inventor and conscience throughout the series: he comes to stand in for the Old Man of La Chapelle-aux-Saints and for the Neanderthals' death at the end of prehistory. The image of Neanderthals that had emerged from the nature-child styles of Forestier and Golding joined with positive stereotypes of Native Americans increasingly common in the 1970s to create a sort of super-Indian—"a

Figure 8.18. A female Neanderthal (note the protruding foreheads) uses a tool to skin and cut an animal in the internationally coproduced TV series *Once Upon a Time . . . Man!* (1978). This was one of the early widespread representations to attribute intention and inventiveness to female Neanderthals.

Figure 8.19. *Quest for Fire* (Jean-Jacques Annaud, 1981).

gentle, gifted, morally superior people with extraordinary tracking abilities," the original Native, as two present-day scholars have
described the stereotype.[45]

Other depictions of Neanderthals remained less generous. Jean
M. Auel's 1980 bestselling novel *Clan of the Cave Bear* (released as
a film starring Darryl Hannah six years later) involved the quasi-
adoption of a Cro-Magnon female by a band of Neanderthals, and it
described in detail the rape and social oppression she suffers at their
hands. The disparaging expression "he's a Neanderthal" became literal. Auel cast her protagonist as a feminist free spirit held down by
a patriarchy dating *that* far back. Jean-Jacques Annaud's 1981 film
Quest for Fire (based on a 1911 novel) placed Neanderthals among
imagined Indigenous peoples and reopened the old question about
the difference between race and species, via scenes of sexual conflict
and transformation. In Annaud's film, the female *sapiens* who travels
with the band of Neanderthals acted as a sex educator—somewhat

like Daenerys straddling Khal Drogo in *Game of Thrones* to introduce him to civilization. Put another way, interbreeding between
the two species was introduced into popular culture as likely (and
violent) almost three decades before Neanderthals were genetically
mapped and the point was confirmed. Violence, especially sexual
violence, came largely from the Neanderthal side, and mostly in an
effort to explain a brutish "backwardness" in *sapiens* men that persists to this day.

SINCE THE EARLY 2000S, THE PERCEPTION OF NEANDERTHALS HAS
changed, along with the evidence base. In the new science on Neanderthals, archaeological analyses of tools and jewelry have gone hand
in hand with advances in mRNA studies and gene mapping. Time and

Figure 8.20. Neanderthal couple confront the viewer—the Atelier Daynès
reconstruction is among the most important recent ones, part of the new
science of Neanderthals.

again, researchers have presented their work as finally free of "erroneous stereotypes."[46] A new wave of representations has emerged alongside the new findings. These depictions are elaborate reflections of contemporary culture, rather than mere "corrections." The "lifelike" Atelier Daynès reconstruction (2008; see Figure 8.20) presents the Neanderthal body as adorned and largely hairless—at any rate no different than *sapiens* bodies. Heavy winter clothing that would have allowed Neanderthals to survive the ice age now replaces the earlier obsession with body hair and signifies industry and intelligence. The stoop, upper-body bulk, and heavy bones are mostly gone. Another reconstruction features a suit-wearing Neanderthal, directly challenging Coon's 1939 vision. Yet another, from a German museum, resembles Rodin's *The Thinker*. The Neanderthal now leans in, pensively looking to the future.

The newer reconstructions are staged quite differently from the earlier dioramas. Neanderthals are now emotionally highly expressive. Even if their actual emotions are not clear to us, nevertheless they are there, irreducibly human. As importantly, these new Neanderthals share one thing: they are all white. Based partly on genetics, this is the most visible, novel, and significant characteristic of the 2000s imagery. In the process of gaining in emotion, they have become newly re-racialized. The whiteness of the Neanderthals intersects with the Neanderthal genocide and "assimilation" hypotheses that arose in the 2000s: even if scholars no longer speak of a Darwinian struggle-for-survival that echoed imperial sensibilities, the genocide idea is once more, at least in part, about white fear.

At its best, this is a fear of culpability for the original genocide. In 2011, the BBC produced *History of the World*, which presents *sapiens* hunting a desperate Neanderthal and pushing him off a cliff—a version of the classic image of the "hunt of the megafauna off a cliff." The BBC presenter Andrew Marr narrates ponderously, staring at the camera: "It's also possible, I regret to report, that we liked to eat them." The genocide hypothesis is saturated with tragedy: like a Greek king, we suffer today for those that "we" killed off. The genocide scenario replaces the earlier struggle-for-existence scenario and participates in debates about the innateness of *sapiens* violence.

At its worst, the genocide theory is about white fear. Even scientific texts aren't free from it: Fred H. Smith's 2012 lecture "The Fate of the Neandertals," published in 2013 in the *Journal of Anthropological Research*, used racialized terminology in discussing Neanderthals and *sapiens*. "The fate of the Neandertals is clear," he writes:

> They helped give rise to Eurasians. . . . Their contribution ["to our gene pool"] was quantitatively small, in large part because there were so few of them relative to the increasing ranks of the modern immigrants. . . . Neandertals were demographically and genetically swamped by the African biological race of *Homo sapiens*. That African biological race comprises all of us alive today and all earlier populations that including the earliest modern Europeans. . . . The indigenous European race was the Neandertals. Although they were partly assimilated into us, the fact is that the indigenous European race no longer exists as such.[47]

Demographic and genetic "swamping" is a euphemism, if there ever was one, for the "white replacement" theories that are prevalent on the Right—theories in which an immigrant "African biological race" (*Homo sapiens*) overwhelms and replaces an "indigenous European" race. Smith hints that these swamped indigenous Neanderthals contributed lighter skin and Aurignacian culture—that they created the Upper Paleolithic.[48] In *New Scientist*, a 2001 article on Neanderthals' role in lighter Eurasian skin pigmentation featured a Neanderthal male captioned: "Who's the daddy?"[49] Really: who fucked whom? The dark web is awash with this question: white supremacists wink that diversity worked poorly for the Neanderthals and look for genes that Neanderthal-derived Eurasians have that Africans do not.

For the extreme Right, Neanderthal extinction is fast becoming the original "white genocide."

Today, anyone writing about the Neanderthals has to engage with the genocide and "assimilation" hypotheses—either accepting one of them or having to sketch out an alternative to explain Neanderthal extinction. Those alternatives include "competitive exclusion"; a "resource competition" that worked out in favor of *sapiens*; and epidemic disease. Each alternative is conceptually problematic and carries a tricky history. The disease hypothesis, for example, plainly pronounces *sapiens* innocent. The "interbreeding" hypothesis has become widely adopted—but it seems to mirror the (often salutary) acceptance of mixed-race partnerships in our time more than anything else.[50] Meanwhile, genetic testing companies like 23andme explicitly advertise "Neanderthal ancestry" on their product description pages. "What percentage of Neanderthal are you? Is your partner more of a Neanderthal than you? Learn with this Christmas gift!" But this is a way to market and normalize genetic data mining. Like the other options on the menu of extinction, "interbreeding" has everything to do with making Neanderthals useful. Each theory involves a decision as to what constitutes violence, what racial assimilation means, and what cultural forms genocide has taken, from social exclusion to wholesale destruction.

This is not to say that the scientific research on Neanderthals is poor—as of this writing, Svante Paäbo has justly won the Nobel Prize for his and his team's genetic decoding of the Neanderthal DNA. But what determines the public life of this research is the speculative nature tied to our dream of projecting onto past millennia a set of concerns and a language that are contemporary. Indeed several clear continuities exist between the 1860s and now. In the 1860s, the debate centered on whether *sapiens* and Neanderthal were separate species; whether Neanderthal skulls indicated a pathological variant of *sapiens*; or whether they were a vanished race. By the mid-1950s, debate raged as to the styles in which they should be depicted:

cannibalistic and beastly? Or capable of communicating intimacy and primordial solidarity? Since the 2000s, the relationship between Neanderthals and "anatomically modern" humans remains tense, even if we no longer look at Neanderthals as brutish or, frankly, as closer to apes than white Europeans. Even today, "our" difference from them lingers somewhere between species and race difference, and the tension has to do with sex and murder.

We still cannot reach the Neanderthal. However much we may "know" about him (and it's mostly a him, not a her, not a them, not an it), he continues to say more about us.

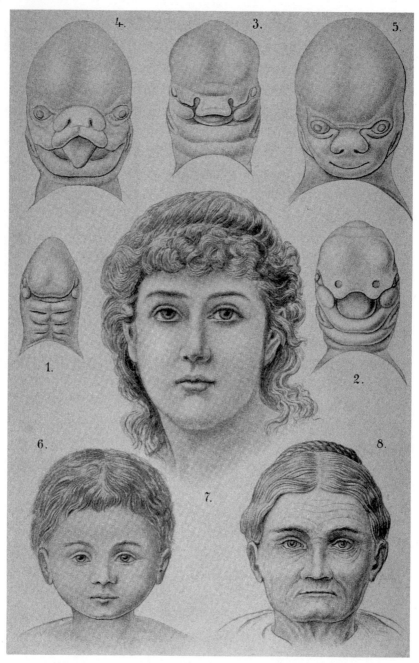

Figure 9.0. "Ontogeny recapitulates phylogeny," with the Apollo Belvedere as the pinnacle of human evolution and individual maturity. Frontispiece of Ernst Haeckel, *The Evolution of Man* (1910).

Chapter 9

THE THIN VENEER

L anguage does not belong to us. As the examples of the thin veneer, the Neanderthal, and the "disappearing natives" show, when we utter an expression, we rarely grasp the range of its meanings and associations. We never weave meaning into a perfect dress out of the whole cloth of language. We stitch, fold, and shape life out of the awkward patches of fabric—the words and images—that we have at our disposal. Even the most benign and boring expressions carry an ocean's worth of implications we do not intend to use. We grind and struggle against these associations as we try to mean what we say. But mostly, we use concepts and expressions merely in passing. We know that concepts do more than we want them to; sometimes they hurt and even kill. And still, we will forever use them partially, inadequately. There's no escape: we give concepts value and texture, we watch them carry unexpected meanings, and we seek, largely unsuccessfully, to build firm knowledge through them.

In matters of human origins, the key expression at the dawn of the twentieth century was the "thin veneer." Or, in its full form, "the savage beneath the thin veneer of civilization." It suggested that civilization, culture, and humanity are an easily breached façade that covers violent, brutish, perhaps instinctual, deeply ingrained thought and life. But the motif has a more complex and interesting history than it appears. It cannot be identified with a single author—a Darwin or Freud or George Bernard Shaw. And while it carried obvious negative meanings, it also became a remarkably productive idea and contributed to the public acceptance of evolutionary theory.

Among scientists, it helped offer answers to questions that seemed unresolvable. It introduced savagery into the unconscious. It filled gaps in theories. Among laypeople it wink-winked that the speaker knew something about the struggle of life. Where does nature end, where does civilization begin? How and why do the vestiges of the deep past survive today? How does one explain the transition from ape to human, or from an early human to a "civilized" one? Why do "civilized" humans have readymade access to a brutal inner self?

Christian thinkers had long used the "thin veneer" phrase to inquire into dubious practices: was the professed Christianity of this thinker here or that people there merely a "thin veneer" over their paganism? Then came the Darwinian upheaval, which raised new questions. Was all of nature merely a struggle for existence? Was human nature therefore merely a dog-eat-dog form of living? As soon as *On the Origin of Species* was published, Darwin confronted the objection that he had brutalized humanity. Was the human story nothing more than an immoral, purposeless, narcissistic struggle? Was morality mere kitsch?[1] Could this masquerade be turned into something productive? Britain frequently criticized other monarchies for thinly veiled brutality. Was anyone really surprised, the story went, if Russia or the Ottoman Empire carried out a brutal massacre? This was, supposedly, their true nature. Other states returned the compliment and accused Britain of savagery in turn. When Arthur Schopenhauer had declared in the 1840s that "Man is at bottom a savage, horrible beast. We know it, if only in the business of taming and restraining him that we call Civilization," he cited American slavery and Portuguese imperial expansion in Africa as the ultimate proof of human predation.[2] Did colonial society—and indeed society in the metropole—do no more than give a pretty face to savage rule? The "thin veneer" also relied increasingly on the image of the man-ape (often called a "troglodyte"). A composite being that dated to racist notions of Africans in the eighteenth century, the man-ape became ubiquitous in the later nineteenth century, particularly in depictions of early humans (see Figures 3.0, 5.1, 8.5, 8.6, 8.8). It now relied on history to explain racial difference. Once all humans were like this, but *we* developed. *They* stayed like that.

Into this matrix of meanings, Darwin inserted a particularly notable formulation with *The Descent of Man* in 1871: "man has emerged from a state of barbarism within a comparatively recent period."[3] He had difficulty explaining how domestication, the "first advance of savages towards civilization," had occurred.[4] So he punted; a year later, he added: "the essence of savagery seems to consist in the retention of a primordial condition."[5] Reason for him—as for so many Victorians—stood "at the summit" of human experience, yet instincts and the "retention of a primordial condition" indicated that an important portion of human experience remained pure savagery.[6]

Other biologists, including Darwin's partisans, tried explaining how this ancient, savage crux of humanity still survived within civilization. The German naturalist (and Darwinian theorist) Ernst Haeckel famously argued that each embryo's development replays the entire history of the human species—and indeed of all the species that preceded it, all the way to the lowest animals (see Figures 9.0 and 9.1). Then the next generation's embryo does the same, and the next. This meant that savagery was a real, inextinguishable part of an "organic memory" borne and reactivated across time. The savage was within. Herbert Spencer translated the basic principles of evolution to phenomena of all sorts, adding that the brain superimposes entire layers of complexity as it evolves. Here too, the savage was deep inside, inescapable, at the core. This savage "within" was a growing threat, all the more so at a time when intellectuals were obsessing about stadial theories, degeneration, and, as Huxley put it, the need for moral and social progress to "check" evolution's unruly, inhuman power.[7] Could the savage ever be overcome?

As a literary device, especially in colonial fiction, the thin veneer did two contrary things at once. It emphasized the difference between savage and civilized, and it pointed to what lurked within "civilization" that explained the capacity of supposed civilized people for violence. In atavistic moments, characters strip away their own civilization to return to their primal, wild nature. They are deeply moral—yet also bear their fangs and kill with impunity.

In *Allan Quatermain* (1885), the white-supremacist sequel to the no less white-supremacist *King Solomon's Mines*, the British colonial adventurer H. Rider Haggard had his protagonist begin the story by lecturing the reader. Out of twenty parts that make us up, he opined, there are "nineteen savage and one civilized." The civilized one shrouds the others. It changes their appearance "as the blacking does a boot, or the veneer a table." In crisis, he goes on, we fall back on those nineteen parts. "Civilization is only savagery silver-gilt."[8] Here, atavism was ideological bleach for visiting brutal violence upon the colonized. A litany of other novels played with the motif—from William Delisle Hay's vile *Three Hun-*

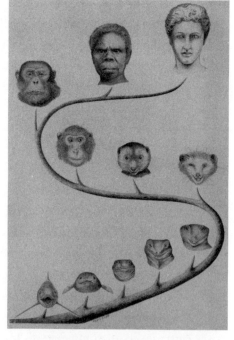

Figure 9.1. Frontispiece of William K. Gregory, *Our Face from Fish to Man* (1929).

dred Years Hence (1884) to American novelist Jack London's *Before Adam* (1906). London turned atavism into a career. In 1906, he wrote "Civilization . . . has spread a veneer over the surface of the soft-shelled animal known as man. It is a very thin veneer; but so wonderfully is man constituted that he squirms on his bit of achievement and believes he is garbed in armor-plate."[9] In his writing, even dogs revert to their wild, undomesticated, powerful truth. Once Buck, the canine protagonist of *Call of the Wild* (1903), finds himself "suddenly jerked from the heart of civilization and flung into the heart of things primordial," he is startled at the ease with which he sheds moral considerations. Tapping "the strain of the primitive," he even recalls ancient memories of a primal scene of domestication by the "hairy man."[10] By 1914, such questions were regularly appearing in literature, from John Buchan and André Gide to Eric Ambler, often in stories of crime or social degeneracy, rather than of colonial encounters. Most famous of all was *Tarzan of the Apes*: the "thin veneer" expression appeared repeatedly in the second volume of the series, *The Return of Tarzan* (1913), as "but a brittle shell, to fall at the least rough usage" and release "the wild beast" within the titular ape-man.[11]

Novelists knew well the Darwinian credo regarding the ape-man. And literature fed off history with as much ease as it fed into politics. In 1899 the US Consul to Hong Kong Rounsevelle Wildman fantasized about meeting James Brooke, an English colonial adventurer who had ruled Borneo as the "White Rajah of Sarawak" until his death in 1868. Brooke had inspired a number of literary characters, so Wildman also enjoyed putting words in Brooke's mouth in order to target the United States' oppression of Native Americans all the while ironizing the British pretense to superiority. "Civilization is only skin deep, and so is barbarism. Had your country never broken its word and been as just as it is powerful, your red men would have been to-day where our brown men are—our equals."[12]

Though they knew better, the British loved to pat themselves on the back that unlike other colonial powers, they created "equals." Where Haggard and London argued that the civilized must grasp their violent inner selves when confronted, Brooke asked how civilization reaches out and transforms the Other. But both approaches were symptoms of colonial anxiety. Were Europeans in the colonies losing their civilized superiority? Would civilization ever reach the hearts and minds of the colonized?[13] Or did they remain fundamentally barbarians?

Artists depicted some of these questions as well. At one end of the spectrum we find paintings and etchings that relate the presumed violence of primitive life—for example Paul Jamin's *Le Rapt à l'age de pierre* (1885; see Figure 9.2). Rape, wife capture, revenge, and horror could feature right on the canvas—the artist had peeked under the veneer. At the other end, the same thin veneer motif allowed

primitivists to indicate that their work allowed the viewer to trek out of her milieu and into primal simplicity, sensuality, and force. There, she could perceive something essential about human beings, idealize a vast realm lurking under her own clothes and culture. To see certain kinds of savagery was to see authenticity.

Depictions of the ancient Greeks began to brim with new meaning. Until the nineteenth century the Greeks were simply the idealized or romanticized non-Biblical ancients. But by 1900, they had become the great capstone of depthless

Figure 9.2. Paul Jamin, *A Stone-Age Kidnapping* (*Le Rapt à l'âge de pierre*, 1888). Jamin's "scenes" from prehistory included this abduction, which approximates McLennan's wife-capture theory (see Chapter 6).

antiquity. Greek gods—especially the Apollo Belvedere (see again Figure 9.0)—came to adorn the images biologists used, as if they had actually escaped nature and reached the sublime. One no longer looked to them as a foil for "the Moderns." Nor as a great culture. One looked back to them to declare that they had trampled over their inherent Dionysian brutality, that they had applied the Apollonian thin veneer, that they had created a sublime human that before them had been unimaginable.

By 1900, the thin veneer carried both promise and threat. Promise: to breach it was an act of art, a gesture toward a deeper reality, more brutal, more authentic. Threat: the Scottish anthropologist James Frazer used the expression in 1900, he made clear that it terrified him. The entire future of humanity depended on "the permanent existence of such a solid layer of savagery beneath the surface of society, and unaffected by the superficial changes of religion and culture." At the time, Frazer was revising his popular book *The Golden Bough* for a second edition, turning it into a massive meta-narrative for myths and legends. His entire purpose, he insisted, was to "plumb the depths" of the deeper world, for it was "a standing menace to civilization. We seem to move on a thin crust which may at any moment be rent by the subterranean forces slumbering below."[14] The point of science was to explain and neutralize them.

WHO BENEFITED MOST FROM THE MOTIF? THE PSYCHOLOGISTS. Psychology in the late nineteenth century was not simply one science among several: it declared itself the science that reached across the final frontier of the soul, the discipline that would be the arbiter of morality, the policeman of the human self, the judge of nations and their character. Its practitioners insisted that their dominion was unlike the simple world of biology and nature: the soul was determined by hidden character and gestures, some secret laws. What could be more welcome to them than the idea that the psyche carried within it the entire history of human savagery and violence, and that the world as we perceive it is but a mask?

To them, the thin veneer intersected with other concepts. "Instinct," for example, now seemed a universal foundation, but only recently it had been the opposite of reason: *we* have reason, *they* have instinct.[15] Spencer's idea of a layered brain also proved useful.[16] Civilization and the will resided in the top layers of the brain, supposedly, and in a crisis, these top layers could suffer a "dissolution." The same could happen to entire cultures.[17]

Some psychologists took particular interest in political passions. It was already common for nationalist authors, in particular, to appeal to the ancient past: Germans to ancient Germans, Greeks to ancient Greece, Zionists to ancient Hebrews. The French intellectual Gustave Le Bon took it a step further. He began as a self-described psychologist studying the "formation of new races,"[18] and (perhaps surprisingly) denounced Europeans in the colonies and especially British traders in India for wholesale thievery (which shows, he wrote, "our thin veneer as civilized men").[19] Having learned this lesson, Le Bon used it to target the French Republic and what he regarded as its false norms. In his famous book *The Crowd* (1895), Le Bon declared that true "sovereign force" of modernity was not democracy, but the crowd in the street. So exciting is the crowd, he wrote, that individuals give up their free will so they can bask in its ecstasy. In a crowd, people are easily hypnotized; they exhibit their instincts, which to him meant their racial inheritance. No truth matters but the idea that sways them; the "best" ideas recall an idealized version of the racial past. The politician who best invokes the racial past through such an image succeeds in moving the crowd, and becomes its expression, its leader. The savage beneath then arises as the true self—against the fake modern politicians who subscribe to reason and the fake society they lord over.[20] (Sound familiar?)

It was Freud who best exploited the motif of the thin veneer. To him, "prehistoric man . . . is still our contemporary," because neurotics and children resembled "primitives."[21] All three groups replayed in their psychology elements of early humanity, elements that civilization had repressed. "Neurotics . . . may be said to have inherited an archaic constitution as an atavistic vestige; the need to compensate for this at the behest of civilization is what drives them to their immense expenditure of mental energy."[22] In other words, the origin and conflict that civilization represses does not disappear. Children are born to a primal state. And those adults who suffer from neurosis find it surging right up and dominating their mental life.

WHEN DID EUROPEANS COME TO BELIEVE THAT THEY ACTUALLY descended from savages? The answer would seem to be during World War I. Not because they began expounding on natural selection in response to the trenches. Something more interesting happened. The thin veneer ceased to be the province of science and colonial literature. Commentators almost stopped using it to refer to Indigenous peoples. In letters, sermons, manuals, lectures, analyses of the war, books about childhood and the future, and other texts, suddenly all sorts of writers and scholars began taking the psychologists' line that everyone carried a savage within their psyche. Some, like Columbia University president Nicholas Butler, bemoaned the brittleness of the veneer, which the war had grated off.[23] Others, like Canadian suffragette Nellie McClung, pointed to the veneer to diagnose fatal flaws in civilization, hierarchies that she thought had become visible with the war.[24] Many, however, simply argued that violent instincts were natural to humanity. As warfare became ever more brutal, the soldier was now called upon to take advantage of his inner beast, his savagery, his raw power.

In 1915, the Serbian government hired the German-Swiss forensic scientist Rodolphe Archibald Reiss to provide a propaganda report on Austro-Hungarian atrocities. He made full use of the thin veneer in diagnosing the reasons why "the army of a people which claimed to be at the head of civilization" instead carried out massacres, rapes, and (he believed) extermination. Habsburg soldiers, he wrote, were frightened by the Serbs "who had always been represented to them as barbarians," and they committed their first cruelties in response to that fear. "At the sight of blood . . . man was transformed into a bloodthirsty brute. . . . A true attack of collective sadism took possession of these troops." Though probably kind and honorable in peacetime, these men now became "victims of the instincts of the wild beast, which slumbers in every human being." Responsibility for atrocities, Reiss concluded, lay with their superiors, who did not keep the beast asleep but "aroused" its instincts.[25]

Reiss was not off the mark. Armies were indeed using the thin veneer both to denounce enemies as savages and to motivate their own soldiers to fight. In a United States Army officers' course around 1910, for just one example, Captain LeRoy Eltinge lectured cadets on military psychology. An acolyte of Le Bon's, Eltinge was a decorated veteran of the Spanish-American War. During World War I, he would be promoted to Brigadier General.

> There is an old saying: 'Scratch a Russian and find a Tartar.' Well may we amplify it and say 'Scratch a civilized man, and find a savage.' Civilization is but skin deep. When the crisis comes, all the outer veneer of civilization is stripped off, and down deep in his being man is responsive only to his emotions.[26]

Eltinge's course was mass-published as an official army pamphlet in 1915, just as the US was debating entry into the war in Europe. The lesson was clear: the enemy is a predator and you too must recapture your primal power to face him.

Physiologists and psychologists agreed. For the American surgeon George Washington Crile, who traveled from Cleveland to France to help the medical effort at the American Ambulance Hospital, the veneer was "astonishingly thin." In his report on the war,

A Mechanistic View of War and Peace, Crile wrote that he saw all around him the "action patterns" of warfare, the bodily structures that mobilized each soldier into action and that dated to the beginnings of humanity.[27] To emphasize the point, he included a frontispiece showing the "most ancient known inhabitant of Britain," the recently "discovered" (in fact forged) Piltdown Man (Figure 9.3).

Crile's archnemesis, the Harvard physiologist Walter B. Cannon, also described atrocities: "when elemental anger, hate and fear prevail, civilized conventions are abandoned and the most savage instincts determine conduct. Homes are looted and burned, women and children are abominably treated, many innocents are murdered outright or starved to death."[28] In his monumental 1915 book on

Figure 9.3. Frontispiece of George W. Crile's *A Mechanistic View of War and Peace* (1915), which used an *Illustrated London News* depiction of Piltdown Man in order to establish the "primitive action patterns of war" at play in World War I.

human emotions, Cannon explained the breach of the veneer by examining the extremes of fear, anger, and pain. Over the long centuries when humans lived in a "wild" state, bodily systems evolved so the body could regulate emotions. For example, prehistoric humans preparing for a hunt could store up adrenal secretions that their bodies could expend during an attack. Modern life involved less physical effort. But it had not eliminated the "ancient pattern of response."[29] Moments of true shock, as occurred in war, drew so much force out of the individual. For these extreme moments, Cannon invented the term "fight-or-flight." Fight-or-flight was the hinge between fear and rage: when the veneer was off, the world was reduced to this binary. A person scrambled for survival.[30] Fight-or-flight remains popular today—and all the way to the 1960s and 1970s, it was a standard reference in popular science accounts of prehistory that promised to explain "How the Savage Lives on in Man."[31]

War, Freud wrote that same year, "strips us of the later accretions of civilization, and lays bare the primal man in each of us." Freud is often credited with this insight—but he was, in fact, utterly unoriginal in claiming that civilization demands the renunciation of our primal desire and violence. As he saw it, civilization pushes murder and death down into the unconscious, hides them there, and replaces them on the surface with the propriety we need to live comfortably alongside one another. But war interferes, once more, by piercing the veneer of consciousness and civilization. "It compels us once more to be heroes who cannot believe in their own death; it stamps strangers as enemies, whose death is to be brought about or desired; it tells us to disregard the death of those we love."[32]

EXPANDING WELL BEYOND ITS ORIGINS IN BIOLOGY, COLONIAL VIOLENCE, and primitivist art and literature, the thin veneer had become a weapon of war. In the 1920s, those who wanted to protect civilization insisted on the need to work, as Henri Bergson put it, "against the deep-rooted war-instinct underlying civilization."[33] Others disagreed with Bergson's (and Freud's) dream of peace. Ernst Jünger, the decorated and widely-read German war veteran, celebrated the war's capacity to arouse the beast that had slept on the refined carpets of civilization. The *Urmensch* or primal man, the cave-dweller in the "unbounded savagery of his unfettered instincts," had arisen from the depths of history to do battle, to build a new world free from bourgeois constraints.[34] For Jünger and many others, fascists and even some on the Left, the trenches had inaugurated a new age, even a "New Man."

By World War II, the motif was again ubiquitous. For example, savage violence was deemed legitimate against the Japanese, whom the British and Americans considered subhuman. The Allied General Thomas Blamey famously declared to his troops: "Beneath the thin veneer of a few generations of civilization he is a sub-human beast, who has brought warfare back to the primeval, who fights by the jungle rule of tooth and claw, who must be beaten by the jungle rule of tooth and claw. . . . Kill him or he will kill you."[35]

The concept had further significance. We like to think that people came to believe in evolutionary theory once they grasped it and saw the evidence. But that is a simplistic view. For modern Europeans to think of themselves as descendants of the earliest humans, other concepts needed to support natural selection. Among such concepts—which included "disappearance," domestication out of savagery, property after matriarchy—the thin veneer was the most important. Whether Java Man or Peking Man was our ancestor, whether natural and sexual selection caused human evolution—these were academic matters. But by 1916, the veneer had become horrible reality, in war and colonization and even in the psychiatrist's office and self-help manuals. Those fighting in the trenches became viscerally aware that they carried the deepest human past within. The easily breached "thin veneer" forced even non-soldiers to see that they lived with antiquity and savagery inside them, much as they lived with their skeletons inside them. Yet the thin veneer, even if it convinced many of evolutionary theory, was another dodge, another justification. It excused the modern savagery of the "civilized" by blaming it on humans' earliest ancestors—with little evidence that those early humans had been savage themselves.

Figure 10.0. Oedipus answers the Sphinx: Freud owned an engraving of Ingres's painting and gradually committed to the importance of its (human) subject for his theory.

Chapter 10

On the Antiquity
of the Psyche

Vienna, fall of 1919. Sigmund Freud was living in a new world. The Austro-Hungarian Empire had been abolished and its ruins kept piling up. The vestigial state of Austria was getting poorer by the day. The republic that had arisen in the aftermath of empire was in turmoil. In May, Vienna had voted in the Social Democrats and gone Red. The Catholic Right was seething. Ex-soldiers and Austrian refugees from former Habsburg provinces were settling in the city, feeding nationalism and antisemitism. Two of Freud's sons had seen battle, and one, Martin, had only just returned from Italian captivity. Early the following year, Freud would mourn his daughter Sophie, a victim of the Spanish flu pandemic.

East and north of his Vienna, the transformations were even more dramatic. The Russian Revolution had devolved into civil war. The 1918–19 German Revolution had been bloodily put down, and the same had happened to the Hungarian Soviet Republic, which only lasted from March until August. Freud's great achievement, psychoanalysis, was itself in flux: already before the war, Freud's fights with his former protégés Carl Gustav Jung and Alfred Adler had shredded the international psychoanalytic networks. The war had thinned out what was left. Now, whole groups of new converts were joining, some distinguished, others just odd.

Freud spent the immediate postwar years revising his theory of the mind. In the fall of 1919, he embarked on a little book about the psychology of mass politics. *Group Psychology and the Analysis of the Ego* presents Freud's bleakest vision of society. With crowds protesting in the streets and pledging themselves to new leaders, with Vladimir Lenin and Gabriele D'Annunzio being celebrated for their wild charisma, Freud now confronted the crowd, the horde in the streets. What did the new political movements mean for the unconscious? They looked as new as "Austria." Were they? No matter what the crowd was, Freud answered, no matter its politics, those in it readily give up their individuality in order to identify with the leader, to imitate him, to be like him. They compete with one another to incorporate him into themselves. Each person in the crowd installs the leader as an "ego-ideal," an idealized version of their self, and they link emotional arms with their fellow believers. Before the crowd's identification and love for the leader, the beliefs in democracy and in a functional or fair state dissolve. Crowds speak in a primal register.

In the book, Freud sometimes sounded like Gustave Le Bon, whom he found perspicacious. But he refused Le Bon's explanations, especially the idea that an individual, in the crowd, could connect with his or her distant racial origin. Le Bon's solution was a cheat, Freud thought, for there is no such thing as a set of original racial images. Nor was the leader self-hypnotized by his own images, as Le Bon claimed. In 1913, in *Totem and Taboo*, Freud had proposed that in modern times the leader replayed the function of the primordial individual—the "primal father." The original horde had obeyed this "primal father"—for a while.

The ominous observation that a primal unconscious was emerging on the streets confirmed that all of history, down to the war-torn present, remained caught up in the dynamics of humanity's birth. Freud contributed to the establishment of another motif about prehistory: what I will call "the world behind the world." Jung, the Swiss psychiatrist who had been his heir apparent and was now his enemy, would offer a far more disturbing and influential version of it.

DID PREHISTORY MATTER TO PSYCHOANALYSIS? EARLY ON, FREUD had been unsure. When he first sketched his theory around 1900, he struggled with the question of how the Unconscious contains both individual memories and deeper structures common to all human life. In *The Interpretation of Dreams* (1899), he resisted history. The Unconscious, this great inner force out of which desire, behavior, and thought emerge, was universal in its structure yet mostly individual in its content. It made its demands, and these demands crashed against the drab, hostile external world. The individual, the ego, was caught in the middle—and it often repressed the events, memories, and wishes that it found unacceptable into the Unconscious.

Freud often complained at the time that scientists were obsessed with heredity and the primeval period at the cost of another, more accessible primeval time: childhood.[1] In a letter, he reported, "In the evenings, I read prehistory, etc., without any serious purpose."[2] Childhood was the more interesting prehistory, and he began using the imagery and concepts of the deep past to talk about his patients. The patients' childhood was a "prehistoric period," to work with them was to take a detour through that prehistory and to dig into the "primitive strata of mental development."[3] As he gradually unfurled his theory, Freud ran into problems of his own making. Did the Unconscious have a history all its own? A past that reaches beyond each person? Does it actually evolve? For as much as he wanted to, Freud just couldn't wave these questions away.

Almost as soon as Freud had published *The Interpretation of Dreams*, he and some of his acolytes regularly undertook to alter it. From the moment that Jung joined the group, he helped introduce a series of "complexes." For Jung, interpreting dreams was a process of finding, through identifiable symbols the patient saw at night, the structures of a collective psyche.[4] Freud came to care about one of these structures, the Oedipus complex, in which children of a certain age exhibit hostile impulses toward their parents (principally the parent of the same gender). In the original edition of the book, Freud had barely mentioned Oedipus, with a nod to Sophocles' *Oedipus Rex*. Around 1908, partly owing to Jung's intervention in the revision process, the boy's desire to replace the father and to love and possess the mother became a "complex." Once Freud started getting sick of Jung, the complexes felt like threats. Not one to capitulate, he rejected them but kept the Oedipus complex. Still, where lay the origin of such a desire for the mother and a rivalry with the father? How could it be universal, an Esperanto of the soul? At the time, Freud was pondering the concept of a "primal fantasy," which he intended to explain shared fantasies and desires, including those arising from the Oedipus complex. If my fantasy is not special to me, is it perhaps deeper than my individual unconscious? Could my desire be powered by structures that predate me by hundreds of thousands of years? Primal fantasies and scenes began to populate Freud's work.

As Freud got more comfortable financially, he also expanded his collection of antiquities. Lying on his couch, a patient would see busts, small sculptures, a Greek relief, a painting or reproduction with satyrs, a Fayyum portrait, a lithograph of the temple of Abu Simbel, a small print of Ingres' *Oedipus and the Sphinx* (see Figure 10.1). Was Freud trying to induce in the patient some deep aesthetic remembrance, to grant her access to primordial grandeur? His interior design declared that the Unconscious travels across history and even that it could stage the past for the patient who contemplated it. Over time, his office and his treatment room became veritable temples, as if the antiquities aroused intimacy with eternity. "My old and dirty gods," he once coyly called his statuettes.[5]

When Freud turned to totemism in 1912, he wasn't just having some fun with speculative ethnology; he was dealing with pressing

Figure 10.1. Sigmund Freud's consulting room in Vienna, with antiquities on the walls and all around the patient lying on the famous couch. The small *Oedipus* engraving (Figure 10.0) is visible above the right end of the couch.

problems in his career. Jung had rejected the primacy of sexuality in favor of a shared unconscious filled with cultural elements—images, myths, figures. Freud had responded by dismissing Jung's theories as unscientific. He doubled down on Oedipus and looked for something more established than Sophocles' play to cast it as a universal experience. He certainly felt that, at the age of fifty-six, with some of his favorite students now hostile, he was being oedipally murdered and dismembered. So he flexed his intellectual muscles, read up on ethnology and religious studies, and made a glorious mess.

Freud was intellectually promiscuous enough to weave into his argument several concepts of prehistory that were current at the time. He tried his hand at stadial theory, working with three stages (animism, religion, and science) as James Frazer had done in *The Golden Bough*. As we noted, he drew a direct line between "neurotics," children, and "contemporary savages." He reported on debates over kinship and marriage, particularly the suggestions of Darwin and John McLennan—the latter of whom Engels had so curtly rejected for his theory of primitive wife-capture. Like so many of his contemporaries, Freud argued that if we have no obvious evidence of the key event that jumpstarted human history, this was because this event had taken place *just before* recorded history: "This earliest state of society has never been an object of observation."[6]

Freud was maneuvering to make psychoanalysis mean something to evolutionary biology, ethnology, and religious studies, which is to say that much of what he presented was not new. The idea, for example, that "savages" should be compared to neurotics dated at least to Théodule Ribot's work in the 1870s.[7] Herbert Spencer, as we have seen, described mental complexity in terms of layers added on over the course of human history: advanced ideas, he argued, were only available to the civilized, those who had advanced. By 1900, philosophers and psychologists around Europe agreed.[8] The idea that Indigenous peoples were akin to children was older still, and Freud found it in Wilhelm Wundt, the founder of experimental psychology in Germany and a rival who had just published his own "psychological history of humankind."[9]

Wundt plainly approved of the march of civilization. "Prehistoric man," Freud argued instead, is not overcome by civilization, but "is still our contemporary. There are men still living who, as we believe, stand very near to primitive man, far nearer than we do, and whom we therefore regard as his direct heirs and representatives."[10] Indigenous peoples, like children and neurotics, did not repress unpleasant feelings and ideas, whereas "civilized" adults do. Civilization forms in each person a dominant conscious ego, which "primitives" ostensibly lack; in turn, this ego imposes more harshly on the libido. Neurotics, Freud now claimed, spend immense psychic energy negotiating their emotional ambivalence not because of some ingrained abnormality but because civilization puts so much pressure on primal psychic conflicts that date to before civilization. These psychic conflicts, he believed, were visible in Indigenous peoples and in children.

As Freud swam back up the stream of time to find human beings who were not repressed, he looked to Darwin, McLennan, and William Robertson Smith. McLennan helped him argue that marriage began with male competition and violence. Robertson Smith had sought to find secular foundations for the Bible (and was tried for heresy before becoming astonishingly influential through his ideas about the secular study of religion). For Robertson Smith, religion, at its foundation, was about community, and each of the earliest communities had centered around a tribal god to whom it raised a totem. The totem animal was the group's sacred protector: it had to be protected in turn. Freud took two points from this: a totem animal, around which the community coalesced, and a ritual sacrifice of that animal, to be shared and ingested at specific rituals that reinforced the communal feeling. This was the *after* Freud sought, the first record in history: the first rituals of civilization, the simple belief, the violent, protective community.

As for the time *before* humanity proper, Freud found it in a paragraph in Darwin: "Primaeval man aboriginally lived in small communities, each with as many wives as he could support and obtain, whom he would have jealously guarded against all other men. Or he may have lived with several wives by himself, like the Gorilla; for all the natives 'agree that but one adult male is seen in a band; when the young male grows up, a contest takes place for mastery, and the strongest, by killing and driving out the others, establishes himself as the head of the community.' "[11]

Now Freud set his sights on the prize—the moment between *before* and *after*, the birth of humanity that would coincide with the birth of the Oedipus complex, thereby marking it as universal. His conceptual choreography was exquisite. The earliest hominids had lived like apes, he agreed; *before* human history, one "father" in each tribe owned and enjoyed all females, leaving his sons out in the cold.

They eventually rose up and killed him. "Cannibal savages as they were . . . they devoured their victim as well as killing him." Now they could take "his" women as spoils and share them among themselves. Which is when the real trouble began. Civilization. Having killed and eaten the primal father, the sons began to feel something new arise within them, namely guilt. "The dead father became stronger than the living one had been." His power, now ingested, expanded into a deafening prohibition. "They revoked their deed by forbidding the killing of the totem, the substitute for their father; and they renounced its fruits by resigning their claim to the women who had now been set free."[12] Had they simply become primal fathers themselves, the prehuman logic would have gone on. But they were no longer Darwin's primal horde. In guilt, they began building totems for protection and ritual sacrifice. Ever since then, civilization has been synonymous with repression and guilt, with the inability to enjoy the deepest desires of the unconscious: to murder the father, to possess the mother.

Guilt, murder, and the renunciation of sex had launched civilization. They had created the Unconscious: thanks to guilt, humans had retained a primal fantasy but also repressed it. In this Unconscious resided a living link to the beginning of humanity: every child renounced what it most desired, and in so doing it really replayed the primal events in its mental life but without carrying out actual murder and incest. In other words, no other Jungian psychological complex could define culture—only Oedipus. The primal murder scene might have even been unreal, merely an unconscious fantasy. But even if so, totems and taboos were testaments to the brute beginning of human history, which was a history of repression.

Freud had also produced a warning. Killing the father gets you nowhere, Dr. Jung, the dead father only gets stronger, I will haunt you from within. If Freud was the "father" of psychoanalysis, a bit like Franz Joseph II was the ancient father of the Austro-Hungarian Empire, then in the aftermath of his reign, the desire for equality and democracy would lead to nothing other than chaotic, guilt-ridden repression.

This was the theory that Freud returned to in 1919–1920 as he wrote *Group Psychology and the Analysis of the Ego*. In the first nine chapters, he gave an impressively straightforward explanation of how love and identification create the threads that bind a crowd to its leader. People in a crowd give up their ideal ego and replace it with the beloved leader. Freud proposed that in fact there is virtually no individual without the group first: an individual is not born, she or he becomes one only by way of submission to a group, starting with the family unit but then expanding to the crowd in the street. Now in the tenth chapter, Freud pivoted to the primal murder. No longer was the murder a matter of oedipal children and "primitives." Instead, the modern crowd was a revived primal horde. The murderous ecstasy of the brothers and the domination by the powerful father were reemerging in the manic crowd and its overpowering leader. What a different meaning the primal scene now took, after war, imperial collapse, and world revolution. How different the claim that Man is "a horde animal, an individual creature in a horde led by a chief."[13]

The primal murder might seem easy to snicker at today after a whole century of chewing and ruminating on every aspect of Freud's thought, from his patriarchal bossiness to his treatment of "hysteric" patients. But our mouths are full of the same cud, and the laughter turns grotesque. The dream of psychoanalysis was for desire to be freed. Was that even possible, if the psyche carried the immense weight of a crime no one can expiate, a guilt no one can put away? If not, should we be surprised at the rise of the crowd, of mass politics, of fascism?[14] Individuality and liberalism were a lovely dream. But no, even today, "the leader of the group is still the dreaded primal father; the group still wishes to be governed by unrestricted force: it has an extreme passion for authority."[15] Between the history of guilt and the violence of the crowd, there was no escape.

⋙

FREUD'S WARNING OFFERS EXTRAORDINARY INSIGHT INTO THE emotional hold of far-Right populism. Some of his politically-minded disciples ditched the prehistoric theory but kept the Oedipus complex for their political theories. Wilhelm Reich, for one, explained fascist mobs as the result of misguided efforts to escape an overbearing family. The more radical interpreters of Freud came to believe that desire is anarchic—capable of shattering the power of the past and of bringing about a better society. Only a truly socialist and truly sexual liberation of desire might break the hold of an age-old patriarchal psychology that fascism rested on.[16]

But regardless of what Freud or the radicals wished, his work also flirted with a more conservative way of interpreting reality: the often conspiratorial idea that a world of more profound agency lies behind the world of mere appearances. In this theory, what looks to us like reality is merely an effect. If only you read the right signs, you could understand a "secret" logic in history, a deeper, "truer" history operating behind simple everyday events.

This way of reasoning was not new. It was half-scientific practice (things have causes) and half-conspiratorial obsession (find the secret cause of everything!). The linguists and the mythologists had led the way when they proclaimed that language is truly foundational to our understanding of the world, that it shapes the myths we grow up with, and that together, language and myth frame our (hidden) symbolic universe. These experts who understood the unfolding of the Indo-European languages and myths thought they could also understand Europe's supposedly deeper history. At their heels, nationalists like Ernest Renan proposed that the nation was a "spirit" that traversed history without being injured: its truths were only partly manifest; its meaning was profound and hidden until people woke up to it. Human history had a secret geometry made of unknown communal laws that scientists could discover.

Figure 10.2. One of Carl G. Jung's paintings in *The Red Book* (1913–1917, reworked until 1930), when Jung revised his approach to the psyche's relationship to myth. From *The Red Book* by C. G. Jung, edited by Sonu Shamdasani.

Carl Jung believed firmly in this universe, much of which he thought had to do with Greek, German, and Indo-European symbols, gods, characters, and myths. Around the time of his break with Freud, he withdrew to reevaluate his theories, and he doubled down on their importance (see Figure 10.2). No existing theory could explain, he thought, the mechanism by which myths and traditions survived. Psychoanalysis *might* do so, provided it recognized how these myths set up patterns inside the self. Jung called these mental patterns "archetypes," and he claimed that they had agency of their own—that they tried to become active in each person's life and to shape that person's character. For him, archetypes derive generally from mythical figures, and they are shared across a culture. Archetypes "can only be explained by assuming them to be deposits of the constantly repeated experiences of humanity."[17] They reside in the "collective unconscious," which he understood as a reservoir of

symbols and myths that is inherited, exceeds the individual's unconscious, and overflows into individual consciousness, underwriting
everyday life.[18] Through the collective unconscious, archetypes structure each individual unconscious. Like orchestra conductors for the
unconscious, archetypes guided individuality and belonging. Jung
promised that if only we followed his particular approach to interpreting the unconscious symbolic universe, then our understanding
of human nature and actions would become infinitely superior. To
get a grip on the archetypes that determined a people's mental life
was to understand how this people held together, how over the centuries they had absorbed their stories into their collective soul. Who
cares if this or that politician is in charge when all of them are mere
marionettes of a world of motives and power that sits out of view?

Some of this was standard fare for psychoanalysis, which is
rooted in the claim that we do not know our true motivations. But
Freud had not bothered so much with myths or symbols. To him,
myths were themselves psychological symptoms; they only had to
be interpreted for a particular patient when this patient gave them
meaning. Thus, even if the Oedipus complex had a common origin and a universal structure, it was still negotiated differently for
each person. For Jung, though, religion and belief carried elements
of a broader, collective psychic truth. They affected each and every
individual. Archetypes sought to express themselves—to escape the
confines of the unconscious and give people the means to chart their
lives. Over the centuries, myth had entered the psyche and now out
it came again: it was a motor of history. As a result, Jung wrote,
"psychoanalytic research into the nature of subliminal processes will
be enormously enriched and deepened by a study of mythology"
(by which he mostly meant "Indo-European" mythology).[19] Jungian
psychoanalysis promised to be the one device that would explain the
psychic survival and meaning of a collective past going all the way
back to the beginnings of civilization.

Freud rejected all of this, though he at times expressed ideas akin to the "collective unconscious." As a result, Jung saw Freud as something worse than just his own personal bully. He saw him as a Jew. Freud the Jew had failed to understand the mixture of race and religion that Jung found expressed in distinct myth cultures. Freud the atheist Jew had failed to *see* the lived reality of religion; he had failed to appreciate the reality of symbols and their hidden meanings in the unconscious. Whereupon it followed that only a psychotherapy that cared about symbols and archetypes could offer a way of living at peace with one's tradition and society. The collective unconscious was the world behind the world, and Jung had sidelined Freud's universal Darwinian origin for an Indo-European or Aryan one.

Why does this all matter? The obvious answer is that the distance from glorious interpretation to conspiracy theory is short indeed. With this "deeper" or "secret" causality, Jung could heap doubt and contempt on scientific and political presentations that lacked this supposed depth. The world behind the world was capable of explaining desires, pressures, heroism, politics. It was racial prehistory made current and permanent. At a time when Protestant churches in Germany were becoming "Aryanized," and when Indo-Europeanism seemed a plausible theory for the origins of language and myth, Jung's archetypes and collective unconscious offered up a psychic location for Aryan myth.

Worse, for Jung the collective unconscious all but affirmed a racial soul. It carried the conditions for the "regeneration" of shared cultural forms and social myths.[20] It was a deeply political concept: it is "self-identical in all Western men and thus constitutes a psychic foundation, superpersonal in its nature, that is present in every one of us."[21] But the French Revolution of 1789, for Jung, had brought about the "end of religion" and had profoundly distorted collective life. For so long, traditional symbols—crucifixion, virgin mother, trinitarian God—had provided structure. After 1789, they could no longer hold sway. Even though the Enlightenment had discarded the gods, Jung claimed, Europeans' "corresponding psychological functioning was by no means disposed of; it lapsed into the unconscious, and men were thereupon poisoned by the surplus of libido that had once been laid up in the cult of divine images."[22] Insofar as the moderns did *not* believe in God in the old Christian way, the collective unconscious had become confused and distraught. Moderns suffered psychically because their frame of reference has become scrambled.

Usually, Jung did not explicitly connect his ideas to politics: only the psychoanalyst could cure the pathology caused by the modern collapse of organized symbolic systems; the patient was a potential hero of his or her own life. Through treatment, the world behind the world could now become conscious reality. But Jung did for a while decide that politics could help, because the collective unconscious was not universal. Jews were outsiders, he chillingly declared in 1934: "The Jew, who is something of a nomad, has never yet created a cultural form of his own and as far as we can see never will . . . The Jewish race as a whole—at least this is my experience—possesses an unconscious which can be compared with the 'Aryan' only with reserve. . . . The 'Aryan' unconscious has a higher potential than the Jewish; that is both the advantage and the disadvantage of a youthfulness not yet fully weaned from barbarism." Using all the classic celebrations of Germanic barbarian authenticity, he resolved that "Jewish" mental categories (that is, Freud's) profoundly misunderstood "the most precious secret of the Germanic peoples—their creative and intuitive depth of soul." National Socialism, he went on, had harnessed "that unparalleled tension and energy" that existed "deep in the Germanic psyche."[23]

After the Second World War, Jung waved away his earlier support for the Nazis. But he continued to present his analyses of myth as ways of unconscious belonging in a community. The death-knells he still rang about Europe's decline came to seem like standard conservative lamentations. By comparison to this style, I am far more sympathetic to Freud's speculation because, even though it too tasked prehistory with explaining everything, it offered no respite, no assurance of racial superiority, no comfort in one's heroism, only guilt and conflict and work without any encouraging resolution.

Jung's work was also taken up by nonanalysts. Mythologist Joseph Campbell, in his influential *The Hero with a Thousand Faces* (1949), refracted a similar approach: "The archetypes to be discovered and assimilated are precisely those that have inspired, throughout the annals of human culture, the basic images of ritual, mythology, and vision. . . ." His hero was Jung's patient, battling "past his personal and local historical limitations . . . The hero has died as a modern man; but as eternal man—perfected unspecific, universal man—he has been reborn."[24] More recently, Jung's masculinism and conservatism have been endorsed by the popular Canadian commentator and former psychology professor Jordan Peterson. In his *Maps of Meaning* (1990), Peterson even reports, without irony, that the mere act of reading Jung on archetypes cured him of his "horrible dreams."[25] It also apparently convinced him of the truth of moral absolutes and allowed him to develop his theory of evil, which he has loudly wielded against such terrifying enemies as they/them pronouns. Where Jung thought that post-1789 culture was confusing old and meaningful archetypes, Peterson sheds tears that contemporary culture ruins young men by confusing their manliness. He is perhaps Campbell's silliest hero, Jung's shallowest disciple.

Part III

—————

THE HORROR, PART I

(FROM 1900 TO THE 1960s)

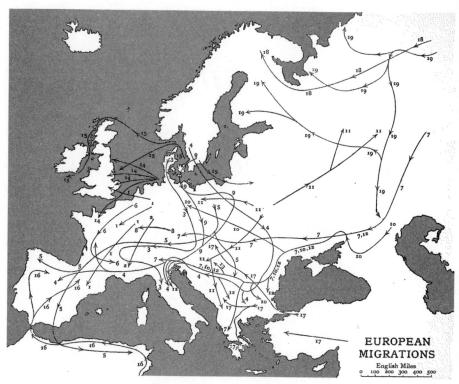

Figure 11.0. Europe as a space of unregulated movement. Right in the middle of a period marked by movements of refugees from war and revolution, stateless people, asylum seekers, and impoverished migrants, Alfred Cort Haddon's 1911 version of the "barbarian invasions" map abstracted "European Migrations" to mere arrows.

Chapter 11

THE HORDES AND THE FLOOD

Metaphors unify facts into a coral reef.

—ALEXANDER KLUGE

In the 2010s, with the alt-right beating its chest, populist leaders in Europe and the United States described refugees and immigrants as "hordes" about to "flood" their countries. Donald Trump's opening salvo for his 2016 presidential campaign bid was a denunciation of Latino immigrants. His notion of "Americans" suffering from an invasion of immigrants was certainly not a new one in the US, where there is a long xenophobic tradition, but it did help normalize the cruelty of wrenching children from their families to put them in cages in camps. Like Trump and Fox News, in Europe the far Right gave new life to an old mixed metaphor: the violent "Asiatic horde" that "floods" the ancient lands of Europe. During the migrant crisis of 2015–19, Hungarian Prime Minister Viktor Orbán callously directed it at Syrian refugees. "Mass migration is a slow stream of water persistently eroding the shores" of Europe, he mused, denouncing "an unprecedented migrant tide."[1] Hungary is landlocked, which somehow made an aquatic metaphor more menacing. In need of enemies after the Brexit referendum had gone his way, the right-wing British populist Nigel Farage denounced refugees as hordes of Islamic State fighters "flooding" in. In Europe, the Right has its litany of invaders to call on, and it reaches very far back in history to find Germans, Huns, Persians, Arabs, Mongol horsemen, Ottomans, and Soviets, all of them salivating, at one point or another, at the chance to sweep away the supposed Natives of these lands. The notion has been systematically deployed by conservatives to denigrate outsiders and dog-whistle at supposed internal threats, Jews and Muslims especially.

How did the mixed metaphor of the flooding hordes emerge? "Why floods, torrents, raging water," asked the German sociologist and writer Klaus Theweleit some forty years ago, "why do they not say . . . 'the Bolshevists advanced like the fourth Ice Age,' or like a 'hurricane,' or an 'Asiatic sandstorm'?"[2] Across Europe, between 1880 and 1940, the metaphor was a favorite expression. It had already been used occasionally for a half-century before then, though who first coined it is impossible to tell.[3] The philosopher G. W. F. Hegel made the point in the 1820s, deriding some sub-Saharan Africans as "terrible hordes" that furiously "poured down" into particular parts of the African continent until their fury abated, and that they had no true history but were "merely destructive and of no cultural significance." He took care to add notes on the geographies through which "the vast hordes of Asia poured into Europe."[4] Scottish economist J. R. McCulloch lamented to the British Parliament in 1826 that "Half-famished hordes . . . are daily pouring in from the great *officina pauperum*" of Ireland, "overflowing upon England." The references were few and far apart until the 1870s–1880s but some trends were already evident. Sometimes the idea expressed British colonial prejudices. Sometimes it abbreviated continental fears about Ottoman, Russian, or East Asian unknowns. Sometimes it legitimized a hatred of "nomadic" Arabs or Jews not bound to a nation state. And eventually, the flooding hordes became a device that translated anxieties about one other to *other* others: depending on the demands of the day, one could use it to attack migrants, refugees, Irish, Jews, socialist revolutionaries, putative Chinese enemies of Europe, then after 1917 Bolsheviks, even Jewish refugees from Bolshevism. It dehumanized the colonized into an imminent threat that would drown Europe's earthy societies with multitudes of aqueous sameness. And by deriding contemporary population movements as mere repetitions of past invasions (whether real or imagined), this prejudice pitilessly refused hospitality to impoverished "outsiders" and presented the deep past as a living enemy.

In the 1880s, Northern European writers began to worry about the colonized wrenching off their shackles and attacking Europe. Adventure writer H. Rider Haggard, author of *King Solomon's Mines*, referred to Shaka, king of the Zulu in the 1810s and 1820s, as an "African Attila"—"a visible Death, the presiding genius of a saturnalia of slaughter" whose "invincible armies . . . had swept north and south, east and west, had slaughtered more than a million human beings. . . . Wherever his warriors went, the blood of men, women, and children was poured out without stay or stint."[5] Haggard tapped into a very 1880s anxiety about the power and savagery of technologically inferior enemies. Britain in the 1820s had been largely unconcerned with the Zulu, but Anglo-Zulu fighting began in the late 1870s, culminating in full-scale war in 1879. The British suffered defeat in their first battle and even the son of the late Emperor of France Louis-Napoléon III managed to get himself killed fighting for them. Haggard's language was revealing: Shaka is like a Hun, his armies solid blocs of power; those armies sweep; blood pours out. Similarly, in 1892, English journalist Edwin Arnold warned that a loss of India would mean "the breaking up and decay of our ancient empire; the eventual spread of Slavonic and Mongolian hordes all over the vacant places and open markets of the world; the world's peace gone."[6]

By the Boxer Rebellion in China (1899–1901), the horde had become a common image in yellow-peril racism. A year before the Boxers, British author M. P. Shiel published, in serialized form, *The*

198   The Invention of Prehistory

Yellow Danger, Or, What Might Happen if the Division of the Chinese Empire Should Estrange all European Countries. Shiel used "flood" to refer first to describe the flight of Europeans to England in advance of a massive Chinese invasion.[7] The conquest of Paris, "the capital of the new Chinese world," meant a *second* flood was ongoing, this time of the invading armies. "At Pekin Yen How . . . knew well that no earthly power save his own strong, present, personal grip, could restrain his wild hordes, fresh from the gleeful slaughter of Europe, from washing their hands in the blood of every human female in Britain."[8] Hordes carried out the killing and the flooding: "The flood behind swept onward the flood before with the inevitableness of the ocean-tide."[9]

In the United States, similar language could be found in Madison Grant's *The Passing of the Great Race* and Jack London's "The Unparalleled Invasion." Lothrop Stoddard even included it in an infamous book title: *The Rising Tide of Color: The Threat against White World Supremacy.*[10] Their primary concern was that a *white* tide was ebbing: the civilizational hierarchy that Stoddard and his ilk deemed natural and appropriate no longer held. In a version known to every high-schooler today, the argument was repeated by Tom Buchanan, Daisy's husband and Gatsby's rival in *The Great Gatsby*, who rhapsodized about Stoddard's *Rising Tide*: "if we don't look out the white race will be—will be utterly submerged."

In Europe, the fear was not simply future-oriented. Biologists no longer believed in the Biblical Flood, but children of course learned about it in school, Sunday school, their family and church. Europeans had spent a century theorizing about barbarians and the end of glorious antiquity, so now they used the flooding hordes they imagined in their own past to smear the colonized of the present, from North Africa to Indochina. Kaiser Wilhelm II directed German troops in 1900 to unleash against Chinese Boxer rebels a violence to rival Attila's. "Should you come upon the enemy, they will be defeated! No quarter shall be given! No prisoners will be taken! . . . Just as a thousand years ago the Huns under their King Attila made a name for themselves . . . so will the German name, through you, be validated a thousand years for now, and no Chinese will ever dare to look askance at a German!"[11] A few years later, the German General Lothar von Trotha used the same motifs of past "invasions" of Europe to justify his extermination order that led to the slaughter of perhaps a hundred thousand Herero in what is now Namibia: this was, he declared, "the beginning of a racial fight" that had to be fought "over there" rather than "back here."[12] Geographical distance meant nothing to subhuman hordes who knew no contracts, only understood raw violence, and did nothing but plot to drown Europe from half a world away.

The racial panic appeared to have a scientific basis, for this was an age obsessed with demography and convinced that the populations of China and Sub-Saharan Africa were vast and utterly unknown. If humanity had emerged in Asia, as Ernst Haeckel had postulated (see Figure 7.0) and as many race theorists and geographers were

Figure 11.1. "Yellow Peril: Everybody Up!" Among the European sovereigns, Wilhelm II of Germany stands up toward the right (because of his appeal to crush the Boxer Rebellion). Tsar Nicholas II of Russia holds the gun (because of the Russo-Japanese War of 1904–5).

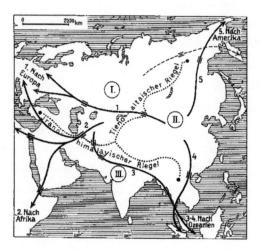

Figure 11.2. The idea that humanity had an Asian origin contributed to yellow-peril racism: proximity to the biological origin supposedly guaranteed Asian backwardness. Geographer Karl Haushofer, who designed this map, also invented the term "geopolitics" and was a Nazi enthusiast. His 1934 map doubled as a warning that the Chinese, having spread all over the earth (from point II on this map), could do so again. "Main Flowlines of Humanity from Asia—I of the White, II of the Yellow, III of the Black Main Stems."

now arguing (see Figure 11.2), who was to stop the hordes that had populated the world from flooding again?

THE MOTIF OF THE FLOOD ALSO FOLLOWED FROM A BROADER ANXIety about movement and from a terror of rootlessness. In their ongoing attempts to cast Arab society as backward, European authors from the later nineteenth century on regularly presented it as essentially nomadic. Some authors—even ones who feared Arab "nomadism"—discovered in the nomad a new noble savage to admire. Scholars resurrected Ibn Khaldūn's celebration of solidarity in nomadic societies and applied the *savages/barbarians/civilized* triad onto nomads.[13] Gustave Le Bon's popular *Arab Civilization* (1884) hailed the nomads' physical, political, and intellectual hardiness, which was due, he declared, to the ostensible fact that "since the most ancient times" the nomads' life had been "invariant."[14] Jew and Arab were kin: Abraham had been "the sheikh of a little nomad tribe."[15] "Nomads and barbarians," Le Bon went on in his *The First Civilizations* (1889), had conquered the Greco-Roman world, before later nomadic invaders like Attila, Genghis Khan, and Timur

(Tamerlane) also destroyed social structures all over Eurasia.[16] Le Bon was using much the same values as his predecessors had used for the German "barbarians:" the nomads knew no technology, no progress, no despotism; they were pure of culture; and in antiquity they had never knelt before the Romans. Houston Stewart Chamberlain, in his super-manifesto of racist Aryanism *The Foundations of the Nineteenth Century* (1889), agreed: Bedouin nomads traveled out of Arabia in migratory "hordes"—Abraham again being one of their number.[17] With them, "a Semitic flood swept once more across the European, Asiatic and African world, a flood such as, but for the destruction of Carthage by Rome, would have swept over Europe a thousand years before, with results which would have been decisive and permanent."[18] For Chamberlain as for Le Bon, these nomads were somehow both Jewish and external to (Judeo-Christian) civilization.[19] They even reinforced the idea that European Jews were nomads. By the 1920s, the idea of the nomad as an indomitable other had become mainstream and made its way even into H. G. Wells's *The Outline of History*. As Wells narrated it, nomadic invasions had "overrun and refreshed" the body and spirit of "primordial civilizations" and had created religion, democracy, and science. "The body of our state is civilization still, but its spirit is the spirit of the nomadic world."[20]

If the socialist Wells could repurpose the fears of nomadism, most contemporaries worried the opposite: that socialism itself was the expression of a "nomadic" other, a "Judeo-Bolshevik" one.[21] The deepest past was nomadic; nomadism was a threat to be feared; worst of all, the socialism of the uprooted presented itself as new but was surely nothing but a profound political regression.[22]

PREMODERN EUROPE ALSO CAME TO BE ASSOCIATED WITH A DEFEN-sive posture against invading armies. The ancient Germans had been treated as a flood; Hegel had written that "like a river, [they] gushed forth over the Roman Empire."[23] But by 1900, the ancient Germans, the destroyers of Rome, were much more strongly identified with the *defense* of Europe against Huns, Arabs, and Mongols.

These were the new old enemies. During World War I, the British and French, later the Americans too, referred to the Germans as roaming, roving, bloodthirsty "Huns." For the Germans, "the Huns" were farther East—in Russia preferably—and Europe was again about to suffer their wrath unless the German army held them back. Europeans could now use maps to create a history in which "they" had long been victims, defenders of their lands against raiders. Some mapmakers simply presented the invasions of Europe with a barrage of arrows, each for a movement or invasion—Europe was forever under threat.[24] On his own dramatic map of "migrating and

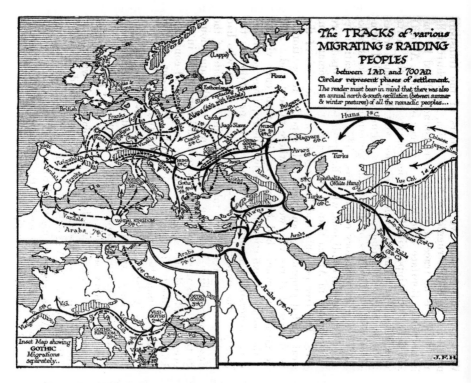

Figure 11.3. H. G. Wells's 1921 version of the barbarian invasions map.

raiding peoples" in *The Outline of History,* Wells presented the feared Huns as the most formidable aggressors (see Figure 11.3). Others highlighted the difference between Germans and Huns: in the 1920s and 1930s, German mapmakers simply removed the Huns from the "wandering of peoples" so as to better celebrate Europe's natives, the Germans.[25]

Popular history books also asked readers to remember terrifying roaming conquerors of old, with the Arabs and Mongols taking up the mantle of the original flooding hordes.[26] The English classicist J. B. Bury made a point in the 1900 edition of Edward Gibbon's *Decline and Fall of the Roman Empire* to correct Gibbon's schema even further so that the Germanic "barbarians" would include even the thirteenth-century Mongols. The galloping horde, Bury insisted, even had a strategic plan.[27] The implication was that a common defense was essential—how else do you defend Europe against a strategic horde?

The interwar was also the time when Ancient Sumer and its Mesopotamian successors came to be decisively identified with the first permanent encampments. Consensus formed that sedentary peoples in Mesopotamia with an improving knowledge of irrigation had jumpstarted the Neolithic period by inventing agriculture and trade. Did this mean that civilization, the building of complex cities and societies, had been threatened since its very beginning by nomadic barbarians at the gates? The contrast between long-established sedentary Natives and roaming (if not galloping) nomads became ever-present, not least because it was at the root of modern antisemitic ideas. Since the 1860s, political theorists like Ernest Renan, artists like Richard Wagner, and Indo-German geographers of language had argued that Semitic peoples were camouflaged by the social fabric but were fundamentally outsiders to the racial formation of Aryan Europe. Once Jews were thought of as non-European stowaways into civilization, they could be lumped with other "Asiatic" threats— Mongols, Turks, Chinese. They need not gallop; they waited like

Trojan horses. The point would become especially forceful in the construction of Judeo-Bolshevism—the identification of Jews with secret agents of communism, which proved so useful to the Nazis.

DURING THE FIRST WORLD WAR, FEARS OF FIFTH COLUMNS WERE consistently directed at Jews and refugees. Refugees were routinely described in liquid metaphors: trickle, flow, flood, deluge, torrent, wave, a "boundless ocean," "riverbanks being broken," lava, avalanche. Such terms, conveying "the familiar, widespread, paralyzing impact of recurrent natural disaster," were used consistently.[28] Even the bureaucrats involved in managing stateless people used this liquid language to describe refugees.[29] The Eastern front shifted often, and it displaced substantial populations. By the start of the Russian Revolution in 1917, the historian Paul Hanebrink shows, "wartime fears of Jewish treason and paranoia about 'floods' of alien Jewish refugees" were everywhere.[30] As Russian, Armenian, Jewish, and other refugees moved around Europe, they attracted the prejudice traditionally directed against nomad "hordes." The image stuck and was repeated endlessly: in Cicely Hamilton's 1922 *Theodore Savage*, "the refugees appeared in their thousands—a horde of human rats driven out of their holes by terror, by fire and by gas."[31] Decades later, during the Spanish Civil War, the far-Right *Action Française* journalist Léon Daudet described the flight of refugees from Spain as "an invasion of barbarians."[32] Refugee movements now recalled arrows on barbarian invasion maps.

By then, the Bolshevik Revolution—and the failed "Sparticist" German Revolution in December 1918—had catalyzed the continental spread of the motif. Nowhere did the flooding hordes play so great a role as in Germany. In the nineteenth century, as we have seen, the "Germans" of Roman times had been projected back to Indo-Europeans ("Indo-Germans," or "Aryans"), those great conquerors and language inventors. Indo-Europeans, scholars agreed, were the real historical "natives" of Germany. From the end of World War I onward, German anticommunist tracts and novels presented Soviet communism as *the* Jewish, non-European, anti-German sub-human wave that threatened Europe, from without and from within. "Undulating and engulfing waves" of communism loomed. In the 1970s, Klaus Theweleit catalogued uses of the expression and found it everywhere, from titles like *Red Flood* (the title of at least two anti-communist novels in the 1930s) to former soldiers' memoirs:

> "The Reds inundated the land;" so says a certain Hartmann, volunteer in the Baltic Army, which "for a time was the only one to stand up to the Red wave" in the Baltic region. "In the east, coming from the Baltic, the Red wave surged onward" (Wal-ter Frank). The Freikorps have rescued Germany "from the Bolshevistic flood . . . and ruination" (Weimar-Borchelshof). In Upper Silesia, German troops struggle to dam up "the rag-ing Polish torrent" (von Osten).[33]

Theweleit's list goes on for pages, to make the point that the motif was omnipresent and also to argue that it stemmed from a psycho-logical terror of the breach of boundaries (political and bodily) that ultimately amounted, in his view, to a terror of women.[34] The flood motif could be applied to other developments deemed problematic: the German jurist (and later Nazi supporter) Carl Schmitt spoke of a "flood" of democracy after 1789 that "no dam" could stop.[35]

The extreme Right did not own the concept. Antifascist intellec-tuals routinely described Nazis as uncultured barbarians. Heinrich Mann in his book *Der Haß* (Hate) referred to Nazis as rapist (*ver-krachter*) barbarian hordes.[36] But to those who insisted that Indo-Germans had been native to North-Central Europe and had moved outward to civilize others, the "galloping Asiatic horde" was the ter-rifying countermovement. If anything, the situation in the modern era was more disastrous: Jews already lived in Germany, so the front against Judeo-Bolshevism had to face both outward and inward (see Figure 11.4).

The National Socialists took advantage, targeting Soviets and Jews in one fell swoop. At the Nuremberg Party rally of 1935, Joseph Goebbels celebrated Hitler and Nazism as having saved Germany from "waves of this vile Asiatic-Jewish flood."[37] *Still!*, he warned time and again, *what if* the Soviets "flooded Germany and the West with their bestial hordes?"[38] In August 1936, in his Secret Memorandum to accompany the Four-Year Plan, Hitler declared that the German economy had to rearm to support his Lebensraum program,[39] because a militarized Germany was the only force that could fend off the Asiatic storm. "Bolshevism's victory over Germany would not lead to a Versailles Treaty but to the final destruction, indeed, the annihilation of the German *Volk* . . . the most gruesome human

Figure 11.4. "Jews Flood Germany—Germans Emigrate." A 1937 teaching poster printed by the Nazi publication house for national literature. It contrasted a "flood" of 80,000 Jews entering Germany with the emigration of 390,000 "valuable" Germans now "lost to the Volk."

tragedy that has been visited upon mankind since the downfall of the states of antiquity."[40]

The fear of the Soviet Union, presented as a defense of Europe, was not exclusive to Germany—in 1939, the French Prime Minister Édouard Daladier dithered about war with Germany because he worried that it would lead to Europe being flooded by Soviet "Cossack and Mongol hordes."[41] Once Germany's Operation Barbarossa—the invasion of the USSR—got underway, and the Allied powers welcomed the USSR into the "Big Three," these anxieties quieted to a whisper. But they became incessant in Nazi Germany: at Hitler's dinner table, in Nazi speeches, in magazines, policy documents, and the military manuals that denounced "barbaric Asiatic methods."[42] On July 16, 1941, three weeks into the invasion of the Soviet Union, Hitler declared the Ural Mountains as the only possible defensive structure against Asian incursion: "Never again must it be possible to create a military power west of the Urals, even if we have to wage war for a hundred years in order to attain this goal."[43] By this point, Hitler was talking obsessively about "Asiatic hordes."[44] And he declared that his ally Mussolini agreed with him.

[The Duce] told me himself that he had no illusions as to the fate of Europe if the motorised hordes of the Russian armies were allowed to sweep unchecked over the Continent, and he is quite convinced that, but for my intervention, the hour of decline was approaching.[45]

All Stalin cared for, Hitler went on, was repeating the ancient Slavic flood. For Stalin, "Bolshevism is only a means, a disguise . . . If we hadn't seized power in 1933, the wave of the Huns would have broken over our heads. All Europe would have been affected, for Germany would have been powerless to stop it."[46] This was not simply a war *now*, but one that drew from history for justification and method.[47] Heinrich Himmler presented the *Generalsiedlungsplan* for the Nazi colonization of the Ukraine as a bulwark against a future "incursion from Asia."[48] Once the Germans began to lose the war, the metaphor took on apocalyptic dimensions.

THE CONCEPT OF THE HORDE/FLOOD MAY NOT SEEM TO HAVE OBVI-ous prehistoric inspiration: one could see it as just a simple form of racism. But, in fact, it was intimately connected with how people thought about belonging, about national and human memory, about their attachments to ancestors in European prehistory. Refugees as primitive nomads disrupted modernity: they intruded *both* as a different race and as savages from the deep past. The threat of barbarians invading yet again raised the specter of a suffering that no one living in the twentieth century had actually experienced. For those groomed in the belief that they were descendants of indigenous Germans or Aryans, attached blood-and-soil to their land, to fight Bolsheviks was to fight Huns, Mongols, Cossacks. It would not be too much of a stretch to say that, saturated with ideas of prehistory, Europeans in the 1920s, 1930s, and 1940s lived at once in several eras: now and back before recorded history.

Yet the motif persisted after the war, in part owing to the new West-East divide and the actual presence of the Red Army in half of Europe. "Civilization," French philosopher Georges Bataille dryly commented after the atomic bombing of Hiroshima, "is no longer a matter of an aristocracy sheltering the order of an empire from the invasion of nomadic peoples."[49] Optimism about the postwar future helped dampen anxieties and changed the tone. The Iron Curtain (and especially the Berlin Wall) blocked most movement from the East. Still, Western Europeans continued to act entitled to their territory and found ever more threats. In France, President de Gaulle and Catholic conservatives roared, in the words of the Catholic novelist and commentator François Mauriac, about "the black, brown, and yellow hordes that would soon invade our shores."[50] (The anticolonial revolutionary Frantz Fanon responded to Mauriac by recalling of how often colonialism involved reducing the colonized to animals: "the *slithery* movements of the yellow race, the odors from the 'native' quarters, the *hordes*, the stink, the *swarming*, the seething, and the gesticulations."[51]) The British worried about Black migrants from former colonies and the capacity of "Anglo-Saxon pallor" to "biologically absorb" them.[52] The British conservative politician Enoch Powell concluded his infamous 1968 speech against immigration with a quote from Virgil that blended the Germanic barbarians motif with the flooding hordes motif. "As I look ahead," he declared, "I am filled with foreboding; like the Roman, I seem to see 'the River Tiber foaming with much blood.' "[53] West Germans did the opposite of "absorbing" when they instituted a guest worker program for Turkish, Greek, Italian, and other *Gastarbeiter* whom they distinguished quite clearly from themselves. For the time being, population movements seemed easier to control, not least because of the "Iron Curtain" that divided Europe into Eastern and Western blocs.

Historians naturalized and depoliticized the concept. In 1959, Henri Baudet presented the history of Europe as "a semi-permanent stream of large-scale invasions and attacks from the east on the western borders. . . . The attackers were not defeated once and for all; the siege was not lifted permanently. . . . Essential to the history of Europe was its defense against the recurrent threat of an Asiatic tidal wave that would engulf the entire continent."[54] In 1941, the mythologist and linguist Georges Dumézil, then a proponent of fascist ideas, posited that the Indo-European motherland lay "between the Hungarian plain and the Baltic" (he couldn't bring himself to name Poland). Between 3000 and 1000 BCE, out of this center, several waves of "hardened riders" who spoke the same language "enslaved all of Northern, Western, Southern, and South-Eastern Europe; ancient inhabitants disappeared, were assimilated, or formed islets that were slowly absorbed."[55] Writing in 1969, long after shushing his earlier political extremism, Dumézil kept the very same language but simply shifted the Indo-European motherland eastward: "Toward the end of the third millennium before Jesus Christ, *migrating horsemen*, who came perhaps across the south of Russia, *submerged in successive waves* the greater part of the European continent."[56] What had been a story of native Indo-German power could now be a tale of ferocious invaders.

Since the 1970s, "ethnopluralism"—a neo-fascist concept that is designed to sound democratic and tolerant but that, in reality, means "each to their own" and, more precisely, "mine to me, immigrants stay away"—has helped the nativist notion of a diluvial foreign threat persist. Owing in large part to right-wing French authors such as Jean Raspail, Alain de Benoist, and later Renaud Camus, but also British conservatives like Powell, the same anti-immigration rhetoric has been mainstreamed, while its fascist past is either winkingly embraced or falls away in the face of indifference. The racist conspiracy theory of the "Great Replacement"—the belief, pioneered by Camus, that Muslims are replacing white Europeans—constitutes the newest incarnation of the flooding horde, once again presenting demographic shifts as existential threats.

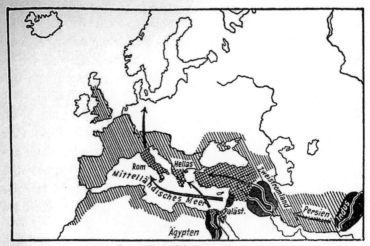

Abb. 4   Das alte Geschichtsbild "Ex oriente lux".

Das Vordringen städtischer Zivilisation in frühgeschichtlicher Zeit.
Hauptträger: westisch-mittelmeerische Rasse.

Schwarz: die angeblichen Ursprungsländer der Kultur: Indien, Mesopotamien, Ägypten und Palästina. — Rechtsschraffiert: jüngere Kulturmittelpunkte des vorderen Orients und Mittelmeergebiets: Persien, Griechenland, Rom. — Linksschraffiert: das Römerreich als politischer Höhepunkt der Mittelmeerkultur.

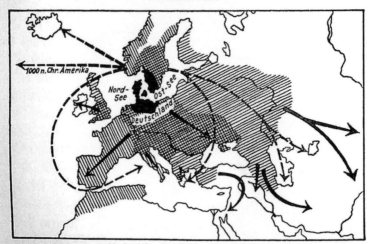

Abb. 5   Das neue Geschichtsbild, Ergebnis vorgeschichtlicher Tatsachenforschung.

Deutschland, das Herz Europas, Ausgangsland von Bauernvölkern nordischer Rasse.

Schwarz: Nordisches Kernland um 5000 v. Chr. — Rechtsschraffiert: Ausbreitung der nordischen Indogermanen bis 2000 v. Chr. — Pfeile: Indogermanische Vorstoßrichtung. — Linksschraffiert: Ausbreitung der Germanen bis 500 n. Chr. — Durchbrochene Pfeillinien: Züge der Wikinger bis 1000 n. Chr. — Politisches Ergebnis: die modernen Nationalstaaten Europas.
(Mit Genehmigung des „Völk. Beobachters". Entwurf: R. Ströbel. Zeichnung: Kurzhauer.)

Figure 12.0. National Socialist map of Indo-German expansions out of Germany, "the heart of Europe." It proposed that "The Old Picture of History: Ex Oriente Lux" that imagined Mesopotamia as the origin of culture had been replaced by "The New Picture of History: The Result of Factual Prehistorical Research."

Chapter 12

NAZIS

The nationalists who came to power across much of Europe after World War I were profoundly invested in history. If the nation was the alpha and omega of politics, then its leaders and sentinels had to match the present to a glorious past. In Portugal, dictator António de Oliveira Salazar's *Estado Novo* (the "New State" that he "founded" in 1933) looked back to medieval glories and insisted on the family as the bastion of long-term continuity. Ioannis Metaxas's Greek dictatorship (1936–41) promised national regeneration and dressed its youth in togas to make the point. Benito Mussolini's fascist Italy (1922–43/45) drew its symbols and dreams directly from ancient Rome, adopting the bundle or "fasces" as the official insignia and portraying its African territories as a restored Empire. It was to be the New Rome.

None of them obsessed over human origins with the promiscuity and ferocity of the National Socialists in Germany. Promiscuity, because from Adolf Hitler at the top and all the way down to local Gauleiters, journalists, and Hitler-Youth leaders, Nazi ideologues grasped at every available moment in German and European prehistory, whether true or imagined. Nazi intellectuals actually varied a great deal as to their particular ideas and beliefs; as a result, they overall give the impression that they went about imagining racial origins almost indiscriminately. But their ferocity was just as important. No regime so emphatically relied on theories of origins as a justification to rule and to kill. Racism was state practice in Britain as in the other empires. This was especially the case in its imperial dominions—where hierarchies of the colonized were a matter of course. Yet in Germany in the 1930s, prehistory itself became a matter of the state. The Nazis worked and acted as though the entire purpose of the state was to put theories of origins into practice for all of Europe. In 1941–42 racial prehistory became a rationale for the extermination of the Jews of Europe. It was not merely used as a vague pretext; it shaped the specifics of the Nazi agenda.

RACE WAS NATIONAL SOCIALISM'S OVERRIDING OBSESSION AND category. But to stop there is to fail to understand what the Nazis' view of race was, and how much their racial ideas relied on a particular vision of human history. Nazi thinkers declared the immutability of race, but they also developed a dizzying series of ideas about origins and an almost unlimited list of examples to cite from. Those ideas and examples legitimized the absurdity of antisemitic declarations and justified the National Socialist vision of the world.

In fact, the Nazis drew on most of the concepts and images we have encountered in this book so far.

- They imagined a version of *survival of the fittest* as nature's command and threat and linked it to their call for racial purity.
- They deployed the *flooding hordes* motif against Jews and Bolsheviks, constructing the invasion of the Soviet Union as a preemptive strike against Asiatic hordes.
- *Ruins*: some, like Hitler, strongly *believed* in ruins, and particularly in the present day as a time of ruin. Hitler

thought that time was a process of degeneration that ruined everything, from Germany's economy to its racial makeup. Already in *Mein Kampf* (1925–26), the now is empty, dissolute, a "fall," a "paralysis," "the depths of our present degradation," a destitution that mirrors the collapse of the German army in 1918, "our present misfortune," "our present impotence."[1] Only a New Man might hold off destruction.[2] Hitler, perhaps along with Albert Speer, his bombastic favorite architect and later Minister of Armaments and War Production, also imagined his regime not just surviving for a thousand years, but eventually collapsing into its own ruins.[3]

- *The ancient Germans*, whom they promoted as Wagnerian Siegfrieds, models of virility and heroism. As new ancient Germans, the Nazis would now destroy not the New Rome of Mussolini but the Rome of the intentional Jewish conspiracy, and by the same token they would defend against the new Judeo-Bolshevik Huns. Some Nazi thinkers, Alfred Rosenberg most notably, gave the theme a distinctly mythical angle, replete with its Wotan and other Nordic gods who acted for him as ersatz archetypes. (Others hated his religiosity.)[4] The regime even put on festivals, with floats and sculptures intended to restage the imagined past and to celebrate ancients and even pagan gods.

- The triad of *civilized/barbarian/savage*: Hitler translated it into Aryans, true creators of culture; those who were mere bearers of culture; and those savage destroyers of culture, the Jews.[5] But Aryans did not fit easily into the "civilized" category, and at times Hitler seemed to advocate that they were indeed barbarians like the ancient Germans. Meanwhile, anti-Nazis in and beyond Germany looked down on the Nazis as barbarians plain and simple. The German politician Hermann Rauschning, who joined the party in 1933 and left it a year later, later denounced Hitler as someone who adopted the position of the barbarian: "Yes, we are barbarians! We want to be barbarians! It's an honorable title. *We* shall rejuvenate the world!"[6] Rauschning—who probably simply invented Hitler's words, was making use of the "barbarian invasions" schema and the belief in nomadic rejuvenation. Hitler, for his part, did indeed identify with the ancient Germans but refused the disparaging connotations of the *Barbaren* motif by insisting on the creativity of the Aryan.

- The *Indo-German* deep past. National Socialism celebrated scholarship on Indo-Germanic origins as a demonstration of Aryan truth. "Indo-Germans," as we have seen, extended the ancient German noble savages backward in time. The Nazi version of this historical synecdoche privileged the ancient Germans as expressions of the creative Aryan "genius" that had supposedly ruled Europe since its earliest Indo-German times. Some insisted on a Northern/Nordic/hyperborean motherland of the Indo-Germans—others saw Germany as the only possible birthplace of Aryans.[7]

- National Socialism turned to ancient Greece as an early pinnacle of civilization, and also (paradoxically) to early Christianity as the birth of a religion aiming to rejuvenate the world against "the Jews." Both of these turns depended on Aryanism. In 1934, for example, Hitler declared the affinity between Greece and Germany on grounds of "Indo-Germanic racial community":

> We are once again looking to the great ancient peoples with wonder, admiring their accomplishments in the field of human culture. . . . They may be far removed from us as nations [*Völker*], but as members of the ancient IndoGermanic racial community, they are eternally close to us.[8]

- The argument about Christianity was more obviously strategic, but Nazism stripped Christianity of its Judaism, often presenting it as part-Aryan. To look back to early Christianity and to Greece was, for them, to rediscover moments of Aryan creation. It even was a way of presenting Hitler as German messiah.[9]

The list could go on. These different, often conflicting concepts did not hold the same importance for every author or speaker, but they all played a part. They allowed for continuity between divergent ideological positions, as well as a rigid opposition to ideas deemed alien, unnatural, Jewish.

IN *MEIN KAMPF*, HITLER HARDENED THE ANTISEMITIC TRADITION that had been developing since the mid-nineteenth century, and he articulated human origins in a polygenistic account, in which only Aryans were a creator culture. Since the dawn of humanity, racial purity had allowed Aryans to build culture and to innovate. All others were condemned to at most carry or else pervert and destroy culture. This distinction between creators/bearers/destroyers required, Hitler went on, that the true people be conscious of its racial heritage and committed to its biological purity. In other words, race was a matter of self-consciousness. One was not an Aryan simply because of one's biology—one had to be profoundly conscious of it and act accordingly. The world was racially ordered but because the "destroyers" denied race, a creator had to actually enforce that order. If the biologically "pure" did not stave off their racial enemies, they would be dominated by their slaves.[10] Hitler committed to this idea all the way to his last text, his "political testament" in 1945. There, he effectively blamed Germany's defeat on the failure of Germans to believe strongly enough or to commit deeply enough to his racism and their own supposed superiority.[11]

Race was National Socialism's foundation, yet it needed the history of Aryan creation, struggle, and ruin. The loop of purity-creation-struggle-ruin-purity activated all the other ideas about prehistory. Further fantasies followed: for example, Heinrich Himmler, the leader of the SS, was obsessed with Tibet and with mysticism, which led him to establish the Ahnenerbe—the SS office committed to Aryan "prehistoric research" that directed expeditions and was only one of several offices involved in "racial research."

One might well ask: why would a whole regime, even a country, believe such a complete jumble of wild ideas? Regardless of what we might imagine, however, even high-ranking officials did not, nor did they have to. Historians have shown that no "100% Nazi" existed and that commitment to Nazism was more a matter of a "disposition" or "posture" than of ideological consistency.[12] A relative diversity of ideological orientations was standard, both within and beyond the state. What mattered was the ethos, "a willingness to adhere to the precepts of the worldview which was vague and indistinct enough to embrace a variety of related perspectives."[13] No one had to go down a checklist and tick boxes but to stand and salute with some version of the broader schema of beliefs. These ideas about origins blended with other beliefs and made this broad, vague schema more plausible: people had less reason to refuse the basic tenets. And anyway, most Europeans already believed in a blend of at least some of these ideas. The same postulates that Hitler made explicit were certainly implicit in academic circles focused on prehistoric archaeology and Indo-European linguistics. A very large percentage of ordinary people mistrusted Jews deeply, celebrated the ancient Greeks and early Christians as creators, liked Hitler, and believed in the originality of their ancients. To these same "ordinary Germans," the Nazis' ideas were convenient and emotionally arousing, especially when they ricocheted off one another: they delegitimized other ways of looking at the world, all the while "proving" and being sustained by antisemitism.

ARTISTS, LINGUISTS, AND POLITICAL THINKERS HAD PAVED THE WAY. Richard Wagner had dreamt of recovering a pure, original Siegfried and restoring him to life against the supposedly Jewish and corrupted music of his time.[14] Ernest Renan, in his early work on Hebrew, had sharpened the opposition between "Indo-European"and "Semitic."[15] Many other scholars spent decades stepping over one another to pronounce on Indo-European origins. In 1878 Heinrich Schliemann had claimed that a Teutonic swastika adorned the pottery he had dug up when he discovered Mycenae.[16] By 1900, archaeologists were increasingly using digs and the artifacts they found in them to stake their claims. Gustaf Kossinna took settlement locations and every-day artifacts as evidence of the spread of Indo-Germans and their continuity with ancient and, therefore, modern Germans. Most lin-guists were convinced that an original "Indo-European nation" had existed.[17] Many, like Antoine Meillet in France, even developed theo-ries about the class status of the original Indo-Europeans. Meillet gradually decided that just as Sanskrit in India belonged to upper castes, the original Indo-European language had been "a language of chiefs and organizers imposed by the prestige of an aristocracy."[18]

The real point of contention concerned where this idealized "Indo-European" language and race had fanned out from. The prehis-toric homeland, *Heimat*, or *patrie* of this "original" Indo-European people—where had it existed? Had its people spread through con-quest, assimilation, or cultural superiority? Three main locations were proposed: somewhere geographically between Iran and North-ern India; somewhere between Denmark and Germany; and some-where near the Caucasus. Kossinna tied archaeological discoveries of Corded-Ware Culture in Northern Europe with Indo-German origins and argued that cultural laws and archeological artifacts located the *Ur-Heimat* in Northern Germany.[19] In the 1920s, the second hypothesis—that the "Aryas" originally emerged in Germany—was gaining ground, and not only among conservatives. Consider one example, the Australian-born, Edinburgh-based archaeologist V. Gordon Childe. In *The Dawn of European Civilization* (1925), he opted for the Central Asian steppe theory. Just a year later, in *The Aryans* (1926), he defected. "Aryans," he now argued, were warmongering neolithic Indo-European conquerors who had come from Northern Germany. Childe added some references to "Nordic blood" and "superiority in physique" for good measure.[20] He soon changed his mind back to the steppe theory, disavowed the book altogether, became a Marxist, and blamed German prehistorians for misleading Science. Their hatred of the Treaty of Versailles after the First World War had swayed them to privilege "the more thoroughly Germanic soil of Saxo-Thuringia!" he declared.[21] Like Childe and with less bad faith, the French sociologist Henri Hubert mocked German linguists for backdating their discoveries so that they could hug their own origins: "With the help of their anthropological bias that prefers blonds to brunets, they search for the origin of the entire language family in the peoples of the North, the Germans, and they have made of the Germans the *Urvolk*, the primitive, excellent people, the first-born branch of the family, the direct successor of the ancient greats who represents the Aryan in his greatest purity."[22]

It wasn't just the humiliation inflicted on Germany by the Treaty of Versailles: as we have seen, German scholars had preferred "Indo-Germanic" (emphasis on Germanic) to "Indo-European" for a century, and by 1900 they routinely postulated a Northern European *Heimat* for their vaunted Indo-Germans.[23] After 1918, nonetheless, this view became politically consequential. Intellectuals continued to track German roots to Tacitus's *Germania* and beyond. The Danish philologist Gudmund Schütte titled a multivolume 1929–1933 book *Our Forefathers: The Gothonic Nations*. Others attempted to produce a singular and consistent interpretation of Indo-Germanic/

Aryan purity by pretending that power was benign, like a gift of technology. Childe, in his ill-fated 1926 book, imagined that "Aryans" succeeded "solely" due to their linguistic superiority.[24] Radicalizing Kossinna's views, archaeologists asserted on minimal evidence that ancient Corded-Ware and Bell-Beaker Cultures had moved from ancient Germany eastward. Culture could never have reached Germany from the uncreative East. Most of them celebrated the coming of Nazism as a realization of their and their teachers' secret hopes.[25] In short, data was used, consciously or not, to bind together the myth of an ancient Nordic-Aryan-Indo-Germanic type that created and spread language and technology.[26]

This ancient ideal(ized) people had developed Europe's linguistic structures, myths, and ideas, in turn influencing life, death, ideas, sacrifice, culture, conflict, and power all the way to the present day. Personal and class "nobility" was based on race and "spirit." Modern history, politics, and society were contingent. For the Nazis, this "world behind the world" paid rich dividends.

FOR THE NAZIS, "ARYAN" ABBREVIATED THESE DREAMS OF HISTORY and belonging. From speeches to laws, from radio broadcasts to the Nazi newspaper the *Völkischer Beobachter* to "völkisch" festivals, antisemitism and Aryanism pervaded popular culture. National Socialist intellectuals spilled an astonishing amount of ink to forge connections between the Nazi worldview, Indo-Germanism, and other discourses on origins. Those connections made Nazi racial categories believable.

Besides the famous figures—Hitler, Rosenberg—many others helped establish an explanation of history that could double as policy. Here, the "wandering of the German peoples" took a back seat to much more aggressive maps of Indo-German expansions that had, in line with Hitler's beliefs, "created" all true cultures. The "Race

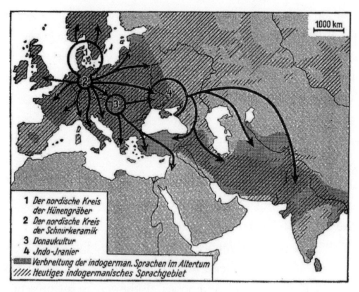

Figure 12.1. Walther Gehl's map of the Indo-Germanic conquests confidently located the Indo-Germanic Fatherland in the Northern German Plain.

Pope" of National Socialism Hans F. K. Günther, in his book *The Nordic Race among the IndoGermans of Asia* (1934), presented a map of the supposed Indo-German conquests of Europe and Asia that decided their original motherland lay in contemporary Belgium.[27] Other widely-published Nazi irredentists like Walther Gehl, Dietrich Klagges, and Karl Buchholz offered very similar maps and prepared teaching posters for this vision. Figure 12.0, with which this chapter began, was Klagges's map; Figure 12.1 Gehl's and 12.2 Buchholz's. For them, Northern Germany was the true birthplace of the Indo-Germans. Many of the same authors presented the Reich as beholden to its obligations to the *Volksdeutsche* (ethnic Germans) who had supposedly suffered under racially foreign Eastern regimes for centuries. Like the ancients who had supposedly traveled eastward from German lands, conquered other peoples in Eastern Europe, the Caucasus, and Central Asia, established languages and founded Germanic communities, so too the Reich had a historical duty: to liberate the Volksdeutsche and realize its racial promise.

Indo-Germanism echoed Hitler's "history" of the race, particularly his conviction that brief periods of racial purity congealed the Aryan spirit and unleashed its creativity to subjugate others. Walther Darré looked to pre-Christian Germans ("and, by the way, IndoGermans," he added) whose "hereditary inequality" served as the basis for the rule of nobility.[28] Architect and race theorist Paul Schultze-Naumburg identified ancient Greece and medieval Germany as highlights of humanity's racial history. His books of photographs of artists and artworks were supposed to substantiate his claim that racially pure art was the only true art—which was also supposed to explain Greece as a version of Germany. Teaching tools or "Training Briefs" (*Schulungsbriefe*) published by the Nazis as soon as they came to power explicitly stated that Greek culture was a product of Indo-Germanic racial prowess (see Figure 12.2), which made it

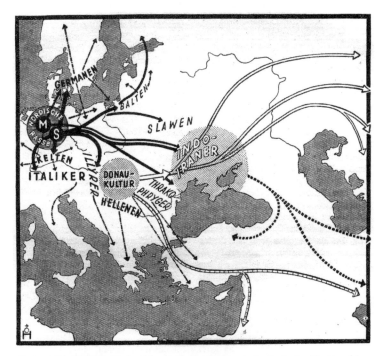

Figure 12.2. Karl Buchholz, "Nordic Racial Destiny in Antiquity"—Nazi teaching tool intended to show the "Nordic" origin of "culture creators" (1934).

ideal for the new Germany. Albert Speer concluded that the "true" architecture for his dream could only be a monumentalizing neo-classicism. Leni Riefenstahl's film *Olympia* (1936) fantasized about the link between ancient Greek male athletic bodies and German Olympic heroes: in its opening sequence, Myron's *Discobolus* comes alive as a German discus thrower (see Figure 12.3). Even intellectual enemies like philosophers Alfred Baeumler and Martin Heidegger could agree that one should look to ancient Greece, with a Nietzs-chean eye, for a high culture specific to Aryans (Baeumler) or a pure thinking unshackled by the constraints of philosophy (Heidegger). By the late 1930s, Himmler's Ahnenerbe office was pursuing "scien-tific" expeditions from Iceland to Tibet.

Figure 12.3. Myron's ancient Discobolus statue comes alive—in the German athlete. From Leni Riefenstahl's *Olympia* (1936/38).

Neither the Nazi state nor National Socialist ideology were monolithic. On the subject of origins, plurality worked in the Nazis' (and Aryanism's) favor. Even advocates of Nazi modernity like Joseph Goebbels (who didn't care for origins) made use of a history broken into thousand-year units, where "our" ancients offered motivation for modern power. It was utterly artificial, but the Nazis were very good at presenting artificiality as natural.[29] Human antiquity was less to be grasped on its own terms than to be brought to life, like Riefenstahl's discus thrower, turned into modern reality. The "modernists" in the regime understood this well. To them, the past did not fully command the state, nor was it the reason for technological developments. Instead, because this past was coming alive again, a new "spirit" was now possible that had not been before.[30] Eugenics, which was favored by so many regimes across the Western world in the first half of the century, acquired a new role in Germany. For the Nazis it could "reveal," at the social and the biological levels, the supposed ancient and original truth of human beings.

At the start of Operation Barbarossa, the multiple explanations of racial inequality and Aryan grandeur were put into terrible practice. With the war on the Eastern front, the massacring of Eastern Europe's Jews, and the establishment of the extermination centers like Auschwitz-Birkenau and Treblinka, the Nazi vision of prehistory became brutal, absolute reality. Against the "Asiatic hordes" and the Jews within, the Nazis claimed to raise a New German Man, durable as granite, who would undo degeneration and end all threats to the race.[31] The same regime that had all the modern means of communication, transportation, war, and annihilation at its disposal also imagined itself as the guardian of the race, as the revelation of the true nature of the world, as the awakening of the deepest human antiquity. The ghettoization and eventually the "final solution" regarding the Jews of Europe was essential to this awakening and supposed racial truth. Reinhard Heydrich's 1939 "Schnellbrief" ordered the rounding up and ghettoization of the Jews of occupied territories in Poland, and it initiated the Aryanization of Jewish property. By putting Heydrich himself in the position of deciding whether Germans or non-Germans should take over this property as owners and workers, the Schnellbrief gave him power to reinstitute the racial and social hierarchy of Eastern Europe of which the Nazis had been dreaming.[32]

The fantasy of the Indo-Germans informed the prosecution of the war. Ideologues like Gehl saw a direct link between ancient grandeur and the task for Germany in the war to come (see Figure 12.4). Figures 12.0, 12.1, and 12.2 earlier, which purport to show the pathways of the Indo-Germans, indicate just as well the hold these images had on the Nazi dream of the future. German conquest routes followed the Indo-German "precedent"—through Ukraine to the Caucasus, aiming to continue to Persia.

Colonization plans were rooted in the same ideas. The *Generalplan Ost* promised that Germany would carry out the expulsions, "depopulations," starvation, and mass murder of between thirty-one million and forty-five million Poles, Jews, and Slavs. The Jews were to be murdered, the others were expendable.[33] SS-Oberführer Konrad Meyer, who authored the *Generalplan Ost* and worked in Himmler's SS office, opened his "memo" with the exclamation, "The German

Figure 12.4. The same Walther Gehl who mapped the Indo-European spread from Germany eastward (Figure 12.2) also offered maps to describe "the German people's soil" in Central Europe and to call for the "liberation" of German communities outside Germany's 1938 borders.

armies have finally won for the Reich the Eastern territories that have been disputed for centuries."[34] What does he mean by "finally?" Or "centuries?" Other Nazi leaders of the SS who were central to the invasion of the USSR and the execution of the Holocaust agreed with his historical horizons. Heydrich compared the new "colonization" to those implemented by the Teutonic Knights of the thirteenth century and Baltic barons of the sixteenth and seventeenth centuries.[35]

At every step, the Third Reich cast its war and conquests as a gigantic return of the racial and spiritual repressed. How could even ordinary, boring, un-Nazified Germans commit brutal massacres with such fervor? Even if they did not strictly follow Nazi ideology, which was itself malleable, they shared a web of ideas that gave metaphysical value to the killing. Germans were fighting both their own war and those of 5,000 years earlier, of the ancient Germans threatened by Rome, of those who defeated Rome, of those who fought Huns and Mongols and Turks. The enemy within and without was a threat going back thousands of years.

With every mass killing, with every train to an extermination camp, the Jew as the racial enemy got a little smaller, the SS officer a little taller. By early 1943, after Stalingrad, the German army leadership understood that the war in the East was a failure. Or was it? Why, it is often asked, did Nazi Germany commit such immense resources to the extermination of the Jews of Europe when it was losing the war against the allies? When it needed every last train to transport matériel and people, and every last pair of hands to pick up a gun? One answer is that the Holocaust was a war aim and, given the immense numbers of Jews that the SS were murdering, it was going brilliantly. Throughout 1943 and 1944, and all the way to the Soviet capture of the extermination camps, the Nazi war against the supposed ancient enemy was being won—and it had to be. Joseph Goebbels, in his famous Sportpalast speech of February 1943, once again presented Bolshevism as the latest expression of longstanding Jewish enmity toward Germany and exhorted his listeners to choose

total war and with it the survival of the *Volk*.[36] If Germany was losing on the military front, it was achieving the world free of the hated
Judeo-Bolshevik nomads.

A few months later, the Nazis' obsession with ancient Germans
culminated in a well-known, astonishing episode. In the fall of 1943,
having lost Southern Italy to the advancing Allies, Heinrich Himmler
assigned an SS detachment the task of discovering and recovering the
earliest existing copy of Tacitus's *Germania* in central Italy. By now,
Germany had lost hundreds of thousands of soldiers in the war. But
this SS detail took the time to ransack old manors in San Marino
in the hope of looting them of the relic. The proof of Aryan origin
could be located, documented, archived, identified; it could be made
current.[37] Even in terminal decline, the regime could not but grasp
desperately at proofs of prehistoric justification.

EVEN SO, ONE MIGHT OBJECT. ISN'T THIS ALL RATHER ACADEMIC?
What is it that mattered—the fantasies of Aryan, Teutonic, and Wagnerian prehistories or the actual slaughter of the Jews, Roma, and
homosexuals in Poland, Russia, and Ukraine, the brutal and violent
internment in ghettos, the meticulously planned transfer of entire
populations to the camps from as far away as Paris and Salonica, and
their annihilation en masse? Do ridiculous notions of racial origins
perhaps trivialize the horrifying reality of the Holocaust?

Primo Levi, for one, did not think so. The Italian-Jewish poet,
chemist, and chronicler of death and survival in Auschwitz did not
simply record the horror of the camp. He painstakingly demonstrated in *If This Is a Man* (better known as *Survival in Auschwitz*)
how the *Lager* (the camp) was designed to reduce its victims to a
state of nature. There, under the eyes of the camp guards, inmates
were forced to struggle against one another, cheating and betraying
to achieve the least and most prominent of all goals: to survive one
more day. By reducing the Jews to nameless subhuman inmates fight-

ing over a shoe or a piece of bread, Nazis could "prove" to themselves their own superiority. When the Nazis spoke of the Jews as vermin, they meant both that they regarded the Jews as vermin *and* that they had to reduce them to vermin.

Who were the civilized? For Levi, it was the cruel and tyrannical man who "understands that if he is not sufficiently so, someone else judged more suitable will take his post."[38] Who were the beasts? Either the violent ones or those with the "divine spark dead within them, already too empty to really suffer."[39] Levi rejected the possibility that he could remain among the civilized. He compared himself to a dignified inmate, Steinlauf, for whom "precisely because the Lager was a great machine to reduce us to beasts, we must not become beasts." But no, this "wisdom and virtue . . . is not enough for me."[40] At another point, he tried mightily to remember the Canto of Ulysses from Dante's *Inferno*—the passages about what it means to be human—only to watch stanzas, and with them his own humanity, wane from his memory.[41] Not for nothing did the camp, that "gigantic biological and social experiment," serve as living proof for the Nazis that Jews were subhuman, alien, uncivilized, destroyers, that they belonged to a different stage of humanity, that they themselves replicated the camp's hierarchies and violence, that they deserved no life. Along with mass death, this was the camp's most brutal, brutish effect, Levi thought, the realization of Nazi fantasies.[42] Ideas of human origins became lived reality before they destroyed and claimed civilization, and became merciless death.

IN 1944, THE GERMAN-JEWISH PHILOSOPHERS AND SOCIOLOGISTS Theodor Adorno and Max Horkheimer noted that Enlightenment "had sought to liberate human beings from fear." In no field does the insight hold true more than the study of human origins. Learning about our origins was supposed to rationalize the world, get rid of superstition, kill off wrathful creator-gods, maybe even explain and end violence. Even today, many still believe that that the pursuit of origins does exactly that, or carries that potential. "Yet," Adorno and Horkheimer rightly continued, "the wholly enlightened earth is radiant with triumphant calamity."[43] A pilot "might be called super-human in comparison to the troglodyte," but to what effect? To spray the earth and "cleanse the last continents of the last free animals?"[44] The super-human being that the Europeans had dreamt might one day emerge, Adorno and Horkheimer argued, but in the shadow of the Holocaust, of world war, of the atomic bomb, this was both unlikely and profoundly undesirable. Advances in knowledge over the caveman had come at immense, unbearable cost.

Figure 13.0. Jules Verne's Robur the Conqueror bombards the "savages" from his airship (1886).

Chapter 13

BOMB THEM BACK
TO THE STONE AGE!

REBEL (*toughly*): My family name: offended; my given name:
humiliated; my profession: rebel; my age: the stone age.

—AIMÉ CÉSAIRE, *AND THE DOGS WERE SILENT*

B omb them back to the Stone Age!" thundered Curtis LeMay,
the storied former US Air Force Chief of Staff, in his 1965
memoir. If the Vietnamese did not "draw in their horns and
stop their aggression," he repeated, "we would shove them back into
the Stone Age."[1] As vice-presidential candidate for the segregationist
George Wallace three years later, LeMay walked back his murderous
fantasy, but only a little: "I only said we had the capability to do it."[2]

LeMay's genocidal swagger was no mere theory. He had overseen
the obliteration of Japanese cities in 1945 and of targets in Korea in
1950–53. It's not clear whether LeMay invented the exact phrase—at
any rate, Western states had long directed overwhelming violence
from a distance, and from the air, against Indigenous peoples. His
boast repeated their old idea of "bombing the savages" and updated
it for a new technological age.

What does it mean to bomb someone *back in time* or, worse,
back into *deep* time? Why think of the bombed in terms of antique

technology? The answer to these questions lies in how aerial bombing first supported, then confused, and eventually transformed
the traditional stadial theories that had been established in politics and international law. In the later nineteenth century, as we
have seen, intellectuals classified societies according to a hierarchy
of *civilized* versus *barbarous*, or else *civilized/barbarous/savage*,
tying this triad to two more: *animism/religion/science* and *Stone/
Bronze/Iron Age*. Bombing intruded into these schemas. At first, it
simply expressed the destruction that the leaders of the new Iron
Age could and did unleash on the colonized (the "savages," the
"animists," those still living in the Stone Age). But by World War
II, bombardment terrified Europeans. Those who claimed to be
"civilized" wondered whether the bombed might be mere victims,
and eventually they largely decided that yes, if you bomb others,
you become uncivilized yourself. In the process, stadial theories
and ideas of progress changed: they became much more dependent
on ideas about technology.

AS OF THE 1870S, STADIAL THEORY SEEPED INTO INTERNATIONAL
law. To enjoy its protections against bombing you needed to be
a sovereign state advanced to the point that you could easily be
declared civilized, which in practice meant Christian and European.
Some states at the margins might be allowed in, but the thrust of
international law was "to justify present European expansion by
making it appear as the fulfillment of the universalist promise in the
origin."[3] James Lorimer, whom we encountered earlier, adopted the
three-stage system of *civilized/barbarous/savage* in the 1880s and
had few qualms about denying savages anything more than "mere
human recognition."[4] Debates on humanitarian law concerned the
question of how much violence could be unleashed upon the uncivilized.[5] John Westlake, the Whewell Professor of International Law
at Cambridge, spoke for his profession in 1913 when he noted that

"All civilized states which are in contact with the outer world are, to their great regret, familiar" with "punitive expeditions, in which the whole population must suffer" in order to repress "inroads or other outrages committed by savages or half-civilized tribes."[6] His words are startling: civilized states "regret," wholesale punishment is necessary, savages commit "outrages." Westlake even made sure his reader would not forget that the Hague Convention only concerned war between civilized states. Article 25, which outlawed terror from the air, "cannot be quoted against the attack or bombardment of a town not having a government sufficient to be the proper object of hostilities."[7] Bombing was necessary, though those doing the bombing would feel bad about it.

Other white Westerners took a more searching approach to the brutality visited upon the colonized. Was not the destruction wielded against the feeblest peoples a sign not of the triumph but of the collapse of justice and humanity?[8] Yet most only saw the actual technological superiority that European states enjoyed, above all in war. Europe's technological arsenal was put in the service of colonization (as was America's), leading to scenes of starkly asymmetric warfare all through the nineteenth century. The British used cannons and artillery throughout the century—starting on Copenhagen in 1807, then Algiers in 1816, then with ever-increasing ease in Burma, Ashante, Sierra Leone, Afghanistan, China, Brazil, Satsuma, Abyssinia, South Africa, and again Afghanistan. By the 1890s they were regularly shelling Native people in battles and "pacification" operations all over Africa, and they continued to do so in the new century. They turned to air-to-surface bombing around World War I, with Somaliland as their test target in 1920.[9] Meanwhile the French repeatedly bombarded forts in Madagascar from warships in the 1880s and 1890s. At the time, the novelist Jules Verne—of *Around the World in 80 Days* fame—fantasized about his hero Robur the Conqueror bombing the savages from his airship (see Figure 13.0 at the opening of this chapter). Until 1900 at least, Europeans distinguished societies principally on the basis of race and culture, not specifically on the basis of technological advance.

Europeans and Americans only began commenting on the savagery of modern technology—as opposed to the "savagery" of the colonized people they used it on—during World War I. The Balkan Wars of the early 1910s were described by observers as unexpectedly "barbaric."[10] Once the Great War broke out in 1914, all parties began denouncing their enemies for their brutal or barbaric violence. Human beings were being turned into war matériel. In 1915, British surgeon Anthony Bowlby lectured on injuries from German shells, listing effects from the splintering and crushing of bones to the mashing of flesh.[11] Now, artillery shells did not reach that far past the front. Aerial bombing did. In Germany already before the war, Count Ferdinand von Zeppelin and strategist Rudolf Martin dreamt of bombing London from the sky. In France, novelist Henri Barbot imagined that a Zeppelin bombing of Paris would fulfill mystical Catholic prophesies.[12] By the later stages of the war, the Germans were indeed bringing the war "home" to the English by dropping bombs on London, albeit from early biplanes rather than blimps.

IN HIS PATHBREAKING BOOK *A HISTORY OF BOMBING* (2002), SWEDISH historian Sven Lindqvist traced the British and, more broadly, European appetite for bombing colonized populations. He too noted a change after World War I. Thanks in part to the shock of German bombs, the British were gripped by fear that colonized peoples or communists would acquire bombs of their own, "flood" into Europe, and bomb its achievements to smithereens.[13] So while Europe continued to rule (by bombs too), beginning with Edward Shanks's novel *People of the Ruins*, bombing entailed paranoia.[14] In the 1920s, it became common to forecast that the next war would be dramatically worse. H. G. Wells had prophesied as much in *The War in the Air* (1908; see Figure 13.1) but at the time he had been relatively nonchalant about destruction. Now, in *The Outline of History* (1920), the "promiscuous bombing of any and every center of population" in World War I, including London, would be "child's play to the bombing of the 'next war.' "[15] Wells believed that religion and education were in rapid decline, and that without a new, socialist trust in humanity and a real commitment to end war, the destruction of civilization was all but guaranteed.[16]

His contemporaries agreed. Cicely Hamilton, in her popular novel *Theodore Savage* (1922, rev. 1928), depicted an England bombed back to a state of nature, with its protagonist rummaging like everyone else for food, humans reduced to scavenging like rats and even to hunting rats. New warring "tribes" would form in the wasteland. T. S. Eliot wrote "The Waste Land" that same year, and in it the same postwar images recur—"I think we are in rats' alley / Where the dead men lost their bones." P. Anderson Graham picked an even more blunt title for his 1923 book: *The Collapse of Homo Sapiens.* "The coloured races by some mysterious accident had mastered the secrets of the west—the mechanical and chemical devices by which the western

Figure 13.1. "As the airships sailed along, they smashed up the city as a child will shatter its cities of brick and card." From the chapter "The War Comes to New York" in H. G. Wells's *The War in the Air* (1908).

sphere had for so long maintained its superiority. They had even discovered a deadlier gas than ours, and explosives of such power that two or three moderate-sized bombs aimed from a moderate height had been enough to wipe London out of existence."[17] Bombs reduced their victims to a state of savagery outside all notions of civilization. Other novels also depicted the bombing of London, from the apocalyptically titled *Ragnarok* (1926) by Irish author Desmond Shaw to *The Poison War* (1933) by the journalist Ladbroke Black. For the British, bombs were especially explosive if their users were savage.

Other Europeans were less fearful. The Germans had spent much of World War I waiting for the promised bombardment of England to win them the war, and they were often more willing to consider air war—even more than war in general—as a way to defeat false Anglo-French "Zivilisation." Their "airmindedness" resulted, too, from the need to escape the naval blockade the nation suffered under in the war.[18] Wasn't all war, they reasoned, highly technological anyway? For Ernst Jünger and other German intellectuals, war renewed the soldier by tapping into his primal instincts. Cries of barbarism were for squeamish Anglo-Saxons, fighter planes and bombers for Germans. Italian intellectuals also gravitated toward the promise and consequences of bombing. For the Italian General Giulio Douhet, one of the early champions of what was euphemistically named "strategic bombing" (bombardment aimed at the enemy's economic capacity and especially the enemy's morale), the plane had effectively ended ground and naval warfare. The reality of war had changed: bombing "ravaged" and "terrified" the enemy.[19] "1,000 tons of explosive, incendiary, and poison-gas bombs dropped on Paris or London could destroy these cities. . . . The air arm gives the means of reaching the most vital of the enemy's centers, and poison gas makes such an offensive as terrifying as it could possibly be."[20] Italian artists—notably,

Futurist painters and authors—followed suit and became obsessed with air power. The tradition of futurist *aeropittura* ("aeropaint-ing," paintings of planes and aerial views) became, historians have argued, "the incarnation of Douhet's 'war in three dimensions.' "[21]

Colonial violence had long elicited domestic criticism, perhaps never more so than in the later nineteenth century. However limited or ineffectual this criticism may have been, after 1918, governments and international lawyers were also seeing benefits to moving the world away from traditional imperial domination. From its founda-tion in 1919, the League of Nations presented itself as the vanguard of a historical movement of progress and civilization. In establish-ing the Mandates system, this first organization for international cooperation (which was heavily weighted toward the interests of European states) proclaimed that "the well-being and development of such peoples form a sacred trust of civilization."[22] In 1922, the American jurist John Bassett Moore headed an international com-mission in The Hague to establish "Rules for Air Warfare." Moore was anguished by how World War I had undercut the distinction between combatants and noncombatants. He insisted that the law of nations that was followed "in the civilized world for more than one hundred years" had been set up in strong opposition to "sav-age" or "barbarous" forms of warfare, and specifically to avoid the killing of noncombatants.[23] "The indiscriminate launching of bombs and projectiles on the non-combatant populations of towns and cities" revolted "the conscience of mankind."[24] The US, Brit-ain, Italy, the Netherlands, France, and Japan—all of them imperial powers—were represented on the commission, which agreed even-tually to prohibit bombing aimed at terror or injury to noncom-batants. But it had difficulty determining some rather key details, for example whether it was fair game to bomb cities that one was already attacking or that harbored military equipment.[25] Of course, the treaty was dead on arrival, for reasons the delegates recognized. No state would hamper its own ability to fight when others might take advantage of a terrible weapon in defiance of a piece of paper. Still, the most powerful states recognized well that the more effective the bombs, the more they undercut international law and reverted *everyone* to barbarism. The real barbarism of the bomber had overtaken the imagined barbarism of the bombed—a twisted kind of "progress" indeed.

The admission that bombing was barbaric broke the hold of the conventional distinction of civilized from barbaric and savage. Bombing also eroded the legitimacy of the League's "sacred trust." In September 1925, American volunteer airmen bombed the defense-less town of Chefchaouen in Morocco; they were fighting on behalf of France, which was supporting Spain's war in Morocco, which all sounds like a very roundabout way of bombing without anyone claiming responsibility. Three weeks later, France bombed Hama in the Syrian Mandate. The French claimed civilization for themselves: they were quelling primitive brigandage. They didn't really convince anyone, but Chefchaouen and Hama were safely outside of Europe.[26] Public outrage (and there was some) could be contained, even though people now understood the power of the latest weapons. Such was the case, too, with the punitive bombing of the village of Co Am that the French carried out in 1930 after the Yên Bái mutiny in Tonkin (in Indochina).

Even Douhet, that booster of aerial bombing, in a 1930 prophesy titled "The War of 19—" recognized the barbarism of the bomber and the fear of the bombed:

> Impressed by the terrible effects of the bombings and by the sight of the enemy planes flying freely and unopposed in their own sky, though they cursed the barbarous methods of their enemy, [immobilized troops] could not help feeling bitter against their own aeronautical authorities, who had not taken enough protective measures against such an eventuality.[27]

Japan—which had joined Moore's commission for "Rules for Aerial Warfare"—left the League of Nations after invading Manchuria (an act that included Japan's aerial bombings of Jinzhou as of October 9, 1931, and Shanghai in January 1932). The League protested. After Japan embarked on its full-scale invasion of China in 1937 and its aerial bombing of cities, it "emerged as the world's leading practitioner of strategic bombing before the outbreak of the Second World War in Europe." The League again expressed its disapproval.[28]

FASCIST ITALY'S 1935 INVASION OF ABYSSINIA (PRESENT-DAY ETHIOpia), which resulted in the killing of an eighth of that nation's population, drove home the terrible barbarity of aerial bombing.[29] The Italians had used the language of "trusteeship" and civilization to justify their invasion, claiming "only to bring law and order to a backwards, warlord-ridden and slave-trading land."[30] F. T. Marinetti, the founder of futurism—also a signatory of the *aeropittura* manifesto and a promoter of its painters—supported the Abyssinian invasion. Futurism had discovered that "War is beautiful because it creates new architectures, like those of armored tanks, geometric squadrons of aircraft, spirals of smoke from burning villages."[31] With *aeropittura*, destruction, like modernity, now traveled by plane. The fascist devotee and aerial painting specialist Alfredo Gauro Ambrosi painted silver planes flying over explosions in a mountainous land in his *Bombardment of East Africa* (see Figure 13.2), while Cesare Andreoni celebrated (this time with planes so silver they mirrored the sky and ground on the painting) the "Mockery" of Addis Ababa, Abyssinia's capital. The painters' distance from the war itself was reflected in the absence of Abyssinian victims and in the celebration of the modernized sky. Geographic distance from death in Abyssinia could now be filled by the glory of the planes that dominated the paintings.

Speaking at the League of Nations after fleeing to exile in 1936, the

Figure 13.2. *Aeropittura* celebrating the bombardment of Abyssinia during fascist Italy's colonial invasion. Alfredo Gauro Ambrosi, *Bombardamento in A.O.*, 1936.

Emperor of Abyssinia Haile Selassie described Italian atrocities, most famously mustard gas sprayed from the air over large areas. Playing on the language used by *both* the theorists of strategic bombing and Moore's commission, Haile Selassie said that "the very refinement of barbarism consisted in carrying ravage and terror to . . . the points farthest removed from the scene of hostilities."[32] In order to associate savagery with Italy, Haile Selassie insisted that *he* had been bringing civilization to his people and pointed out that both Italy and Abyssinia were states recognized by the League. It did not help the Italian claim to civilization that Mussolini's son, a pilot, was quoted boasting about the Abyssinians he had bombed:

The bombs hardly touched the earth before they burst out into white smoke and an enormous flame and the dry grass began to burn. I thought of the animals: God, how they ran . . . After the bomb-racks were emptied I began throwing bombs by hand. . . . It was most amusing: a big Zariba surrounded by tall trees was not easy to hit. I had to aim carefully at the straw roof and only succeeded at the third shot. The wretches who were inside, seeing their roof burning, jumped out and ran off like mad. Surrounded by a circle of fire, about five thousand Abyssinians came to a sticky end. It was like hell.[33]

The German-Jewish philosopher Walter Benjamin took note of Marinetti and Mussolini's boasts: Marinetti was the poster boy for fascism's credo about the "beauty of war." In his essay on art losing its "aura" in modernity, Benjamin treated aerial warfare as characteristic of current dehumanizing trends:

Imperialist war is an uprising on the part of technology, which demands repayment in "human material" for the natural material society has denied it. Instead of draining rivers, society directs a human stream into a bed of trenches; instead of dropping seeds from airplanes, it drops incendiary bombs over cities; and in gas warfare it has found a new means of abolishing the aura.[34]

By the time Benjamin's essay appeared, the emblematic aerial bombing had taken place. On Hitler's orders, the Luftwaffe, which had studied the Japanese attacks on Chinese cities, bombed Guernica during the Spanish Civil War, on April 26, 1937.[35] The anxiety about being bombed was no longer confined to fantastical novels fed by racial frenzy or Wells-style sci-fi. A year later, Benjamin jotted down some theses on history—an essay that turned out to be his requiem: "There is no document of civilization which is not at the same time a document of barbarism."[36] Did he have international law and the "sacred trust of civilization" in mind? Was not his famous tale about the Angel of History an allusion to bombing? "Where a chain of events appears before us, he sees one single catastrophe, which keeps piling wreckage upon wreckage and hurls it at his feet. The angel would like to stay, awaken the dead, and make whole what has been smashed. But a storm is blowing from Paradise and has got caught in his wings; it is so strong that the angel can no longer close them."[37]

ON SEPTEMBER 1, 1939, AS GERMAN TANKS ROLLED INTO POLAND with the Luftwaffe overhead, and as the British and French debated going to war, American President Franklin D. Roosevelt appealed to them all (plus Italy) to shun the "ruthless" aerial bombardment that had indiscriminately maimed and killed defenseless civilian populations. This "inhuman barbarism," as he called it, "has sickened the hearts of every *civilized* man and woman, and has profoundly shocked the *conscience of humanity.*"[38] In the Second World War, however, it was the American and British air forces that carried out the most devastating bombing raids. From early in the war, the Nazis were perceived as "barbarians" and, because of the Luftwaffe's (failed) attempt to subdue Britain in 1940, as the aggressors in the air. In the Pacific theater, American propaganda made every effort to describe the Japanese as subhuman. During Roosevelt's own presidency, Allied bombing became normalized. It devastated, at times even razed, many cities and towns in Germany (most famously Hamburg and Dresden) as well as most major Japanese cities—not least because the Germans and Japanese had carried out such shocking atrocities of their own. After the US used atomic bombs to annihilate Hiroshima and Nagasaki, The Bomb's apocalyptic quality finally assigned blanket innocence to the victims. (Needless to say, it did not do so immediately—as Figure 13.3 shows).

It was no longer states that should be protected from bombs, as the Hague Convention had originally understood it, but human beings. The older adage about the civilized, barbarians, and savages

Figure 13.3. "This Is the Atomic Bombing Hour!" *Los Angeles Times* (August 7, 1945). The US had dropped the first atom bomb on Hiroshima a day earlier. The cartoon presents it as having a geological effect, blasting Mount Fuji out of the atmosphere. But no human victims.

was more an anachronism than ever. It had relied on the belief that the complexity of a culture determines its status and its character; now the surviving distinction was technology itself. One no longer bombed "savages," nor did one bomb others "into savagery." Rather, one bombed the barbarous enemy back, as LeMay suggested, to an earlier technological stage.

Debates among both Anglo-Americans and Germans during World War II concerned the *target* of the bombing. Should "area bombing" and widespread urban devastation be employed to kill wantonly and undermine the enemy's will to continue fighting? Or should air attacks be focused on military and infrastructural targets alone? The German Minister of Armaments and War Production Albert Speer later considered the British insistence on area bombing to have been a mistake that emboldened and united Germans without, for the most part, breaking up the key centers of industry and arms production.[39] The US Strategic Bombing Survey did not agree: it quoted Speer's own March 15, 1945, report to Hitler that the German economy was "heading for an inevitable collapse within

4–8 weeks" and it commented that regardless of the situation on the battlefield, Anglo-American bombing would have forced the German army to "cease fighting by June or July."[40] Bombing, the Strategic Bombing Survey rejoiced, had won the war. Both Speer and Anglo-American authorities had an agenda in making these claims, of course. The Strategic Bombing Survey would soon turn its attention to Korea.

RECONSTRUCTION AFTER THE WAR—IN GERMANY BUT ELSEWHERE too—contended with often thoroughly destroyed urban and industrial landscapes. On a continent ruined already by Nazi occupation, then pockmarked with craters and buried under rubble that still concealed rotting body parts for months, "technology" meant the rebuilding of a smoothly functioning infrastructure. Could new technologies promise a world better than the one that had been reduced to food, shelter, and maybe transportation? Meanwhile, the postwar fields of developmental economics and modernization theory also recovered the logic of technological ages: from Egypt to India, from the Sahel to West Africa, as new states were established, a new triad and hierarchy replaced the old: *developed/developing/underdeveloped*. W. W. Rostow's 1959 theory of economic development opted for five stages, beginning with "the traditional society" and continuing through economic "take-off" toward economic "maturity" and mass consumption. Technology, infrastructure, and participation in the world economy became the primary markers of a country's status. In 1961, US President John F. Kennedy made it clear that he wasn't interested in a war over the Western Papuans, a people "living, as it were, in the stone age."[41]

The shift from culture to technology as the primary rubric of cultural status had precedents before the war—after the war it prevailed. V. Gordon Childe, the doyen of European prehistoric archaeology, had shifted to technological comparisons already in the 1930s, using pottery-making and the evolution of agricultural tools to study ancient societies.[42] After the war, his work in technologically grounded archaeology became the standard. And children in Europe and America began to be exposed to tools and instruments from supposed earlier technological eras—for example in music class, where the Orff Approach exploded in international popularity. German musical theorist Carl Orff, though politically compromised by his successful career in Nazi Germany, became an international sensation when he directed pupils to begin with "the most ancient musical instruments of mankind" before they moved on to more complex instruments.[43] The Orff Approach is used in schools to this day. Far from the sites of bombing, Orff and other educators and intellectuals contributed, little by little, to the veneration of technological development.

Why did Europeans, given that they thought that savages lived in the past anyway, bother to actually bomb them "to the Stone Age?" To bomb someone back into the Stone Age was to destroy their place and potential in the modern era. The punishment lay not in the killing, but in the destruction of infrastructure. In the massive US aerial bombing of North Korea during the Korean War, the term "Stone Age" now indicated the destruction of buildings and bridges, roads and factories. It was another euphemism, like "area bombing" or "strategic bombing" had become, a convenient way to avoid speaking of mass death. And it signaled that the bombed could only be saved by new economic ideas that would drag them forward into modernity on Western terms. In Vietnam, the US proposed to bring the country "into the twentieth century" or else abandon it decimated in its own backwardness. For LeMay, one could not show superiority and technological power without embracing barbarism— the beast under the thin veneer—when that seemed necessary. Technology was superiority, and the Vietnamese deserved to remain in the Stone Age. But for most of the people who heard his quip, LeMay was exactly the barbarian that had to be rejected or, at least, disavowed.

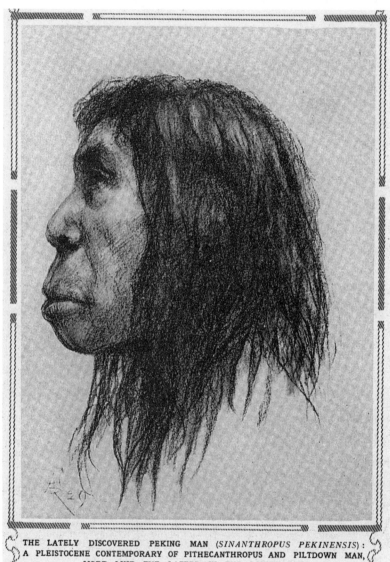

THE LATELY DISCOVERED PEKING MAN (*SINANTHROPUS PEKINENSIS*):
A PLEISTOCENE CONTEMPORARY OF PITHECANTHROPUS AND PILTDOWN MAN,
MORE LIKE THE LATTER IN THE LOWER JAW.

Figure 14.0. Illustration of Peking Man (*Sinanthropus pekinensis*), August 14, 1930.

Chapter 14

THE MANCHURIAN CATHOLIC AND
THE FUTURE OF HUMANITY

In late 1939, when Poland was overrun by German and Soviet
armies, the French Jesuit priest Pierre Teilhard de Chardin was
in Zhoukoudian, outside Beijing, writing his magnum opus *The
Human Phenomenon*. Already under Japanese occupation for two
years, Zhoukoudian was famous as the site where the remains of
"Peking Man" (*Sinanthropus pekinensis*) had been progressively
unearthed since the early 1920s. Teilhard had worked at the site for
almost fifteen years. In letters to friends in France, few of which ever
arrived, Teilhard reported on his shock that "the West was ablaze,"
horrified as he already was about a "Far East flooded by nature and
laid waste by an insidious invasion."[1] Still, he would not be deterred.
His grand theory posited that humans had emerged (even *emanated*)
out of the earth's biosphere, had become ever-more complex, and had
transformed the planet by forming a new mental and spiritual layer
around it. "No, a thousand times no!" Teilhard roared, "however
tragic the present conflict may be, it contains nothing that should
shake the foundations of our faith in the future."[2] The future had to
be wrested back from a "barbaric" authoritarianism of the strongest.
Christianity and ostensibly all the world dreamt of the convergence

into one humanity.[3] This humanity would ultimately become one with God, in a mystical union he called Omega. Teilhard stayed in China, finished his book, and after the war, submitted it for approval to Jesuit authorities in Rome, hoping against hope that they would allow him to publish it.

In the drama of modern paleontology, few cut a figure as singular as Teilhard—so pious and anguished, so committed to an evolutionary theory of humanity's origins and future, so distrusted by the Church. Protestant pastors, concerned after the 1925 Scopes Trial in the United States with the spread of evolution in secular education, laid the foundations for the creationism that is today the archnemesis to evolutionary theory. For its part, the Vatican rejected evolution, though it was a bit squeamish about engaging in conflicts with scientists. Teilhard seemed to walk alone: a frocked scientist who believed as firmly in evolution as in Catholic theology. He and his supporters choreographed his public persona and shared typescripts of his writings.[4] In the years after his death in 1955, he became a posthumous superstar, celebrated among anticolonial intellectuals, ecologists, scientists, and counterculture gurus.

PIERRE TEILHARD DE CHARDIN WAS BORN IN 1881 IN AUVERGNE. HIS family had been ennobled in the eighteenth century. After first studying physics and chemistry, he turned to geology and paleontology; he spent two years in Egypt carrying out geological research. When he returned, he became a star student of the paleontologist Marcellin Boule at the Museum of Natural History in Paris. Once a jawbone and skull were "discovered"—in reality forged—in 1913 at Piltdown (in England), Boule sent Teilhard to the dig as his representative, where he found a planted "ancient" tooth.[5]

Until it was revealed as a fraud in the early 1950s, Piltdown had held great value. Before it, European scholars had two "missing links" in the chain from ape to human: Java Man or *Pithecanthro-*

pus, discovered in Indonesia, and Cro-Magnon Man, an early *sapiens*, discovered in France. Neanderthals and "Heidelberg Man" were not believed to be precursors of *sapiens*. So "Piltdown Man" offered British prehistorians the golden chance to stand up to the dominant theory that East Asia was *the* site of human origins.[6] The hominid was discovered less in that quarry in Sussex than in a very well-tilled conceptual ground. Some scientists rejected it as a fantastical chimera: the skull and jaw, they argued, could not belong to the same species.[7] But all most could see was that right here, in England, the missing link had been recovered.

After Piltdown, Teilhard went to dig at Santander in Spain, where he wrote beautiful reports to his parents on the excavation: "we go up to the cave, in our outlandish rig-out, and stay there till six in the evening, in the open air and wonderful sunshine, with a magnificent view in front of us . . . Seeing these traces of a mankind

Figure 14.1. Painting by John Cooke (1915) of scientists discussing the Piltdown cranium. Note Grafton Elliot Smith pointing to it, Arthur Keith examining it, and Charles Dawson just in front of Darwin's painting. Dawson is generally "credited" with the forgery.

earlier than any known civilization really gave us something to think about; it's wonderful to stand in front of it, alone, in an absolute silence that is broken only by the sound of water dripping from the stalactites."[8] His romantic description moves between the beauty of the landscape and the labor of science: he grasps the world with eyes and hands. Evolution, Teilhard decided around this time, should not be understood as a material matter. It is a spiritual scene.

Teilhard was then drafted into the French army in World War I, serving as a stretcher-bearer. The carnage did not change his mind. In astonishing passages reminiscent of soldier-mystics like Franz Rosenzweig and Ernst Jünger, Teilhard described the terrible war as a step forward for human transformation. He proposed that the front was not no-man's-land, "but the 'front of the wave' carrying the world of man towards its new destiny . . . you seem to feel that you're at the final boundary between what has already been achieved and what is struggling to emerge."[9] With his unit fighting in the Battle of Verdun to recapture Douaumont, a quaint town with a half-destroyed medieval fort, Teilhard declared ecstatically that retaking Douaumont would "mark and symbolize a definitive advance of the world in the liberation of souls." He committed to his work "with all my soul"—a mad bravery that bordered on adoration of war. "If I am destined not to return from those heights I would like my body to remain there, molded into the clay of the fortifications, like a living cement thrown by God into the stone-work of the New City."[10] The human rise toward such ecstasy pointed toward God, destiny, the "New City": human transformation was now to be his central scientific concern.

In May 1918, six months before the war ended, Teilhard took his vows and retreated to the theologically innovative Jesuit seminary in Jersey, the English island off the coast of France.[11] He began writing his own (now well-known) prayers. In them, the grandeur of heaven dovetails with a sensuous planet and the toil of humanity. Consider this passage, part prayer and part essay, where the blessing shifts from the earth to evolution to God.

Blessed be you, harsh matter, barren soil, stubborn rock: you who yield only to violence, you who force us to work if we would eat. Blessed be you, perilous matter, violent sea, untamable passion: you who unless we fetter you will devour us. Blessed be you, mighty matter, irresistible march of evolution, reality ever new-born; you who, by constantly shattering our mental categories, force us to go ever further and further in our pursuit of the truth ... Blessed be you, universal matter ... you who by overflowing and dissolving our narrow standard of measurement reveal to us the dimensions of God.[12]

Across the rhythmical invocation of matter—harsh, perilous, mighty, universal—Teilhard glides through scenes of violence and the creativity of evolution as an ineluctable force. The spiritual unity of everything, minerals included, unfolds as an attempt to redeem the world.

"Agreement will be reached quite naturally," Teilhard insisted, between Catholicism and evolution. Science and dogma would join "in the burning field of human origins. In the meantime, let us take care not to reject the least ray of light from any side. Faith has need of all the truth."[13] In 1919, he visited China for the first time, and in 1922 he defended his dissertation on mammals in the lower Eocene. He began to teach at the *Institut Catholique* in Paris—a major site of theological instruction. He shared his writing privately, but texts leaked and controversy followed. With the Vatican already hostile to the Jesuits, the shock couldn't have been a surprise. Teilhard was removed from his teaching position, banned from publishing in theology, and threatened with further sanction. In May 1925, he wrote: "it's done—I am being moved from Paris, and the most I can hope for is to be left here for another six months to finish the work on hand, and get ready for going back to China."[14] It's hard not to sympathize when he begs a friend: "Help me a little, I've been keeping up appearances, but inside me there's something like real agony. It is essential that I should just show by my example that even if my ideas appear an innovation, they still make me as faithful as any man to the old attitude."[15] From here on, Teilhard would wear the tight Roman collar of the isolated truth-seeker.

This image of Teilhard was only partly accurate. His friends circulated his sermons and texts and treated them as fetish objects, the possession of which signaled proximity to revelation. His well-positioned contacts conveyed to the Vatican that a defrocking or a full-on refusal of evolution would be tantamount to a new Galileo affair.[16]

TEILHARD RETURNED TO ZHOUKOUDIAN. FOR THE JESUITS, CHINA had once been centrally important; now it was distant enough to neutralize Teilhard's theological challenge and the political danger of his work. What was more, it was a great site for a paleontologist. From 1921 on, and especially in the late 1920s and 1930s, Davidson Black, Weng Wenghao, and others discovered skulls, bones, and fragments of *Sinanthropus pekinensis*. The site became a goldmine for paleontology—and a frustrating prompt for international disagreement. Clashes proliferated among Europeans, the better-funded but disliked Americans, and the Chinese, who were in actual control and used the site to promote Chinese science internationally.[17] Teilhard concentrated on geology, animal remains, and tools. He also participated in the "Yellow Cruise" (see Figure 14.2), a PR stunt for the automaker Citroën, where one group of tank-like modified cars drove east from the Mediterranean to meet another—Teilhard's group of cars—that traveled from Beijing southwest, through the Himalayas, and then northwest toward the rendezvous.

For Teilhard, the advertising adventure was the first of many half-returns. His focus, for the time being, was to use Zhoukoudian to fit human prehistory with the development of nature. The origin of

Figure 14.2. "Yellow Cruise" picture of Pierre Teilhard de Chardin with Dr. Yang Zhongjian, his colleague at the Cenozoic Research Laboratory of the Geological Survey of China. They stand next to the Citroën travelling westward from Beijing. Ürümqi, China, July 1931.

the world and the origin of humanity were ultimately one: an opening of spirit. Teilhard began asking about the possibility of a higher unity—the "Omega point." He published geological, paleozoological, and paleotechnological essays—some three thousand pages in total!—through the printing press of the Cenozoic Research Laboratory of Peking. In 1937 he received a first major award, from the Paleontological Congress in Philadelphia. Throughout, he remained an important if not always a leading member of the lab in Zhoukoudian, both during the directorship of Davidson Black and that of his successor, Franz Weidenreich. Acknowledged for his help in a colleague's article, he scribbled acidly on his copy "I believe that I have more than helped."[18]

Why was *Sinanthropus* significant? Coming after the discoveries of Java Man, Cro-Magnon, and Piltdown Man, and before the post-WWII acceptance of *Australopithecus africanus* (the Taung Child

Figure 14.3. North side of the Zhoukoudian excavation site (where *Sinanthropus* remains were found), China, 1931.

discovered in South Africa in 1924), Peking Man seemed a crucial early link—the first hominid that could be called more human than ape. Teilhard and his colleagues had "never any serious doubt that we are studying the vestiges of members of our own race."[19] *Sinanthropus* helped to establish continuities and differences that undercut the notion of the "linear" movement from *Pithecanthropus* (Java Man, perceived as more ape than human) to *Homo sapiens*.[20] Teilhard saw the vanished hominids as just that—alternatives, paths not taken (see the arrows moving outward from the main stem in Figure 14.4).[21] *Sinanthropus*, like Neanderthal, did not seem a direct progenitor of *sapiens*. But it made tools—a whole bone and antler "industry."[22] It was thus an animating force in Teilhard's transition from the *biosphere* to the *noosphere*. If the biosphere was the space where all living beings exist, then how did something in nature come to *think*?

The concepts of the biosphere and the noosphere that Teilhard was using had been reinvented by the Russian geochemist Vladimir Vernadsky. For Vernadsky, we are wrong to conceive of the earth starting from the bottom up, from its physicochemical structure. Oxygen and the atmosphere make no sense "before" plants: it is living beings that give shape to matter. The earth exists only because of the layer of life that surrounds it. Teilhard thought much the same about the emergence of thought out of life, the noosphere: "it affects life itself in its organic totality, and consequently it marks a transformation affecting the state of the entire planet."[23] Nature became conscious of itself somewhere between *Pithecanthropus* and *Sinanthropus*.[24] Intelligence, latent till that point, now slowly bloomed over the millennia.

If these arguments seem un-Christian, Teilhard was relying on the work of two other thinkers to remain theologically relevant. The first was Henri Bergson, who by 1920 had become the Catholics' favorite secular philosopher (and their favorite Jew).[25] Bergson accepted evolution, but he found it baffling that the complexity of life had been perfected simply thanks to natural selection. Could the eye have evolved without vision as its purpose? (The answer—though not Bergson's—was and is, of course, yes.) Instead, Bergson insisted that life was purposive, that it had a goal. His view proved key for Teilhard's *The Human Phenomenon*. Humanity exists because its purpose had been latent.[26] Even during the long eras when there were only plants and premammalian fish, humanity hung like a ghost over the earth. And so too were the future redemption of humanity and even of the planet itself latent in the present time.

For this redemption, and to attempt to give new purpose to the Church, Teilhard turned to his friend Henri de Lubac, a Jesuit critic of Catholic conservatism and the leading intellectual of the "new theology."[27] De Lubac's theology of the "mystical body" of the Church granted it a new role—to raise all of humanity to a closer relation to God. Intended as a rejection of fascist ideas that subordinated the individual to the collective, De Lubac's beliefs became influential in Catholic ecumenism at the Second Vatican Council in 1962–65. Teilhard's paleontology relied on the same vision: that he, like the Church, was meant to fuse the universe together and lift it up.

FROM ITS VERY BEGINNINGS, TEILHARD ARGUED, THE UNIVERSE WAS suffused with religious and spiritual force.[28] In this distant past, even if no "us" existed to see it, "our" possibility was already embryonic.[29] Teilhard asks the reader to appreciate how life emerged, how animals of the species *Homo sapiens* interacted with other animals, how thought arose out of the biosphere. The beings that stood at the threshold of humanity, *Pithecanthropus* and *Sinanthropus*, allow us to "glimpse a whole wave of mankind."[30] Hominization involved a long list of qualities that Teilhard saw stretching from nature through humanity, and always upward toward God: "sexual attraction . . . the inclination to struggle for survival . . . the need for nourishment, with the accompanying taste for seizing and devouring; curiosity . . . the attraction of joining others to live in society."[31] But now that humans were on the planet, the planet itself changed, as human involvement (the noosphere) wove the natural world back out of the human one. With *Homo sapiens*, with the noosphere, the earth is no longer just a living unity, but a thinking one: "the earth 'gets a new skin'. Better still, it finds its soul. . . ." This birth of thought "is the only thing comparable in order of importance to . . . the advent of life itself."[32]

Human beings are not merely participants in the natural world. They observe it, experience it, enjoy it, and raise it to a new unity. Humans emerged "silently," in the deepest past, without knowing it; now that they are conscious, they are the guardians of the earth's consciousness, they make it into a spiritual earth that can have a relationship with God.

The overall movement of nature reaches thus all the way up to

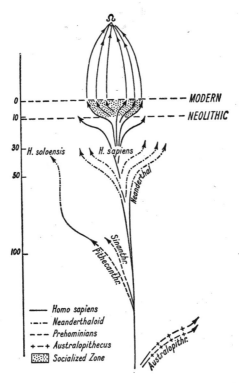

Figure 14.4. "The development of the Human Layer. . . . The
hypothetical zone of convergence on the point Omega is
obviously not to scale." Diagram prepared for Teilhard
de Chardin's *Le phénomène humain/The Phenomenon of
Man* (1956).

the grand complexity of thought and spiritual experience to Teilhard's "Omega point." In the most famous representation of this
idea, included here as Figure 14.4, different human species rise from
nature toward a mystical height, an end of history. *Pithecanthropus, Australopithecus, Sinanthropus,* and Neanderthals move outward and are gradually lost. "Modern" humans—meaning *sapiens*
in general but also Teilhard's contemporaries—lead the way to this
future of convergence and unity. We are all being sucked forward
by evolution toward the Omega point, Teilhard thought, where we
will overcome our limitations, be ourselves, and achieve a mystical
union with God.[33]

AFTER WORLD WAR I, THE HUMANIST DREAMS OF PROGRESS, INDI-
viduality, and dignity were treated with deep suspicion. If you thought
that the world was getting better, how could you explain the mass
death and horror of the war? Or the glaring poverty and dramatic
inequality? Or the violence in empire and capitalism? How about
the rise of fascism? Had humanism failed? Or had it succeeded—
and been complicit?[34] Many Catholics like Jacques Maritain mocked
humanism by insisting that a real, "integral" humanism required
humanity's attachment to (the Christian) God. Teilhard concurred:
his ideas were a promise of "ultra-hominization," ultra in the sense
of "ultraviolet . . . a better organized, more 'adult' version of humanity."[35] Progress was a matter of the universe, not of science or the self.
It would lead to a future "civilization of the universal."

The belief in cosmic progress wasn't specific to Teilhard. Soviet
scientists, their fellow-travelers in Europe, like J. B. S. Haldane, and
some progress-oriented liberals like Julian Huxley also espoused a
nonreligious version of the idea. Huxley came to a position strikingly
close to Teilhard's. He had spent the 1920s and 1930s bridging his
liberalism with his eugenicism. He eventually recognized the threat
of Nazi eugenics and opposed its racism. After the war, Huxley was
appointed the first Director-General of UNESCO, and he advocated
for what he called "transhumanism." Seeing Teilhard as a kindred
soul, he proposed that humanity now could control evolution and
the fate of the universe. "As a result of a thousand million years of
evolution, the Universe is becoming conscious of itself, able to understand something of its past history and its possible future."[36] Teilhard
similarly called the universe a "collector and custodian of consciousness."[37] As humans are conscious of themselves, Huxley went on, so
too the universe can be said to have become, through them, conscious
of itself. Humans are the "business managers" of evolution, the ones
who can know, control, transform it all. "The cosmic office," Hux-

ley called it: "a vast new world of uncharted possibilities awaits its Columbus."[38] Unlike the advocates of humanism in the 1920s and 1930s, Huxley and Teilhard argued that it was essential to mobilize humanity to transform itself into something bigger.

Huxley was ambivalent about "humanism" alone, sometimes speaking instead of "scientific humanism" or "evolutionary humanism."[39] His ideas never veered very far from eugenic fantasies and the scientific management of humanity, even if he was careful not to say as much in public. He and Teilhard developed a complicated relationship: Huxley invited Teilhard to participate in his Idea-Systems Group, a think tank for "geniuses" to influence humanity. But then Huxley pushed him out, deeming him too religious and idiosyncratic. Teilhard, for his part, took the rejection a bit too graciously, at least in public—he replied warmly and with forgiveness to Huxley, but was privately quite scathing toward him.[40]

AFTER WORLD WAR II, THE VATICAN CONTINUED TO SANCTION TEILhard. He suffered from the bans, but also gained stature and notoriety. In 1948, he was offered a prestigious professorship at the Collège de France in Paris. "I went to Rome—where the head of my Order was perfectly nice with me, but finally stopped everything (for practical or semi-political reasons): publication of a book [*The Human Phenomenon*], Professorship at the Collège de France, lectures in the States—nothing was granted."[41] To his critics, Teilhard offered an empty Christianity without Jesus, belief, the Church, or revelation. In 1950, Pope Pius XII further denounced Teilhard in his encyclical *Humani Generis*: "Some imprudently and indiscreetly hold that evolution, which has not been fully proven, even in the domain of natural sciences, explains the origin of all things, and they audaciously support a monistic, pantheistic opinion that the world is in continual evolution."[42]

As with the "Yellow Cruise," there could be no definitive return

to Europe for Teilhard, only half-trips. Yet he was becoming celebrated, both in some secular circles as well as in some religious ones.
With China closed to foreign researchers after the triumph of the
Communist Party in 1949, he moved to New York City. He acknowledged that humanity emerged in Africa and developed a following
across that continent as well.[43] The Christian world was changing,
and his career and thought expressed the need to manipulate traditional orthodoxies so as to negotiate with secular ideas. Teilhard's
religious writings—including *The Human Phenomenon*—continued
to be privately printed and circulated, now often with his clear blessing.[44] He acquired the aura of a closely-guarded secret: a deeply religious and truthful scientist (and conversely, a deeply scientific and
truthful religious man). Photographs of him circulated and also supported the aura: his photogenic piety and gentleness didn't hurt.

On Easter Sunday, 1955, Teilhard died of a heart attack. Over
the following two years, several of his works were officially published and widely translated. While the official Catholic reception
in Europe was at best ambivalent, Teilhard was publicly acclaimed
by a surprising number of visible scientists.[45] One UNESCO conference, "Science and Synthesis," celebrated him as a figure on par with
Einstein. Huxley now wanted a piece of his eminence—and wrote a
preface to *The Human Phenomenon* for its posthumous publication
in English in 1959. Teilhard also became influential among anticolonial intellectuals in both Africa and the Caribbean. Most significant
among them was the leading Négritude poet Léopold Sédar Senghor,
who was elected the first president of the newly independent Republic
of Senegal in 1960. Senghor found in Teilhard a way to reconcile his
anticolonialism, Marxism, and deep Catholicism. The same year as
he became president, he wrote on Teilhard, explaining how the late
Jesuit enabled him to develop his own prayers of human unity and
material equality. Négritude would help create "a new humanism,
more human because it will have reunited in their totality the contributions of all continents, of all races, of all nations."[46]

While the details of his paleontological work faded into oblivion, Teilhard was becoming a key inspiration for theologians from the developing world including those associated with liberation theology—a socialistic and revolutionary form of Catholicism that pressed the Church to tend to the poor of the Global South—and ecologically minded Christians who sought ways to imbue the earth with spiritual meaning.[47] To Oprah Winfrey and New Age authors who proffer spiritual unity (for example, Rhonda Byrne, author of *The Secret* and a series of sequels), Teilhard has been an ideal reference point.[48] And today, the Vatican has reappropriated Teilhard. Popes John Paul II and Benedict XVI presented his consecration of the world as (per Benedict) "the great vision of Teilhard de Chardin: in the end we shall achieve a true cosmic liturgy, where the cosmos becomes a living host."[49] Benedict had been a supporter of Teilhard's since his earliest days as priest and theologian. The trend toward acceptance, if not fervent embrace, has continued under Pope Francis I.[50] One recent graphic novel even presented Francis's election as the result of a conspiracy by renegade Teilhardian intellectuals.[51]

Why all this advocacy? Teilhard developed an evolutionary theory that avoids the pitfalls and antiscientific crudeness of "Intelligent Design" theories popular among evangelicals and that allows Catholic leaders to proclaim themselves warm to *some* version of evolutionary theory. So long as Teilhard's approach remains metaphorical, a "great vision" in Benedict XVI's words, it remains possible to glide over its details, to treat its idiosyncrasy as mystical vision, to find profound meaning in his care for the earth, for the transformation of humanity, for the "cosmic liturgy." Thus, while Teilhard's variant of evolutionary theory has justifiably remained very controversial among scientists, it has been thoroughly enabling to Catholic thinkers cognizant that without it, the Church stands far apart from visions of humanity's genesis.

Amid all the celebrations of Teilhard since his death, two quieter citations of his Omega stand out. These are Chris Marker's cult film *La Jetée* (1960) and, more famously, Stanley Kubrick's *2001: A Space Odyssey* (1968). Both include sequences that represent the Omega point. Chris Marker had begun as a Catholic-influenced leftist thinker who wrote for the Catholic journal *Esprit*, worked occasionally for *Présence Africaine*, published photo-books, and then moved sharply to the left, criticizing colonialism in the film *Les statues meurent aussi* (1953). In *La Jetée*, European civilization is destroyed in a nuclear war and humans go underground. The surviving scientist-leaders first send the imprisoned protagonist back in time to learn the lessons of the past (he spends time at the Museum of Natural History in Paris). They then ship him forward in time so he can be received and humanity can be redeemed by the all-knowing humans of the future. Just before he meets them, he passes through a barrier resembling a petri dish, as if crossing through the beginnings of life. The all-knowing future humans await him at this new Omega point. In Kubrick's *2001: A Space Odyssey*, after the astronaut Dave defeats the supercomputer HAL-9000 and learns the purpose of the mission he and the other crew had been sent on, he enters into the so-called "stargate sequence," which is a perfect visualization of Teilhardian self-transcendence. Having brought technology back under human control, Dave overcomes himself. Lying on his deathbed in an unearthly room, an aged Dave stares at a mysterious black monolith and is promptly replaced by a fetus in a transparent ball. The camera gaze moves through the monolith to reveal the same cosmic fetus confronting the planet. The noosphere of the future stands free of the earth altogether.

Part IV

===

THE NEW SCIENTIFIC IDEOLOGIES; OR THE HORROR, PART II

(SINCE 1930, AND STILL ONGOING)

VOLUME III — N° 6 — 7   PRICE : 10 Cents (U. S.), 6 Pence (U. K.), or 20 Francs (FRANCE).   JULY-AUGUST 1950

Courier

PUBLICATION OF THE UNITED NATIONS EDUCATIONAL   UNESCO   SCIENTIFIC AND CULTURAL ORGANIZATION

FALLACIES OF RACISM EXPOSED

UNESCO PUBLISHES DECLARATION BY WORLD'S SCIENTISTS

MORE than fifteen years ago, men and women of goodwill proposed to publish an international declaration which would expose "racial" discrimination and "racial" hatred as unscientific and false, as well as ugly and inhuman. The world at that time was running downhill toward World War II, and so-called "practical" considerations prevented publication of the statement — even if they could not prevent the war.

False myths and superstitions about race contributed directly to the war, and to the murder of peoples which became known as genocide — but victims of the war were of all colours and of all "races". Despite the universality of this agony and destruction, the myths and superstitions still survive — and still threaten the whole of mankind. The need for a sound unchangeable statement of the facts, to counter this continuing threat, is a matter of urgency.

Accordingly, Unesco has called together a group of the world's most noted scientists, in the fields of biology, genetics, psychology, sociology and anthropology. These scientists have prepared a historic declaration of the known facts about human race, which is reprinted in this issue of the Courier.

Unesco offers this declaration as a weapon — and a practical weapon — to all men and women of goodwill who are engaged in the good fight for human brotherhood. Here is an official summary of the conclusions reached in the declaration :

● In matters of race, the only characteristics which anthropologists can effectively use as a basis for classifications are physical and physiological.

● According to present knowledge, there is no proof that the groups of mankind differ in their innate mental characteristics, whether in respect of intelligence or temperament. The scientific evidence indicates that the range of mental capacities in all ethnic groups is much the same.

● Historical and sociological studies support the view that the genetic differences are not of importance in determining the social and cultural differences between different groups of **Homo sapiens** and that the social and cultural changes in different groups have, in the main, been independent of changes in inborn constitution. Vast social changes have occurred which were not in any way connected with changes in racial type.

● There is no evidence that race mixture as such produces bad results from the biological point of view. The social results of race mixture, whether for good or ill, are to be traced to social factors.

● All normal human beings are capable of learning to share in a common life, to understand the nature of mutual service and reciprocity, and to respect social obligations and contracts. Such biological differences as exist between members of different ethnic groups have no relevance to problems of social and political organization, moral life and communication between human beings.

Lastly, biological studies lend support to the ethic of universal brotherhood ; for man is born with drives toward co-operation, and unless these drives are satisfied, men and nations alike fall ill. Man is born a social being, who can reach his fullest development only through interaction with his fellows. The denial at any point of this social bond between man and man brings with it disintegration. In this sense, every man is his brother's keeper. For every man is a piece of the continent, a part of the main, because he is involved in mankind.

(See pages 8 and 9 of this issue for the full text of the important statement on race, published by Unesco on July 18th, together with an article, "Race and Civilization", written by Dr. Alfred METRAUX, the well-known American anthropologist.)

Figure 15.0. "Fallacies of Racism Exposed." Cover of *UNESCO Courier* (July–August 1950).

Chapter 15

DARWIN IN THE
AGE OF UNESCO

It took until the 1930s for someone to frontally attack the idea
that Indigenous peoples are literally "primitive." This someone
was the American anthropologist Ruth Benedict. In her book
Patterns of Culture (1934), she bluntly noted that "man everywhere
has an equally long history behind him. Some primitive tribes may
have held relatively closer to primordial forms of behaviour than
civilized man, but . . . our guesses are as likely to be wrong as right.
There is no justification for identifying some one contemporary
primitive custom with the original type of human behaviour."[1] The
book would become a landmark in antiracist anthropology, with
a profound influence on the way scholars conceive difference and
human origins. Benedict offered an alternative reason for seeking
out Indigenous peoples: the more remote from Western society they
were, the more likely it was that characteristics other groups shared
with them were universal.

Benedict died in 1948, but two years later, her French colleague
Claude Lévi-Strauss wielded her hammer in a pamphlet, *Race and
History*, that he wrote for UNESCO. In a single hard downward
swoop, Lévi-Strauss exploded several longstanding assumptions. He

denied that Indigenous peoples, especially the poorest among them, lived in humanity's infancy: "There are no peoples still in their childhood; all are adult, even those who have not kept a diary of their childhood and adolescence." No, they were not "less developed," nor did "stages" of development exist, like steps on the staircase of progress. No, there are no peoples without history—it's just that most people are not aware of other peoples' histories. Each Indigenous history "is and will always be unknown to us." His point, so close to Benedict's, was deceptively simple. "All human societies have behind them a past of approximately equal length. . . . For tens and even hundreds of millenaries, men there loved, hated, suffered, invented and fought as others did."[2] Terms like "primitive" would no longer do.

Today, when we have seen so many efforts at antiracism fall short, it is difficult to grasp the effect Lévi-Strauss meant for these sentences to cause. So it is worth recovering the moment when, after the fall of the Third Reich, the politics of human origins emerged in the biggest possible forum. With international support, social anthropologists tried to design a new standard: humanity had emerged as one, it was one, and its differences were not racial.

At the time, intellectuals who had fled Europe before or during the war were often so appalled by what they saw on their return as to immediately behave like ethnographers. Many turned to metaphors from nature and prehistory. Alfred Métraux, a Swiss-born ethnographer, was working for the US Strategic Bombing Survey in 1945 when he witnessed "this massive obliteration of urban life, of all that makes our civilization." Germany "has been smashed like a nest of termites . . . Are we really going to witness a total cultural regression?"[3] Two years later, the German writer Alfred Döblin compared Berliners—whose wild, bustling lives he had written about in *Berlin Alexanderplatz* (1929)—to mute cave dwellers (the very definition of troglodytes).[4] The annihilation of the Jews of Europe was harder still to fit within stories of the ascent of humanity. The Holocaust was slowly coming to be recognized as absolute, almost unintelligible brutality. Already in 1946, Hannah Arendt called it the most difficult story to tell, noting how the extermination camps forced Jewish victims down to "the lowest common denominator of organic life itself, plunged into the darkest and deepest abyss of primal equality, like cattle, like matter, like things that had neither body nor soul, nor even a physiognomy upon which death could stamp its seal."[5]

In that moment, Ruth Benedict's call for anthropology to "make the world safe for human differences" drew a great deal of attention. The limits of "humanity" were manifest—and the peoples of Europe could make almost no claim on civilization.[6] If Europe was so capable of savage war and barbaric rule—as non-Europeans had long known, experienced, and said—what was left to gloat about but a whole world in ruin?

Such feelings were particularly strong among anthropologists, and they found expression most notably in UNESCO, one of whose founding tasks was to develop a world without racism.

THE UNITED NATIONS EDUCATIONAL, SCIENTIFIC AND CULTURAL Organization (UNESCO) was born of the Second World War and the internationalist impulses of the Allies. Proposed in the San Francisco meeting where the UN Constitution was drafted, UNESCO had a constitution in November 1945 and started its work in earnest a year later. The preamble of its constitution declared that World War II was "made possible by the denial of the democratic principles of the dignity, equality and mutual respect of men, and by the propagation, in their place, through ignorance and prejudice, of the doctrine of the inequality of men and races."[7] But then UNESCO's first Director-General was none other than Julian Huxley, who still dreamt eugenic dreams and imagined humanity as the new "managing director of the biggest business of all, the business of evolution."[8] Huxley pushed UNESCO in the direction of "One World

Culture," a sense of humanity as unified, as "Man." Its headquarters was established in Paris, and the organization began working on several major projects, including illiteracy, "International Tensions," the "children of Europe," and "the field of less-developed peoples."[9] Early on it was decided that ending racism would be a central aim of the organization.

UNESCO hired social scientists for key positions and recruited even more of them for the committees that pursued its goals. Huxley, for his part, passed the baton to Jaime Torres Bodet, formerly Mexico's Foreign Minister, who was more invested in differences between cultures than in a single humanity. By 1949, UNESCO was playing host to a cascade of scientific committees, but it was also armed with a serious, perhaps unprecedented public relations operation.

The first committee to draft a "Statement on Race" met in December 1949; it was made up of little-known authors from India and New Zealand to Brazil, Mexico, and France. Among them was Juan Comas, a Spanish-born Mexican anthropologist who had taught children orphaned by the Spanish Civil War and had sought to develop a physical anthropology that would not be held hostage by race. American anthropologist Ashley Montagu had trained under Ruth Benedict and in his early work had denounced "ruthless cruelty, injustice, dispossession, and wholesale murder" of aboriginal populations like the Arrernte. He had then published several works on anthropological method.[10] Reviewers were often negative—criticizing him for over-reaching in his activist fervor and undermining his own case.[11] And then there was Lévi-Strauss, a French Jew who would soon become the most famous anthropologist of his generation. The whole enterprise was supervised by Alva Myrdal, head of the Social Science division of UNESCO, who counted among her deputies the anthropologist Alfred Métraux (the anthropologist who compared Germany to a destroyed termite nest) and the sociologist Robert Angell.

It is not entirely clear how the Statement was prepared. Outside the committee, Montagu was believed to have taken advantage of his position as *rapporteur* to write it basically on his own. What he produced carried hints of Huxley's "One World Culture" vision, even though he was no longer the Director-General. The Statement opened by announcing a scientific consensus that humanity "is one" and that "all men are probably derived from the same common stock."[12] Race had caused "untold suffering," a heavy toll, an "enormous amount of human and social damage." Any biological foundation in race "should be disregarded from the standpoint of social acceptance and social action. The unity of mankind from both the biological and social viewpoints is the main thing."[13]

Fighting words, and wonderful ones indeed. They caused a world of trouble. While the Statement was still in draft form, Angell shared it to be "checked by a few outstanding scholars."[14] Among them: the Swedish economist Gunnar Myrdal, husband of Alva and author of the landmark work on racial inequality *An American Dilemma: The Negro Problem and Modern Democracy* (1944) and the Canadian psychologist Otto Klineberg, who was the public face for statistics that showed there was no innate difference in intelligence or psychological superiority between Caucasian and African Americans. (Klineberg would become a star witness in *Brown v. Board of Education*, the 1954 Supreme Court decision that rejected segregation and that cited both his work and Myrdal's.) They supported the Statement. Huxley was colder: he liked its spirit, he wrote, but "would not like my name to appear on the document" unless certain amendments were made.[15] Geneticist Theodosius Dobzhansky at Columbia University lodged objections to the refusal of race, on the grounds that biologists used the term, though not in the common sense. Still, he too signed the statement.

Montagu's sentiment in response to their criticisms was captured in a phrase from Darwin that he had included in the Statement and now began obsessively defending. We've seen it before, at the end of Chapter 5, expressing Darwin's belief that civilization leads to a universal humanity:

> As man advances in civilization, and small tribes are united into larger communities, the simplest reason would tell each individual that he ought to extend his social instincts and sympathies to all the members of the same nation, though personally unknown to him. This point being once reached, there is only an artificial barrier to prevent his sympathies extending to the men of all nations and races.[16]

The passage may seem innocuous. Did Montagu quote these sentences out of a sincere belief in the promise of international agreement? Or to make his political argument against "tribalism" in the Euro-American context? Or was it an attempt to outflank biologists by appealing to Darwin himself? Today, we can easily recognize Montagu's blindness: he plainly failed to think through what these sentences implied about the "small tribes" that are not "united into larger communities." According to the passage, the small tribes are "less advanced," they have no sympathy for others, they have not learned civilization. Montagu's well-meaning universalism had not really advanced beyond the "civilizing missions" of European empire. The passage shows both the grandeur and the poverty of the Statement.

BY THE SPRING OF 1950, THE STATEMENT WAS NEARING PUBLICATION. Métraux was now reassigned to handle it and he ran without realizing it into a dense minefield. Montagu and other participants in the Statement's creation were still arguing, and Métraux had to spend his time mediating. Huxley disliked the Darwin passage, but Montagu retorted that Darwin "belongs to the world" and should "under no circumstances" be omitted.[17] For his part, Métraux hated the translations of the Statement into French and Spanish, and he reworked them himself. And he worried constantly about its reception. Eventually, UNESCO went ahead and published the

Statement with only small changes to the original draft, also issuing a triumphalist article titled in all-caps: "UNESCO LAUNCHES MAJOR CAMPAIGN AGAINST RACIAL DISCRIMINATION, BRANDS RACE A 'SOCIAL MYTH.'" (The accompanying issue of *The UNESCO Courier* agreed, as Figure 15.0 shows: "Fallacies of Racism Exposed.")[18]

Protests immediately poured in: from the Royal Anthropological Institute in the UK, whose objections also appeared in *The Times*; from Dutch anthropologists; from American critics. To brand race a "social myth" proved a step too far when several disciplines, including genetics and physical anthropology, used it (rightly or wrongly) as a classification device. *Eugenics Review* printed an editorial that it was forcefully opposed.[19] So did *Man*, the journal of the Royal Anthropological Institute, whose authors took to scoffing at "The Ashley Montagu Statement."[20] The UK Ministry of Education wrote in that it "deplores the fact that UNESCO has issued, and given wide publicity to a document on Race which does not command the support of leading physical anthropologists."[21]

It would be convenient to characterize the opponents of the statement as incorrigible racists. Some of them were: Métraux worried that apartheid South Africa would, like the British, exploit the physical anthropologists' objections and attack UNESCO as a whole. But many were reacting to other matters. Geneticists and physical anthropologists were offended about being left out. They thought that they needed the category of race, that they didn't need to be lectured, and certainly that they did not need to be taught antiracism by an upstart American anthropologist who had finagled international backing for his ideas. They also felt that they had gotten rid of the race problem when they developed the "modern synthesis" of evolutionary biology and genetics. Dobzhansky had been profoundly involved in establishing the modern synthesis, and Huxley in publicizing it. The idea was that the modern synthesis had already undercut racism because it showed that populations were characterized by astonishing genetic diversity. Given the sheer extent of this diversity, scientists needed to be able to look at subpopulations and to use classifiers (like race) for subgroups so that particular kinds of variation could be appreciated, rather than lost to a population-wide generality. Dobzhansky, like many others, deeply resented being suddenly implicated by the Statement in Hitlerite racism or segregationism on the grounds that he didn't reject the word "race."

One of the criticisms of the Statement is particularly interesting, because it points to a very real limitation. The geneticists and physical anthropologists asked: If you claim that humanity is simply one and separated only by minor differences, don't you in fact over-standardize the human species? Don't you end up celebrating as normal what is in fact an invented average that doesn't reflect any human group? Isn't "one humanity" too normative, at once bland and violent toward human differences?

Montagu and the other authors thought that they had managed a way around this objection—dismissing "races" in favor of "human groups" or "ethnic groups." The latter terminology dated to the 1930s. Huxley had proposed it together with Alfred Cort Haddon, an English anthropologist who was famous for leading the Torres Strait Expedition but who had also been obsessed with race science throughout his long career at Cambridge.[22] Huxley and Haddon had dismissed racial measurements as an inexact, arbitrary, poorly used method "susceptible of a variety of interpretations" and reliant on poor field research.[23] Now in 1951, critics of the Statement asked if "ethnic groups" were actually any different from "races." Wasn't the category of ethnicity just as vague as "race"? In keeping with anthropological consensus, the Statement also made broad generalizations about "Caucasian," "African," and "Mongolian" ethnic groups. Wait—how were these "ethnic groups" not "races"? Geneticists wondered how UNESCO could in one fell swoop change the word, ignore advances in biology, and insinuate that biologists were racists. Meanwhile, critics on the Left (like the communist biologist

J. B. S. Haldane), insisted that the Statement was altogether benign and boring in its criticism of racial prejudice and should have gone much further.

By the spring of 1951, the situation had become embarrassing for UNESCO, as some states (often with vested racist interests) were openly objecting to the Statement, and Jaime Torres Bodet, the Director-General, convened a new committee for that June that would revise the Statement or write a new one. Métraux played a double game, pressing for support for the original Statement all the while diplomatically appeasing critics with polite letters and mea culpas. He tried to avoid inviting Montagu to the June meeting. Understandably aghast at the vitriol he was receiving, Montagu invited himself anyway, emphasizing his passion for the project and his status as scientist and "not bigot."[24] Métraux was exasperated. "I have seldom been so exhausted," he confessed to the leading anthropologist Margaret Mead (who had also been Ruth Benedict's close friend) in a letter where he berated Montagu's "exhibitionistic demeanors."[25] To Montagu, whom he now started officiously calling Dr. Montagu, he responded, "I do not think we can present the former Statement as 'The Truth.'" The anger in his terseness is palpable: you messed it all up, Montagu, I'm left picking up the pieces, I'm now obliged to hand over control to those damn biologists, who think it's all about their disciplines and are anxious about race "mixtures."

The second Statement that the June meeting produced was less confident and provocative than the first.[26] It succeeded quite brilliantly in satisfying no one. Mountains of further correspondence poured in from the experts, specialists, and authorities to whom Métraux wrote for their further opinions. The UNESCO office sought to present the second Statement as a complement to the first, in that it represented biologists and not social scientists, and in that it testified to the complex diversity of modern science. Everyone knew this was euphemistic language and no satisfactory Statement on Race could be agreed on.

By October, when Métraux was traveling to the United States,

Alva Myrdal brusquely recommended that he take a vacation after the fiasco.[27] Even the Director-General was fed up. Myrdal informed Métraux that Torres Bodet had summarily rejected her request for funds to publish more about the "Race Question" or push for a race resolution in UNESCO's 1953 General Meeting. Torres Bodet recognized the political danger of UNESCO overpromising and underdelivering. He also worried that "we would not be able to steer a middle course between metropolitan interests and those of the freedom-seeking peoples." Decolonization had finally reached UNESCO: Who would the organization stand for? The two Statements had meant well but had done nothing more than express competing and confused metropolitan interests.

AT THE BEGINNING OF HIS TENURE IN 1950, MÉTRAUX HAD COMMISsioned four pamphlets. One, by the French memoirist-anthropologist Michel Leiris, used Ruth Benedict's work to articulate his anticolonial sentiments and echo those of his close friend, the Négritude poet and anticolonial activist Aimé Césaire.[28] Another, *Racial Myths*, by the Spanish-Mexican anthropologist Juan Comas, a member of the committee to draft the first Statement, made clear that social Darwinism amounted to a genocidal politics. Tactfully (if wrongly), Comas left Darwin blameless. It was the colonial powers that had used his work to legitimize the idea "that slavery or death brought to 'inferior' human groups by European rifles and machine-guns was no more than the implementation of the theory of the replacement of an inferior by a superior human society."[29] The other two pamphlets handled race prejudice and the relationship between race and biology. When everything blew up over the first Statement on Race, Métraux took advantage of the crisis to ask Lévi-Strauss for a fifth pamphlet too.

This was the pamphlet in which Lévi-Strauss lifted Ruth Benedict's hammer against Western beliefs about "primitivism." From his very first paragraph, Lévi-Strauss jeered at the original Statement, supporting some of its spirit while attacking its famous points.[30] The racism UNESCO thought it had chucked out by declaring race a social myth had in fact remained right at the Statement's core. Lévi-Strauss rejected its division of humanity into three large "divisions"— Caucasian, African, and "Mongolian." He refused the conceptual quagmire of the idea that "races" had offered "contributions to humanity." How did any of this undercut racism? And if the first Statement was a mess, the second was beneath contempt and altogether undeserving of his time. A plague on both their humanisms.

For the most part, his argument was that of the many younger, more progressive anthropologists. He reaffirmed Benedict in writing that all of humanity had a past approximately equal in length. He concurred with Métraux, who regularly used platitudes about Australia as a young continent with archaic animals and peoples—but in order to argue for the *complexity* of Aboriginal societies. If they resemble "our prehistoric ancestors . . . these primitive people have developed a social organization and a system of relationship of such complexity and refinement that it requires an able brain and a degree of mathematical ability to unravel all its intricacies."[31] "Our" social organization is crude, by contrast.[32]

Lévi-Strauss went one further. Whereas the two Statements on Race had mostly looked at the relations between large ethnic groups (or "races"), he argued that the real targets of racism were Indigenous peoples. They were not prehistoric or primitive; on the contrary. The deepest past could not and would not be accessed through them. One could not do battle with the biologists on their own terrain. But one could hack at their presuppositions and starve them of ideological sustenance.

One of his unnamed foils was Wilfrid Le Gros Clark, an Oxford anatomist and primatologist. He had contributed to the revision of the UNESCO Statement, averring his distaste for racism but nonetheless protecting categories of race. In *The Antecedents of Man* (1959), Le Gros Clark depicted an Aboriginal Australian in a visual continuity, even a little scene, with primates (Figure 15.1). This, for Lévi-Strauss, was a misguided idea of continuity, a racial belief in a "childhood of Man" that linked *some* humans, easily depicted as crude and animalistic, to humanity's prehistory. Against such reasoning and imagery, he played down evolutionary continuity, turning instead to human difference.

Lévi-Strauss had long hated social evolutionism, the "absurd

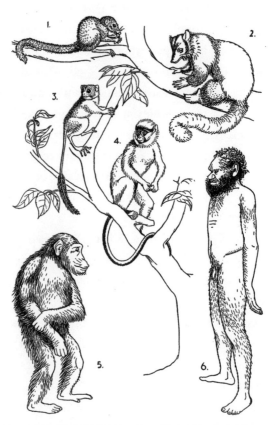

Figure 15.1. From Wilfrid Le Gros Clark, *The Antecedents of Man* (1959): Note the visual contrast between the aboriginal man and the rather genteel chimpanzee who faces him and seems to be the focus of most gazes. Note also the play on gazes: the chimpanzee seems their focus, and we, standing outside the little scene, look in to examine them all.

idea" that by studying social institutions "across tribes and peoples, we can reconstitute the chain of evolution on the social sphere."[33] Now, after the second UNESCO Statement, Lévi-Strauss became far angrier about it. Denying others humanity by treating them as barbarians undermined humanity itself: "The barbarian is, first and foremost, the man who believes in barbarism."[34] The disintegration of native societies was not caused by evolution, nor by colonial violence alone. Western civilization as a whole was to blame:

> Western civilization has stationed its soldiers, trading posts, plantations and missionaries throughout the world; directly or indirectly it has intervened in the lives of the coloured peoples; it has caused a revolutionary upheaval in their traditional way of life, either by imposing its own customs, or by creating such conditions as to cause the collapse of the existing native patterns without putting anything else in their place. The subjugated and disorganized peoples have therefore had no choice but to accept the substitute solutions offered them or . . . to seek to imitate Western ways sufficiently to be able to fight them on their own ground.[35]

Racism had to be understood not on the terms of the first Statement on Race, not on the abstract grounds determined by someone who declared what race is, but as a consequence of the plain destruction of native societies. This argument on social decomposition became the key figure of Lévi-Strauss's work; it also had the peculiar quality of diminishing the effects of specific acts of colonial violence. In his bestselling memoir *Tristes tropiques* (1955), as in his career through the 1960s, Lévi-Strauss argued for the value of non-Western forms of life.[36] Without them, humanity was "in the process of creating a mass civilization, [the same way] as beetroot is grown in the mass." Instead of safeguarding multiplicity, the West had taken to traveling and tourism. Europeans believed they were creating equality when they were, in fact, destroying complexity and difference. Polynesian islands, he wrote, "have been smothered in concrete and turned into aircraft carriers solidly anchored in the southern seas . . . the whole of Asia is beginning to look like a dingy suburb . . . shantytowns are spreading across Africa . . . civil and military aircraft blight the primeval innocence of the American or Melanesian forests even before destroying their virginity."[37] UNESCO, by working to convince an educated Western audience, had failed miserably to understand the difference of others—that "men can coexist on condition that they recognize each other as being *equally*, though *differently*, human."[38]

At the time, Montagu and Klineberg were becoming the public faces of antiracism in America. Lévi-Strauss was trying to outdo them and UNESCO both. Though blunt, he appealed to the same concerns that motivated UNESCO and other public projects—the immensely successful exhibition *The Family of Man*, for instance. First staged at the Museum of Modern Art in New York in 1955, it toured the world with funding from the US Marshall Plan and the Agency for International Development, and it was eventually viewed by some nine million people.[39] The exhibit celebrated a purportedly universal human experience bridging love, work, learning, funeral rituals, children, conversation, and, overall, rather benign forms of pain. The ideology it represented, and some of the links it drew between Europeans and non-Westerners struggling for their civil rights—including lynched African Americans, including Indonesians—*was* new. Quotes from Shakespeare, Sophocles, and Jefferson accompanied the photographs, but so did Sioux, Kwakiutl, Navajo, and African proverbs as well as passages from the *Bhagavad Gita*. The exhibit offered precisely the escapist humanism that Lévi-Strauss detested and against which he invoked the suffering and irreducible otherness of Native peoples. Yet it did not stand so far from his own thinking. *The Family of Man*, Levi-Strauss's *Race and History*, and UNESCO's two Statements all drew on human origins to link diversity with the story of humanity. National Socialism had shattered human history into hierarchy and racial annihilation. "Restoring" the unity of humanity was the express goal of cultural organizations, and as Lévi-Strauss argued, an often-misguided ideology that failed to understand the value of difference.

Figure 16.0. Fernand Windels, photograph of the "Dead Man and Bison" scene in the shaft of the Lascaux cave.

Chapter 16

A History of Cave Painting

In the depths of the Lascaux Cave in Southwest France, a man lies splayed out, gored by a bison. His palms face down. He has a bird's head and beak. The bison too is coded as dead, speared and with its entrails hanging out. The bison appears stylized, wrathful, textured, beautiful. Its head is still twisted to gouge the man.

The bird-man, figured here on the opposite page, is the only human drawn on the cave's walls. But unlike the bison, he's a simplistic outline, a mere husk of a figure. Has he already fallen or is he still in motion, still falling? His avian face does not belong only to him: it also appears on a bird that stands on a stick drawn just underneath. He has but one distinguishing characteristic: his erect phallus.

For all his poverty, this obscure, humble dead man has been forced to speak in response to all the questions put to caves like his: the caves with the oldest art known today.

THERE'S SOMETHING ABOUT CAVES. WE ENTER THROUGH THEIR "mouth," their "yawning," and the sky melts away, the earth too, as we move into total dark. The stillness of the enclosure becomes overwhelming, the lair a damp, disorienting, near-blind matrix against which a visitor whispers, sighs, and sings. Caves lure us, the philosopher Gaston Bachelard recalls, from a world of vision to a world of sound: caves breathe and speak, "they sigh and murmur."[1] To experience the caves, to think about them, is to situate them in a history of the imagination. It is to "catch mythology in the act."[2] It's hard to imagine other landscapes where our perception is so influenced by the literary and religious tradition, so invested with "imaginary voices."[3] Caves mix eroticism, terror, regeneration.

In ancient Greece they were sites of worship for cults of Demeter, Dionysus, and Cybele. In the *Odyssey*, the Cyclops's cave is a putrid mess, a barbaric dump. Can we be surprised that such a terrible guardian, a member of a primal *genos* that "hold no councils, have no common laws," gives no quarter to Odysseus and his fellows?[4] Aeschylus in the *Oresteia* had Athena pacify the Furies by offering them a cave by the Acropolis, from where they would protect the city and guarantee its fertility. Their earthen hospice calmed their ancient rage, but it is a poisoned gift, a threat to the fabric of Athenian life. Should you *ever* opt for civil war, Athena warns the Athenians, the Furies would burst out of the cave, arise again as their ancient murderous selves, and gorge on the city.[5] More famous still is the seventh book of the *Republic*, where Plato imagined human beings chained to the floor of a cave, with light projecting from a fire behind them to the wall in front. They only see the shadows of movements taking place behind them and imagine these outlines as the entirety of reality.

Since around 1900, Plato's Cave has served as an allegory of cinema—the encavement of humans sitting by one another in a mass hypnosis. But it has also hinted at a different authenticity specific to the earth. How beautifully the outlines projected by firelight recall the paleolithic cave paintings discovered starting in the 1880s.

After Plato, the cave hardly lost its promise for the imagination.

Figure 16.1. Peter Paul Rubens, *The Entombment*
(about 1612).

The Christian tradition is replete with images of Jesus entombed and
the Angel, beside the tomb, announcing the Resurrection. In one
version of Rubens's *Entombment* (see Figure 16.1), Mary gazes up,
beseechingly, but a shadow has already started to fall across her face,
and the darkness of the looming cave seems to close off any access
to the divine. Despair reigns. Even more intense is Hans Holbein
the Younger's *The Body of the Dead Christ in the Tomb* (1520–22,
Figure 16.2): Jesus spread wide, bloody palm facing down, his skin
sinking into the corpse inside a flat ossuary. Jesus's tomb stands in
for the site most distant from the Resurrection itself—an abyss where
a humanity bereft of life resides, awaiting its redemption. What follows is of course the Resurrection—the breakout from the cave, the

Figure 16.2. Hans Holbein the Younger, *The Body of the Dead Christ in the Tomb* (1520–1522).

overcoming of the terrestrial realm, the triumph of the spirit. Sometimes paintings depict the angel sitting on the tomb's rock (as in Pieter Brueghel the Elder's *The Resurrection of Christ*, ca.1560), sometimes Jesus himself bursting out (as in El Greco's *Resurrection*, 1600).

By the 1500s, when most of the paintings just mentioned were made, Europeans were fascinated by Roman catacombs. First, these were seen as scandalous pagan sites.[6] Then, in the later Reformation, the Vatican appropriated the catacombs for its war against Protestantism and presented them as sites of early Christian purity against Roman persecution.[7] The story went that Mass had been established in the darkness of the catacombs, in the authenticity of the hidden community of martyrs. Leaving the cave was akin to transcending the old earth.[8]

The Romantics obsessed just as much about the caves. William Blake took advantage of the scary warmth of the earth and the blindness of the cave dweller to set up his poetic prophesies. Some seventy years later, Nietzsche found the cave an excellent headquarters for Zarathustra to plan how he would transform all values. And literary moderns from Victor Hugo to Virginia Woolf and D. H. Lawrence concurred: caves are depths for mystical experience, places of reprieve and regeneration.[9]

Caves were not just the stuff of the imagination and of religion and art. The discovery of Brixham Cave (today Windmill Hill Cavern) in Southwest England in 1858, just months before Darwin published *On the Origin of Species*, sparked a rush to find more caves with ancient bones and flint tools. The later nineteenth century was the new age of mining—if manmade industrial digging was a dirty, grotesque business, what could be more appealing than primordial, pristine caves? Who had been their occupants? How had they lived? What significance might these ancients hold for modern conceptions of the human?

In the fervor for ancient caves, we get to catch mythmaking in the act, starting with the first discovery of a cave full of paintings in 1879: the Altamira Cave in Spain.

Only two such caves had been found before 1879, in the Pyrenees, and even those caves were ignored. Why just two? Did people simply avoid them? Had other caves been destroyed as supposed sites of witchcraft? The earliest comment on cave art dates to 1878, by the prehistorian Léopold Chiron at the Anthropological Society of Lyon. But his description went unnoticed, partly because the carvings he had found were not especially striking. In 1879, as the story goes, the agronomist-archeologist Don Marcelino Sanz de Sautuola was searching for portable paleolithic art—bones or stones with scratch marks, shaped like the replicas of paleolithic objects that he had seen at the 1878 Universal Exhibition in Paris. His eight-and-a half-year-old daughter Maria ventured into passages that led deeper into the cave, then emerged, like Alice crawling back out of Wonderland, shouting: "*Toros! Toros!*"[10] She had looked up at the ceiling, at an angle too steep for the grown-up to try, and seen bulls and aurochs. This story would be told, turned into fable, celebrated.[11] Sautuola and his daughter came to represent amateur ingenuity, especially because the authorities refused to recognize the authenticity of the cave or the truth of this provincial Galileo.

How could they have? It wasn't only that Sautuola was an amateur, or worse, that his aide-de-camp was a girl. Evolutionary theory was regnant, and Herbert Spencer's variation dictated that art and symbolic/religious ideas could only have emerged very late in human history. The doyen of French prehistory, Gabriel de Mortillet, was appalled by Sautuola's argument: for art, one needed religion, and for religion one needed evidence of burial practices. None had been found, ergo paleolithic spirituality was an oxymoron.[12] Stone-Age art did exist, but it was limited to fetishes and other animal-modeled animistic objects from the Reindeer Age (the late Upper Paleolithic when, thanks to retracting glaciers, reindeer became plentiful in parts of Europe). In an engraving from 1870 (see Figure 16.3), the artist Émile Bayard dated ancient art to a later time, the Bronze Age. For Mortillet and others, religion and the consciousness of death could only have arisen in the Neolithic period. Sautuola must be a fantasist to question evolutionary theory—or a forger. By the late 1890s, Altamira was less suspicious. But no less nettlesome.

The conundrum was resolved by 1903, five years after Mortillet's death.[13] By this point, eight caves had been found, in south-

Figure 16.3. Emile Bayard, "The arts of design and sculpture in the Bronze Age," in Louis Figuier, *L'Homme primitif* (1870). Like the leading prehistorians of their day, Bayard and Figuier could not imagine art emerging as early as it had.

west France and northern Spain, and Émile Cartailhac—a leading archeologist and prehistorian who had been dismissive of Sautuola—visited Altamira, accepted its authenticity, apologized to Maria, and penned an article together with a precocious Catholic priest named Henri Breuil. Cartailhac had been Mortillet's leading student, while the abbé Breuil, aged just twenty-six, was already an authority on the late ice age, and had discovered two more caves (Combarelles and Font-de-Gaume) together with colleagues in 1900–1901. Breuil began copying the paintings—see, for example, his gorgeous reproduction of the "kissing" or "licking reindeer" from Font-de-Gaume, (Figure 16.4). Their article was received as an official prehistorians' communication, an actual authentication—and suddenly a lot began to happen. The French-Jewish art historian Salomon Reinach linked the paintings to early religion even before Cartailhac and Breuil's article was out. (He was already communicating with Breuil.) Reinach was Vice-Chairman of the Alliance Israélite Universelle. He was deeply affected by the virulent antisemitism that was then bursting into political and academic affairs in the Dreyfus Affair. Interested in scholarly attempts to universalize religion, he adapted E. B. Tylor's and James Frazer's claims about animism to the problem of cave painting.[14] Reinach argued that totemism—the celebration of a totem animal by a community—was designed "to assure, by magic means, the multiplication of the animal" on the totem.[15] In other words, for Reinach, people depict what they worship because they don't understand the fundamental difference between the animal and its representation. They think (he claimed) that the totem will intercede so they can successfully hunt or control the animals themselves. The "troglodytes" of ancient Southwest France had believed that to paint bulls was to command real ones. "The image of a being or an object offers a grasp on this being or this object; the author or possessor of an image can influence what it represents."[16] This belief preceded religion.

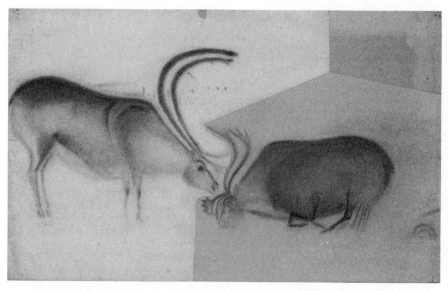

Figure 16.4. Henri Breuil's painting of the "kissing reindeer" at the Font-de-Gaume cave.

The "sympathetic magic" idea made an impression. Breuil and some of his colleagues concurred with Reinach that influencing animals was the goal of the paintings.[17] The venerated Frazer himself was now studying homeopathic or imitative magic and the "savage theory of telepathy in war." He even compiled a list of "savage" nations that used magic to enhance warriors' virility.[18] In his 1904 book *Apollo*, Reinach turned the caves into the first chapter of the history of art. The "realistic" paintings showed that religion and art shared the same origins.[19] Reinach marveled at the effort to render movement in the paintings: the galloping horses, with their many legs, resembled Étienne-Jules Marey's chrono-photographed animals. In his view, sympathetic magic in the caves was a sort of proto-cinema, and it gave birth to universal spirituality. The motif survives: the animated film *Ice Age* (2002) includes a scene where paleolithic paintings are viewed by the animals themselves; the paintings come alive, and Manny the mammoth watches his kind being hunted by the humans who made them.[20]

Figure 16.5. Paul Jamin, *A Decorative Painter* (1903). Jamin was probably the first to copy the "kissing reindeer" from Breuil (Figure 16.4) into his painting. Many other artists followed suit.

The year 1903 witnessed an early attempt to represent the scene of painting. Paul Jamin staged a sexualized fantasy of the prehistoric painter's life: the broad-shouldered painter at the center receives instructions from an older sage, while bare-breasted women and attentive pupils ooh and aah before his creation.[21] On the walls of his cave we see copies of Breuil's early reproductions of bison from Altamira and the deer encounter from Font-de-Gaume (see Figure 16.5). In modernity, Reinach argued, the "magic of art is not meant literally"— but "it once was rigorously true, at least in the belief of artists."[22] That was Jamin's dream: to take for himself the magic art of the caves and unleash it on his own spectators.[23] Other, less sexualized depictions by the artist at work retained the key point: power rested with the artist.

IN SUBSEQUENT DECADES, ENTHUSIASTS AND SCHOLARS DISCOV-
ered many caves, both in southwest France and in northern and

eastern Spain. They unveiled a landscape unlike any seen since the early moderns got excited about Roman catacombs. Caves added human depth to the two sides of the Pyrenees and grafted humanity back onto the glacial world of the last ice age. They also produced yet another picture of glorious Europe by supporting the belief that Piltdown Man and Cro-Magnon Man had been superior to other early humans. Some cave paintings and portable art would be found elsewhere in Europe, and large rock paintings were discovered in Scandinavia, Western Spain, Jordan, Australia, and especially across Africa.[24] But nothing approximated the stylistic complexity of the animals represented in caves in Northern Spain and South-West France.

Or so it seemed, thanks to the work of the abbé Breuil. Breuil had been an energetic participant in prehistory disciplines since about 1900 and became a professor at the Collège de France in 1929. He was brilliant at leveraging his gaze, expertise, and art to raise considerable funds for projects. He built a large bourgeois audience and evaded criticism from the Vatican. For a good sixty years, he was an international star. In cave after cave, he made copies of the paintings and published gorgeous reproductions in massive, lush volumes that were funded by the principality of Monaco and later, more creepily, by diamond ore companies operating in Africa. Breuil was more responsible than anyone else for familiarizing international audiences with the caves and for the view that the paintings represented the birth of spirituality and not some random superstition. This was natural religion, and it began a pathway that, for Catholics but also theorists of art, led inexorably to Christian belief.

One site would become synonymous with Breuil's career. It was discovered on September 8, 1940, only a couple of months after Hitler toured Paris to humiliate France after his armies had defeated it. Once more, the discovery was the work of the young. Three teenagers were playing with their dog near Montignac, a small town in the Dordogne region of South-West France. Like Timmy in Enid Blyton's *Famous Five* novels, the dog vanished into a hole, which turned out to be a cave. They descended, searched, rescued the puppy, and reported their discovery to their teacher, who called in Breuil to authenticate the paintings inside the cave. The boys' story became a cause célèbre amid the occupation, and a source of French pride. For 50 years, Lascaux, 17,500 years old, eclipsed all other caves. Even after it was closed to visitors in the 1950s, it would remain the most famous, photographed, and significant cave in the world. It was nicknamed the "Sistine Chapel of Prehistory," a term now routinely applied to each new impressive discovery of ancient art around the globe.[25]

It was Breuil's reproductions that made the caves famous. Small, awkwardly angled photographs did no justice to the huge, lush, startlingly beautiful aurochs, bison, and horses. In most cases, the irregular, narrow corridors and walls simply didn't offer enough depth for a camera. The paintings in Lascaux "defy the camera," the lament went.[26] But Breuil also had an interesting response to the problem of copying the paintings. How do you put to paper works that were painted and often also scratched into an uneven three-dimensional surface? Breuil isolated the grand animals from each other, removed the interactions between them, and inserted the dynamism of the scene into each figure. On the bison from the Altamira ceiling, for example, the cave painter(s) had used a hump to convey its massive shoulder. The shape of the wall becomes part of the body of the animal; if caves breathe and sigh, surely this wall and bison breathe together. Compare Figures 16.6 and 16.7, Breuil's copy of the bison. Here, as in Breuil's other copies, the wall's texture, cracks, and bulges disappear. He would outline the paintings with great clarity, then abstract them into the two dimensions of lavish burgundys, browns, and grays. Often he worked his way through several drafts, each

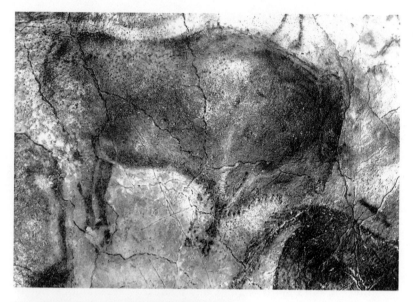

Figure 16.6 (*above*). Hugo Obermaier or Fernand Windels, Photograph of a bison in Altamira, 1920s. First published in Breuil and Hugo Obermaier, *The Cave of Altamira at Santillana del Mar, Spain* (1935).

Figure 16.7 (*left*). Henri Breuil, painting of bison at Altamira. The painting was published along with the photograph (Figure 16.6) so that readers could compare the two.

abstracting from the prior. The bison he devised seems sharper, more compact, bulbous. By centering individual figures, his romanticized work made them easily and immediately recognizable. Thanks to Breuil, the paintings became symbols.

Readers had little choice but to follow his interpretations. In *400 Centuries of Parietal Art* (1952), Breuil dated the invention of cave painting to the moment after the arrival of Cro-Magnons in Europe, roughly 40,000 years ago. Over his career, he had spear-headed the scholarly consensus that styles could be divided into four

Figure 16.8. Charles Knight, *Cro-Magnon Artists of Southern France*, 1920.

particular styles, Aurignacian, Perigordian, Solutrean, and Magdalenian. Each was for a different period and followed the chronologi-
cal division of the Upper Paleolithic devised in the 1880s after areas
where sites with tools and bones had been discovered. Four hundred
centuries ago, art ceased to be some individual talent or "caprice"
and became a social and "spiritual unity . . . [even] an orthodoxy,
which presupposes the existence of a sort of institution that would
rule over artistic development."[27] Breuil echoed Reinach's point
from half a century earlier: paintings confirmed human spiritual
unity. He added, however, the idea about institutions that judged
and trained new artists: an idea that, much as in Christian painting,
knowledge and conventions had been handed down.[28] (This idea
influenced other paleoartists, for example Charles R. Knight who,
as in Figure 16.8, always depicted cave artists in conversation or in
an engagement with their fellows.) What kind of society had generated that kind of institution? What role did the paintings play in
that society? In a slender popular book with amateurish drawings
he made of scenes "beyond the bounds of history," Breuil went a
step beyond Reinach's animistic "spiritual unity" and presented the
art as enabling a shamanistic ritual.[29] In one scene (see Figure 16.9),
the human actors are covered in animal parts; other animals, bears

Figure 16.9. This child-like drawing, published by the abbé Breuil in 1949, almost single-handedly established shamanism as the leading interpretation of cave art and its social function.

for example, have been speared and lie dying. An audience raptly observes the participants. This "hunting magic" would see the lead figure take over the animal. Notice also the sketch on the wall—what Breuil would call the God (at other times the Horned Sorcerer) of the Les-Trois-Frères cave.

BREUIL'S CONTRIBUTION TO THE BELIEF THAT THE ARTIST WAS A shaman had rather little to do with prehistory and everything to do with the place of art and the artist in the twentieth century. In her brilliant book *Transfixed by Prehistory* (2022), Maria Stavrinaki accounts for modern artists' response to the geological, cave painting, and tool discoveries.[30] In 1937, the Museum of Modern Art (MoMA) in New York staged a large exhibit of copies of paleolithic paintings from Leo Frobenius's Institute in Frankfurt. MoMA juxtaposed them to modernist works of art by Pablo Picasso, Joan Miró, and others. On one floor the museum installed prehistoric images, on another the modernist paintings that supposedly bore witness to their importance and impact. This produced a kind of continuity of art across human life.[31] The MoMA's director declared that "200 centuries BC" an antediluvian Adam "drew the animals before he named them." Adam's imagination had survived for all these thousands of years and returned now to influence modern art itself.[32] It was hardly an original argument: over at *Cahiers d'art*, for example, Christian Zervos was juxtaposing Giacometti's sculptures with Cycladic art of the third and second millennium BCE. After the war, other exhibits proposed a link between prehistory and modernity. Lascaux was "liberated" along with the rest of France in 1944 and opened to the public in 1948. The novelist André Malraux visited while serving as France's Minister of Culture and later professed that it had "certainly been sacred, and it still is."[33] At the Institute of Contemporary Arts in London in 1947, a modernist exhibit placed the prehistoric paleolithic works at the very end. You would travel through modern art, then arrive at its true origin.[34] Picasso flamboyantly declared that art had created nothing new since Lascaux.

It isn't hard to see why that idea was appealing, and how it made the artist into a shaman of sorts. Lascaux stayed open for fifteen years. Visitors marveled at the spectacular paintings—all the while the CO_2 they exhaled built up mold all over the walls. Above all, they asked, how were the paintings made? They would have required the painter to work with torches, in near-darkness, and on some sort of scaffolding. The caves were too damp, too difficult to traverse to be real dwellings, but for the same reason, they seemed to be spaces appropriate as sanctuaries. Some of the works were made at such depths as to be almost inaccessible. The art must have been powerful

and sacred to be so special, to require "the participation or approval
of a whole society."[35]

Also: What were the paintings for? A new answer dominated
the postwar period, and it was linked to Breuil's argument: cave
paintings were not merely representations of the hope that a hunter
could control animals (that is, sympathetic magic or animism), but
rather they surrounded priestly scenes of shamanistic ritual. Breuil
and others instead conceived primordial communities as centered
around a shamanic figure who mediates with the world of spirit,
controls or is possessed by the animal, and unites the human world
and the animal world. The paintings were involved in rituals where

Figure 16.10. Rudolph Zallinger's "A Puberty Ceremony" (1961) replays the essential
elements of Breuil's shamanistic painting of 1949 (Figure 16.9). The shaman wears deer
antlers, three males dance in a circle, cave paintings (including the "kissing reindeer" and
Altamira bison, Figures 16.4 and 16.7) adorn the ceiling above.

the shaman *becomes animal* to bind a community together. Breuil, in his 1949 sketch of the Les-Trois-Frères cave, offered an early version of the shamanistic idea. It spread quickly. Like other artists, the Czech illustrator Zdeněk Burian and the American paleo-artist Rudolph Zallinger copied Breuil's reproductions into their own art (see Figure 16.10).[36] Burian in particular loved the shamanism theme. Their shamans are clad in animal headgear, including deer antlers and fur. In a trance before a captive audience, the shaman pulls the community into his ecstasy: he crosses a threshold from his human nature into that of an animal, wearing the eyes of the animal that he fuses with on his mask.[37]

Some of the shamanist imagery used in the twentieth century dated back hundreds of years. The substitute second face with deer antlers is a classic image in shamanistic descriptions going back to the 1690s, when the term "shaman" was first coined by Nicolaes Witsen, a Dutch cartographer and later mayor of Amsterdam, who had traveled in Siberia. His Tungus shaman (Figure 16.11) was the prototype: with his antler headgear, his second pair of eyes, his ani-

Figure 16.11. The first depiction of a "Tungus shaman." Nicolaes Witsen, *Noord en Oost Tartarye* (Amsterdam, 1692).

mal legs and feet, his skin and the fur clothing seamlessly fading into one another.

Despite being undercut and criticized over and over, which it has been ever since, the shamanistic interpretation persists to this day.[38] Why did it prevail? The dream of the artist guiding humanity was powerful—especially among artists and critics. Then, as Nazism, Italian fascism, and Stalinism were blamed on charismatic leaders, the violence of these political ideologies was explained away as an ecstasy that tyrants alone could use to turn their supposedly passive followers into killers. That explanation is wrong and far too easy, but at the time it had purchase. We are rational, they are hypnotized Nazis—and, similarly, our most ancient progenitors were shamanists, which is to say politically controlled by priests and irrational beliefs. What was more, because Burian, Zallinger, and other artists relied on Breuil, the shamanism idea strengthened with each new depiction. Who could paint it best and transport the viewer most vividly to this other world?

Scholars also built an echo chamber. In 1952 the German prehistorian Horst Kirchner published "An Archeological Contribution to the Prehistory of Shamanism," linking contemporary shamanic practices to Witsen's Tungus Shaman and then directly back to prehistory.[39] Everyone talking about cave art and primitive religion suddenly had to deal with Kirchner's article, which so conveniently linked past and present.[40] Romanian-born scholar (and former fascist sympathizer) Mircea Eliade was perhaps even more important, owing to his enormously influential book *Shamanism*. Eliade defined shamanism as a technique of ecstasy, a movement beyond the self. Surrounded by his community, the shaman leaps into a realm between human and animal, bridging the worlds. After first publishing in French in 1951, Eliade settled at the University of Chicago and expanded *Shamanism* for its American edition, now using Kirchner to dilate his own argument all the way back to prehistory.[41] Eliade insisted that the communal force of early religious life must have been especially intense, that it survived in the magical and religious beliefs that followed in later years, and that the shaman's power was the most likely explanation for the intensity.[42] And he thought shamanism had been near-global. He spared Indo-Europeans, but otherwise he claimed it began with prehistory, became dominant in Siberia and central Asia, and was present from the Americas to Australia and southern Africa.[43]

What motivated Kirchner's argument was less the bird-headed "Dead Shaman" in Lascaux than the strange figure from Les-Trois-Frères cave. Breuil called it The God (or "Horned Sorcerer") of the Les-Trois-Frères, and he too used it to insist on the universality of religion and its origin in shamanism. In Breuil's drawing (see Figure 16.12), the God/Sorcerer of Les-Trois-Frères has horns, telltale hypnotizing eyes, well-defined muscles almost transmuted into animal body parts, a tail, and a prominent phallus. On the wall of the cave itself, things are not quite as clear, because the work is partly etched into the wall and only partly drawn. Breuil's flattening technique allowed scratches and paint to congeal into his desired two-dimensional interpretation. But what guarantees the unity of

Figure 16.12. Henri Breuil, The "God" or "Horned Sorcerer" of the Les-Trois-Frères Cave.

the scratch and the painting on the cave wall? Or that they referred to a priest, rather than a grotesque? Or even that they were made at the same time?[44]

By Breuil's death in 1961, the concept of shamanism aligned an idea of universal early religion with the eminence of the painter, the beauty of the cave art, the violence of the imagined ritual, and the political influence of charismatic leaders. It explained the painter as a shaman too. Like Picasso, Breuil enjoyed what this meant for himself: the painter saw and moved where others could not, and like the shaman he plumbed the animal depths and made them accessible to everyone.[45] The Renaissance of cave painting in the twentieth century was built on this myth. Contemporary artists, confronted with an unpleasant, disenchanted world in their own time, couldn't resist.

NUCLEAR WARFARE SUPPLIED THE CAVES WITH NEW MEANING. FOR Pierre Teilhard de Chardin, the atomic tests in Arizona had transformed the very surface of the earth.[46] Fifteen years later, in *Dr. Strangelove*, Stanley Kubrick imagined a post-apocalyptic, fascist future that involved a return to caves "to preserve a nucleus of human specimens. It would be quite easy . . . at the bottom of some of our deeper mineshafts. The radioactivity would never penetrate a mine some thousands of feet deep." French philosopher, sociologist, and pornographer Georges Bataille agreed. He used the potential of atomic extinction to guide his readers back to the beginning of humanity. "Light is being shed on our birth at the very moment when we confront the notion of our death."[47] Bataille vaporized history so as to teleport back to the very beginning and make the present meaningful again.[48] In the "miracle" of Lascaux, the paintings still "look as though they were painted yesterday," Bataille noted in *Prehistoric Painting* (1955), perhaps the first book to include large color photographs of the works.[49] Worried as he was about atomic war,

Bataille was in no mood to celebrate that eternity. The caves were just sites of reprieve from the bleeding of the earth: "The Earth has been disemboweled, yet from the inside of her stomach, what men have extracted is above all iron and fire, with which they never cease to disembowel one another." Lascaux allowed him to swat away conventional attitudes to human history and art. To close the book on the long-standing fetishization of the Greeks. To avoid the language of "genius" that Picasso, Malraux, and others loved and linked to Altamira.[50] To leave the evolutionary ladder aside. To get rid of shaman theory.[51] And he finally tripped over a good response to the philosophers, too: history had not begun with masters and slaves, it hadn't ended with Hegel and Napoleon and Stalin. The humanity that was about to get pulverized in atomic war had begun on the cave walls, in a proto-ecological drama about humans' separation from nature. Humanity was not about a species or about philosophy, it was about *self-recognition* and *death*.[52]

Cave paintings, Bataille argued, recorded how "Man" had first announced himself "clad in the glory of the beast."[53] The ancient human killed animals, then mastered them by painting and deifying them, as though to make amends.[54] Sure, early humans had dreamt of the future, but they had also looked backward to a moment in time when they were still animals. "An animal is in the world like water in water," Bataille proposed in 1948.[55] A beautiful, enigmatic sentence: animals flow in the world, without consciousness separating them. Animals "await for nothing, and death does not surprise them. Death in some way eludes the animal."[56] What had ended the harmony of animals with the world was human consciousness and representation. Consciousness was consciousness of death. We might flatter ourselves that we represent an improvement on animals. But all we gained was the knowledge of death, an incurable wound in our intimacy with the world. Humans might hunt other animals, but now they were conscious that they too will die. Bataille exclaimed that painting had commemorated the particular moment when humans first recognized that the harmony and continuity had ended. "These men made tangible for us the fact that they were becoming men . . . by leaving us images of the very animality from which they escaped."[57] The birth of the human, the birth of representation, confirmed the end of paradise, that earlier time when humans too had been like water in water. Cave painting mourned the loss of a past dispensation, before the destructive human economy was born.

While the painted animals on the cave walls are exquisite, "extraordinarily more detailed," Bataille continued, the painted humans are pitiful, childish stick figures. Or else, like the Horned Sorcerer, they are distorted assemblages of animal parts. Bataille was consumed by the Man-Bison pairing in "the depths of the holiest of holies in the Lascaux cave," and by the dead man's erection in particular. No surprise, perhaps, because the painting had been made as if for Bataille alone. In his infamous pornographic novel *Story of the Eye* (1927), Bataille had staged two bullfights, one leading to the bull's death, one to the matador's. At Lascaux, both were dead. Bataille called the Man-Bison painting "a measure of this world; it is even *the* measure of this world."[58]

Death and arousal: the mutual killing on the wall, the bison-human dance, brought both loss and eroticism. Disguised as an animal, the Lascaux man defined himself as "the king of animals" and then "concealed his humanity behind an animal mask."[59] Humanity was nothing better than a stick figure trying hard to reembrace the animal itself. "The human predator asks forgiveness for treating the animal as a thing, so that he will be able to accomplish without any remorse what he has already apologized for doing."[60] He is, as in Figure 16.0, artful but indecent and insufficient before the animal. Erect, he dies. Which meant that paleolithic art was not really about art, not really about religion either: it was the dream of a lost Eden, a world with humans "like water in water," wishing to retreat back into the animal kingdom. Human-animal relations remained crucial

to later scientific and popular interpretations of the paintings, including in recent ecologically minded works, like those of the artist Trevor
Paglen or the filmmaker Werner Herzog.[61]

A MORE SYSTEMATIC APPROACH TO THE CAVE PAINTINGS WOULD
dominate the next era of interpretation.[62] The star of the new attitude was Annette Laming-Emperaire, who already in her days as
an archaeology student in the late 1940s expressed deep skepticism
toward shibboleths of cave research like Breuil's dating conceit that
referred to the earlier paintings as "Aurignacian," and the later as
"Magdalenian."[63] A trained philosopher who had participated in the
French resistance and only turned to archaeology with the liberation of France, Laming-Emperaire wanted archaeology to become
a hard science. She pioneered a completely different approach that
owed much to Claude Lévi-Strauss's structuralist account of social
organization. Laming-Emperaire was utterly indifferent to existing
theories, except she spent no fewer than 150 pages of her dissertation debunking them.[64] Her predecessors' obsession with anatomy
and ritual, their belief that one could easily make sense of particular signs on the wall, their focus on individual images: all this was
meaningless, amateurish.[65] Instead, the paintings had to be read
together, within each cave and across them all. They were a system
of representation.

To reconstruct this system, Laming-Emperaire mapped, tabulated, and compared details of the works: the length and twists of the
animal horns, the markings on their bodies, the stylistic touches on
the necks, the engravings, the quasi-geometric symbols.[66] The individual paintings lost their value.[67] For example, she noted that the
dead and erect Bird-Man of Lascaux was anything but special: semi-
human figures proliferated in other caves like Pech-Merle, Altamira,
Couffignac, and elsewhere, and semi-humans who resembled birds
were "always in a difficult situation, injured or beaten to death."[68] We

Figure 16.13. Horse and buffalo painted atop one another (1962): Annette Laming-Emperaire abandoned Breuil's approach that separated the animals and she focused on the superimpositions (and gender) of animals depicted.

can't even really consider him, Laming-Emperaire insisted, unless we look first at conventions, techniques of fabrication, even traditions of artistic practice. A millennia-long, infinitesimal development of styles across hundreds of paintings linked to each other and to particular spaces—this was her research domain. Laming-Emperaire also explained the gendered social structure of the painting.[69] Fascinated by the paintings that superimposed bison and horses (as in Figure 16.13), she eventually concluded that the key opposition in the paintings was between bovine/male and equine/female.[70]

This was such original work that even her dissertation advisor André Leroi-Gourhan, an anthropologist-prehistorian to whom we will return, started over, the better to follow in her footsteps. He published his own project, a massive, field-defining book, a decade after hers, and he drove Laming-Emperaire's technique to the extreme, using 2,000+ variables, treating the exact topography of each cave as meaningful, and reading the datasets as revealing an origin for all human metaphysics. He described "a cycle of life's renewal, the actors in which form two parallel and complementary series: man/horse/spear, and woman/bison/wound."[71] Curiously, while they agreed that the cave paintings encoded gender and social structures, Laming-Emperaire and Leroi-Gourhan came to the exact opposite conclusion.[72]

Their shared idea about equines being female and bovines male (and vice-versa) was dead on arrival. Archaeologists laughed at it even as they celebrated and adopted their techniques for the new "processual archaeology" that arose in the 1960s. Even the two of them wondered how they could have reached the exact opposite conclusion.[73] Laming-Emperaire shifted her focus to Chile and Brazil, where she made important discoveries but also supported the theory, widely scorned at the time, that humans had populated South America by traveling on boats across Polynesia (and not across the Bering Strait). She died tragically in 1977.

The most important and interesting of her critics was Margaret Conkey, a leading feminist in prehistoric archaeology, who celebrated Laming-Emperaire's and Leroi-Gourhan's work as massive achievements and yet criticized Leroi-Gourhan especially (and Laming-Emperaire implicitly) of failing to understand the context of the paintings' production.[74] Conkey objected that their structural approaches had focused on a visual "language"—the paintings, sculptures, and inscriptions in the caves—at the expense of studying the societies that had made them. At stake was the role of women's labor. Archaeologists loved projectiles, handaxes, hunting, and other activities coded as male.[75] This came at a cost—on top of scientists presuming the masculinity of these tasks and ignoring others, there was a symbolic cost too, having to do for example with the way gender was imagined and then reinterpreted off the cave walls. Leroi-Gourhan's and Laming-Emperaire's theories still relied on the painter being male, on the celebration of the hunt, the symbolic encoding of traits (recall Leroi-Gourhan's easy attachment to "female/bison/wound"), and so on. Conkey instead turned to the process by which symbols were generated: the relation of human to animal populations at and near the caves, the remnants of flora and fauna, and the purposes of the tools discovered at the sites.[76] Making symbols is a social process and "we need to know how human needs and wants are manifest symbolically in the landscape."[77] Modern interpretations of the caves had suppressed women's labor; as importantly, those interpretations had ignored the kind of social life that would have occurred at the sites themselves. Most archaeologists since Conkey have sought to reconstruct that social life and the process by which early humans made symbols.

More recent discoveries—notably of the Chauvet Cave in Southwest France in 1994—have reprised some of the same questions about humanity and animality. Some cave art has been controversially attributed to Neanderthals.[78] More attention is being paid to non-European rock and cave art. But the theories have not improved by much. Shamanism has returned, notably in the popular work of Jean Clottes, who proposes that an "altered consciousness" was essential to the making of the paintings.[79] In the 1990s the pendulum swung to evolutionary psychology, then to neuroscience. I prefer Bataille's and Laming-Emperaire's myths—but largely because they recognized that they could not offer a full account, that they were interpreting paintings unwilling to be interpreted and were even deceitful. Only Mel Brooks offered an irony superior to Bataille's in his movie *History of the World, Part I* (1981): "And of course with the birth of the artist came the inevitable afterbirth: the critic." Werner Herzog, in his gorgeous film *Cave of Forgotten Dreams* (2010), observes a wolf's and a child's footprints in a recess of the cave and can't help but ask: "Did the wolf stalk the boy? Or did they walk together as friends?"[80]

Herzog longs for an answer. And yet the paintings and the caves themselves refuse that answer. Who can fail to be moved by this terrible elegance that refuses meaning, by the scenes the paintings conjure, by how they force us to think about centuries of human descent, by the way the animals in the paintings now play against the backdrop of a century of modern art, nuclear threat, and ecological catastrophe?

Figure 17.0. "The Dawn of Man": in a bone-smashing frenzy, the "killer ape" australopithecine learns to use the femur as a bludgeon. *2001: A Space Odyssey* (Stanley Kubrick, 1968).

Chapter 17

KILLER APES FOR AN
AGE OF DECOLONIZATION

An ape encounters a black granite slab. Under its enigmatic influence, he begins to hammer a heavy humerus or femur against other bones. With Richard Strauss's *Also Sprach Zarathustra* playing, he fashions a bludgeon. He remembers the felling of a tapir and suddenly, furiously conscious of his bone weapon, he begins to rise toward a bone-smashing ecstasy (Figure 17.0, on the opposite page). He returns to a battleground in contested territory and lures an adversary. In this amphitheater, with their respective bands as viewers, he murders the enemy with his new contrivance. The territory now belongs to his group.

This is the "Dawn of Man," the opening scene of Stanley Kubrick's *2001: A Space Odyssey* and one of the most famous sequences in the history of cinema. The granite monolith was based on Arthur C. Clarke's short stories. But otherwise, for this "Dawn," Clarke and Kubrick relied on an idea developed by a screenwriter turned nonfiction author, Robert Ardrey, who in turn drew on the work of Australian-born anthropologist Raymond Dart.[1] Dart and Ardrey presented humanity as emerging out of australopithecines that were distinguished by one thing: violence.

In 1925, Dart had made a discovery, known as the "Taung Child" or *Australopithecus africanus*. Dart lived in South Africa and had already published on other finds, notably *Proconsul, Dryopithecus,* and the "Boskop skull." He was spending most of his time on racial questions related to cranial divergences. In his 1925 article in *Nature* reporting on the Taung Child, he argued that it was a hominid, a missing link.[2] "*Australopithecus* was not an ape-man like *Pithecanthropus,* but a man-ape. He was a creature who emerged just before the dawn of man."[3] The Czech-born American anthropologist Aleš Hrdlička, based at the Smithsonian Institution, was elated.[4] But in England, Dart's claims were perceived as going several steps too far. His mentors, Arthur Keith and Grafton Elliot Smith, wrote about the Taung Child, as they had done in reaction to his previous work (Elliot Smith even waxed enthusiastic), but they did not endorse his conclusions.[5] Skulls already were to the scientist like diamonds to a miner, only it took great effort to confirm what seemed so blinding to the eye.

Dart had not dug out the skull himself, he had been handed it, so he lacked some key details, like its position in the rock, which would have revealed something about its age. It was a single, somewhat damaged skull, and that of a child. How it related to other fossils discovered in South Africa and the protectorate of Rhodesia (now Zambia) was unclear, because those fossils were surrounded by unanswered questions, too. Plus, Dart had the bad taste to go over his mentors' heads and appeal directly to Darwin, who had speculated—unusually for his time—that humans had originated in Africa. What if Dart's baby was an earlier ape, like his other discoveries? What if the most important characteristic that indicated its humanity, namely the vertical *foramen magnum* (the hole that links the skull to the spine), was distorted like the rest of the skull—and thus an abnormality?

British scientists looked either to Java Man and East Asia, or else to Neanderthals and the forged Piltdown Man in Europe, to identify human origins. Piltdown had all the trappings of a recent human: a large braincase (hence a large brain) and a very English location. In China, the Americans and the French were expending considerable sums on research, and *Sinanthropus* seemed more promising than Dart's find (see again Figures 14.0 and 14.3). After further *Australopithecus* discoveries were made in the late 1930s by Robert Broom, Dart felt vindicated. Still, only once the "Out of Europe" and "Out of Asia" paradigms declined—once the *Sinanthropus* remains were lost during World War II, once Piltdown was proven a forgery, once a generation of researchers had died out—only then did Southern Africa become compelling as the cradle of humanity.[6] Maps from the 1940s, like the one in Figure 17.1, still imagined the "probable origin of man" to be in Asia.[7]

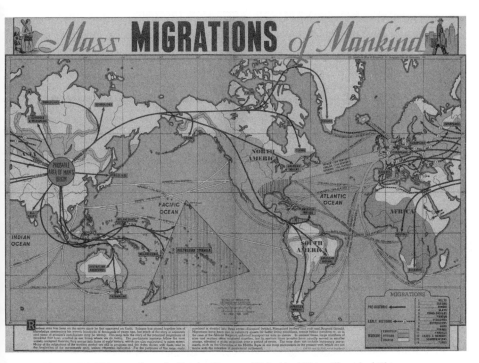

Figure 17.1. This map from the *Sunday News*, published in 1944, still marked China as the origin of humanity (see the large circle), and presented humanity as a long history of migrations.

Around 1950, Dart's 1925 *Nature* article was resurrected and declared prescient. So were many of his other claims. Dart had proposed that "our troglodytic forefathers" had benefited from "open veldt country where competition was keener between swiftness and stealth." That "fierce and bitter" environment was the ideal "laboratory" for humanity.[8] Now, in the postwar era, Dart was doubling down on the centrality of violence. He invented the notion of the broken thigh bone that served as both bludgeon and dagger, named his theory "osteodontokeratic," and declared that the "transition from ape to man" had been predatory.[9] Dart published his theory just as Piltdown Man was coming under scrutiny, only to be publicly exposed as a hoax in 1953. (The forger is mostly believed to have been Charles Dawson, who "discovered" the cranium and mandible, and who had habitually forged other discoveries, though the controversy has snared several others over the years.) More importantly, these were the years immediately following the establishment of apartheid in Dart's adoptive country. He ominously spoke of the "sanguinary pursuits and carnivorous habits of proto-men" in the "blood-bespattered, slaughter-gutted archives of human history." He was fully aware that the South African government was turning the dominating screw, associating Black Africans with primitiveness and violence, expanding what had been British colonial "reserves" into *bantustans*, and forcibly moving "coloured" populations into them. If anything, Dart strengthened the government's rationale. His research had always been about racial difference and the evolution of living bodies out of ancestral races.[10] He even participated in court trials about racial classification.[11] His argument certainly hinted that the brutality of Africans' origins influenced Africans in the present day. He supported the accepted history of the African south, which absolved whites of violence by presenting the San ("bushmen") people as having been devastated not by European colonization but by Bantu and Zulu peoples. And his emphasis on the weapons of *Australopithecus* was new. These implements were either "wielded and propelled to kill during hunting or systematically applied to the cracking of bones and the scraping of meat." His prose was straight-up goth:

> Either these Procrustean proto-human folk tore the battered bodies of their quarries apart limb from limb and slaked their thirst with blood, consuming the flesh raw like every other carnivorous beast; or, like early man, some of them understood the advantages of fire as well as the use of missiles and clubs.[12]

These "killer apes" were the foundation of world history, cannibals who bore into the skulls of their victims to scoop out their brains for eating. Dart compared World War II with what he considered native customs: "the world-wide scalping, head-hunting, body-mutilating and necrophiliac practices of mankind." Hominids differed from other primates owing to "this common blood lust differentiator, this predaceous habit, this mark of Cain."[13] It had all begun with the bludgeon doubling as a dagger, Kubrick's ecstatic ape.

In Robert Ardrey's hands, Dart's discovery would become the first chapter in the "Out of Africa" story. Which it was not.

FROM 1905 TO 1925, SEVERAL DIFFERENT VERSIONS OF THE OUT OF Africa thesis had been proposed. They differed from the version that has now become accepted is that they were culturally, not biologically, grounded. One schema saw origins in West Africa—then in German hands—and another in Southern Egypt—then in indirect British control.

The first was the brainchild of Leo Frobenius, a self-taught German anthropologist and adventurer who first arrived in West Africa in the late 1890s and gradually rose, thanks to the dramas he wove into the material he so persistently collected, to become an internationally respected expert on African culture. He kept "one eye on

primitive vitality and the other on civilized apocalypse."[14] In addition
to cataloguing tales, myths, and art, Frobenius pursued research on
migrations inside Africa, assuming that the expansion of the Sahara
Desert had pushed prehistoric African peoples south. By 1905, he
was half-seriously proposing that Atlantis, the original civilization,
had actually existed—and he proceeded to seek out its ruins.[15]

In England, Dart's discovery of the Taung Child drew him into a
controversy for which Grafton Elliot Smith was directly responsible,
and whose main claim today seems wild. Having spent a decade in
Egypt studying mummified remains, Elliot Smith became obsessed
with the idea that all culture derived from Southern Egypt.[16] Obsessed
may be an understatement, though: together with his friend William
J. Perry, Elliot Smith presented what might be the mother of all conspiracy theories, in which every last sign from early human history
must be tied to every other, all of them pointing back to a single cultural movement out of Africa (see Figure 17.2). Elliot Smith played
a minor role in contemporary Egyptomania, examining mummies,

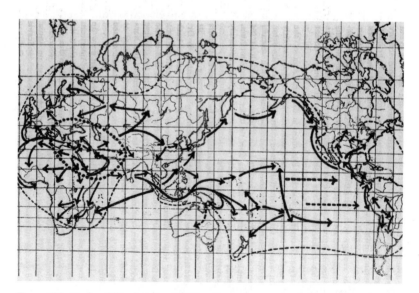

Figure 17.2. Grafton Elliot Smith and William James Perry, map of "The Reality of
Diffusion" (1934), with all culture diffusing out of Egypt.

debating myths, publishing on Tutankhamen's newly-discovered tomb.[17] But his was Egyptomania on steroids. How, he asked, could the savage Papuans be using mummification techniques that it took the civilized Egyptians 3,000 years to develop? How could boats used in the Pacific or by Indigenous peoples in Latin America have the same sun design on their bow? How could megalithic culture spread if not from a single origin?

The only explanation, Perry and Elliot Smith concluded, was that culture had diffused out of Egypt. With some well-placed supporters, they started tracking the entirety of human culture as a series of technology-bearing one-directional migrations. America was colonized, they insisted, not along the Bering Strait, but across the South Pacific. It is hard to appreciate today how much this argument both made sense in some strange, airless way *and* was relentlessly mocked by anyone who did not agree. Friendships were made and broken, theories of the brain were built, data in some corner of the world was interpreted on the basis of its fit into the notion of diffusion from Egypt. For the Polish-born, Vienna-trained anthropologist Bronisław Malinowski, who was based at the London School of Economics and who established participant-observer fieldwork as the foundation of anthropology, Elliot Smith was patching together random minor elements of culture to build an all-encompassing theory. The French sociologist Marcel Mauss thought that his good friend W. H. R. Rivers—who had done pioneering ethnographic work in the south-west Pacific but then supported diffusion—had gone mad. Still, if you had caught the bug, everything suddenly made sense.

And why not? The idea that culture had spread out of an original center was a standard one, for example, in Indo-European linguistics. Much as sixth-graders today learn about Mesopotamia first, Phoenicians and Egypt second, then Greece and Rome, for most of the twentieth century, schoolchildren in Europe learned about the origins of Indo-European languages, the Dorian invasions into Greece, and so on.[18] Many archaeologists were diffusionists, work-

ing for example to track the drifts of ancient pottery, domestication, and animal husbandry, usually outward from Mesopotamia
and the Mediterranean. Better to assume technological diffusion,
archaeologist V. Gordon Childe argued, than to always posit population movements—that is, ancient racial expansions.[19] If language
diffused out of an original site, wouldn't technology, myths, and
cultural objects do so too?

This was a generation after James Frazer. His *The Golden Bough*
had enthralled and horrified readers by piling myths of renewal upon
tales of kingly sacrifice. Everyone was scraping around for the material and cultural foundations of myth. For social Darwinists, the
ground was ultimately evolutionary or racial. But many anthropologists thought that social Darwinism offered only a crude explanation
of widespread myths, habits, and forms of social organization. The
more they demonstrated errors in Darwinism, especially Herbert
Spencer's, the more they looked for alternative explanations.[20] Plus,
diffusion was not domination. The British Empire of course insisted
during the interwar years that its presence in the Middle East was
benign, that it was merely involved in improving other societies, that
Egypt was not its protectorate: Elliot Smith and Perry transposed this
ideology of "diffusion, not domination," to the ancient Egyptians
who, to them, spearheaded human civilization. The theory also differed a bit (but only a bit) from the more blatantly racist "Hamitic
hypothesis," argued at the time by Giuseppe Sergi and Charles G.
Seligman, which insisted that culture in Africa was the product of
quasi-Caucasian Mediterranean or "Hamitic" cultures—and that
Black Africa's cultures were derivative at best.[21]

Dart was a believer. He embraced racially weighted forms of diffusionism and race typology early on, and he remained a diffusionist
to the end.[22] But consider what it meant for him and Taung to saunter
into this debate in 1925. Respected in osteology and physical anthropology as Grafton Elliot Smith was, he was also cranky and disliked
as something of a covetous crackpot—and he knew it. His critics,

who by now included the respected anatomist-anthropologist Arthur Keith, had had enough of his banter about Africa. Why should Elliot Smith (or other professional archeologists and anthropologists) go with Dart's version? After all, Dart was making clear that if culture diffused from North Africa, the species did so from his own private front yard in the South. On top of everything else, Elliot Smith had the habit of using the peremptory rhetoric of discovery and true science—he was, in his own eyes, the seeker who followed where evidence led him. Dart was filching his brio to claim that he was the real investigator, looking to a more primal period and insisting that his skull was no ape, no intermediary, but the true missing link.

AMONG SCIENTISTS, CULTURAL DIFFUSIONISM WAS TROUNCED. Malinowski, for example, showed in a public debate with Elliot Smith in 1928 that mutual cultural borrowing explained Elliott Smith's own data better than arrows radiating across a map.[23] Still, Elliot Smith's and Perry's idea that the Americas had been reached across the South Pacific was shared in unexpected quarters, including Paul Rivet, the socialist deputy and director of the Musée de l'Homme in Paris.[24] In 1948, the Norwegian explorer Thor Heyerdahl traveled with five companions on a raft across the Pacific to prove the point. He called it the Kon-Tiki Expedition, won an Oscar for the documentary they filmed, and became a media star, a precursor to Jacques-Yves Cousteau and David Attenborough.

Kon-Tiki was a rather neutral variety of diffusionism—others were more racially minded. A revealing example is the interpretation of a rock painting in the South-African-administered mandate of Namibia proposed by the abbé Henri Breuil. While South Africa's prime minister Jan Smuts was fighting for his political survival in 1948, Breuil presented him, over dinner, with his copy of a San painting that he first called "Our Lady of the Brandberg" and then, softening the Christian enthusiasm, "The White Lady of the Brandberg" (see Figure 17.3). He told Smuts, and then the world, that the main figure of the complex procession was the "White Lady," its leader; that most accompanying figures were "originally" white-skinned and were later blackened; that a Black Death haunted the White Lady. This was not the work of a bushman: "certainly no primitive, uncultured people would depict such ceremonies or give such a central position to a woman."[25] In a part-"Hamitic," part-diffusionist hypothesis, he insisted that the painting had been made by white Mediterraneans, perhaps Minoans, who had gone on an elaborate odyssey all the way to Namibia.[26] They had left nothing else, but this painting was definitely theirs. The Black figures, if they hadn't originally been white, were surely her enemies.

Breuil's account thundered as a warning against miscegenation and degeneration just when South African settlers were obsessed with white replacement theories.[27] Smuts, one of the key figures in the development of apartheid in South Africa, described himself as moved. Still, everything about Breuil's interpretation of the work was soon ridiculed as fantastical. The painting was not ancient

Figure 17.3. Mary Boyle and the abbé Henri Breuil stand over his copy of a rock painting in Namibia that he called "The White Lady of the Brandberg."

but at most a few hundred years old, the central character was not racially white but wore makeup, perhaps for a ritual, and best of all, "she" had a phallus.[28] Breuil's white, Christian dynamic was entirely ideological.

Diffusionist ideas also contributed to an opposing set of political claims: Pan-Africanism and decolonization. Already well before World War II, William H. Ferris had advocated for an African origin of all culture, and Frobenius had maintained a correspondence with W. E. B. Du Bois and other Black intellectuals.[29] In the 1930s, African anticolonial intellectuals began to celebrate Frobenius for asserting the autochthony and authenticity of African culture. Senegalese poet and statesman Léopold Sédar Senghor later described an entire generation of 1930s Black students as viscerally affected by Frobenius's "sacred" books. Senghor, his colleagues in the Négritude movement Aimé Césaire and Léon-Gantran Damas, and other African readers "still carry, in our mind and soul, the master's marks, like tattoos executed in an initiation ceremony in the sacred wood."[30] How could Frobenius be the master in a ritual of intellectual liberation? Because of how utterly rare his view was that Africa was deep and original and had a spirit all its own. As President of the newly-independent Senegal, Senghor declared that Frobenius "has given us back our dignity!"[31] Aimé Césaire included Frobenius in even his most radical texts: in the famous *Discourse on Colonialism* (1950), he invites a white man to relearn Africa through Frobenius: "Say, you know who he was, Frobenius? And we read together: 'Civilized to the marrow of their bones! The idea of the barbaric Negro is a European invention.'"[32] That quote from Frobenius became proof of African authenticity. Other anticolonial politicians and radicals, including Frantz Fanon, followed suit: reading Frobenius was a conversion experience.[33] Kwame Nkrumah, President of Ghana, quoted Frobenius in a big speech on the defense of African monuments and memory at the first meeting of the Editorial Board of the *Encyclopaedia Africana* on September 24, 1964, and cited him regularly in his speeches when he decried the violence that Portuguese conquistadors had inflicted on the West African coast.[34]

This framework enabled a brilliant kidnapping of Frobenius's legacy, which was by no means unambiguous. Diffusionism could now serve what came to be known as Afrocentrism. Senegalese historian Cheikh Anta Diop pointedly advanced diffusionism as a response to the popular denigration of Africa in Europe and America. As opposed to a "Hamitic" attack, Diop invoked Frobenius and especially Elliot Smith to argue—as his book called it—for *The African Origin of Civilization*. "Civilized to the marrow of their bones!" he repeated, turning a whole set of colonial ideas on their head.[35]

THE EARLIER CULTURAL DIFFUSIONISM HID IN DART'S OUT OF Africa/killer ape model. And Dart's postwar reception should be seen in this context—one in which the turn to Africa had multiple political meanings, with some commentators committed to the continuation of colonial power, some to its destruction.

Charmed by Dart, and well aware of the effect of his claims about violence on his audience, Robert Ardrey, then a Hollywood screenwriter, began to write nonfiction, to think of himself as a naturalist, and to propagate Out of Africa. In the decade between 1950 and the publication of his *African Genesis* in 1961, Africa actually gained relatively wide acceptance as the cradle of humanity (see Figure 17.4). Scholars in London and the United States pronounced themselves convinced. In Tanzania, Mary and Louis Leakey discovered *Zinjanthropus* in 1959, adding to the picture sketched by Dart and Broom. Meanwhile, *Sinanthropus* had been turned into a national ancestor in China, but elsewhere it was no longer granted the same importance, not least because China was mostly closed now to outsiders.[36] In international scientific discussions under UNESCO's aegis, only the Soviets were doubtful about Dart's *Australopithecus*

Figure 17.4. As early as 1952, the Czech painter and paleo-artist Zdeněk Burian depicted a band of *Australopithecus africanus* in a field of bones, with the lead figure pensively fondling a bone that could double as a weapon or as a spoon. A similar painting he made figured the centerpiece male holding the same bone like a knife for stabbing.

and the Out of Africa scenario in general, and were so perhaps for geopolitical reasons.[37]

Nevertheless, Ardrey marketed himself and Dart as vanguards of a "new Enlightenment" that had been repressed, and that was now "seeking light under darkest cover." The slights Dart had experienced became the basis for fatuous praise of his search for truth. With this approach, as Erika L. Milam argues, Ardrey's book rode a new medium—the cheap paperback—to bestselling fame and treasure.[38] He spread the killer ape idea far and wide, turning it into a popular story—if not the most popular story at the time—about human origins and "who we are."

"Killer apes, our immediate forebears. . . ." So began his

search, beneath the thin veneer, for the origins of atomic destruction.[39] Africa was the origin, so distant yet so "immediate," of the now-dawning atomic and space age. Africa was the tail of the serpent of technical power and destruction. Ardrey was good enough a stylist to let the reader surmise that Hiroshima, concentration camps, and the Cold War were no surprise for a species born of killers. Violence was the motor of history, as he made clear—African history especially.

While proclaiming himself an antiracist, and (inaccurately) casting Dart as marginalized by racists, Ardrey made his own racism clear: "The African independence movements are rapidly converting a continent into something approaching a political state of nature, where primitive human behavior may be observed not as we should wish it to be, but as it is."[40] Decolonization in Africa represented not the overthrow of colonial oppression, but Africa's return to a deplorable state of nature: to killer apes. The year 1961, when Ardrey wrote this sentence, was above all a year of violence *against* decolonization. Ardrey mimicked the rhetoric of South Africa's government, which had just banned the African National Congress in 1960; the rhetoric of Belgium, whose continued meddling in the Congo after its independence led to the assassination of prime minister Patrice Lumumba in January 1961; the rhetoric of the French generals who, that April, attempted a putsch against Charles de Gaulle to stop Algeria from gaining independence from France. Seventeen countries had declared independence in 1960, the British and French Empires were crumbling, and even the British government—which had just defeated the eight-year-long Mau Mau uprising in Kenya—was acknowledging that freedom movements could not be stopped.

Ardrey's readers were vaguely aware of the meaning of decolonization in Africa, but the truth of the Holocaust was also just entering public consciousness. Was violence universal, or was it especially relevant to the so-called dark continent? The gory vision appeared at just the right historical moment to catch on. Ardrey spent weeks "at the heart of the contemporary revolution"—revolution *in science*, he meant, not revolutions against colonial rule—to answer the question. This "heart" he talked about was the triangle between Rwanda, the Congo, and Tanzania. Louis and Mary Leakey, who had been peripherally involved in the British colonial government's suppression of the Mau Mau rebellion, were now scouring the Olduvai (or Oldupai) Gorge for hominid remains.[41] Jane Goodall had just reached the Gombe Stream for the first time, having encountered Belgian troops and refugees leaving the Congo. She and her mother prepared 2,000 spam sandwiches for them, in an episode she soon mythologized as an *emptying* of colonialism from these lands before she could arrive at the pure preserve. ("When can a virgin white woman best represent Man?" Donna Haraway later asked about Goodall. When she replaces Man's violence toward the colonized with a "feminine" attachment to the ape, the naturalized Black man.[42]) To get his own answers on violence, Ardrey spent his time studying (African) animals and insects and stuck to his haughty view of the "laboratory" of humanity's supposed return to a Hobbesian state of nature—"not as we should wish it to be, but as it is."[43]

He elaborated on this picture in *The Territorial Imperative* (1966), which claimed that human territoriality was innate, evolution-driven. He linked his work to the studies of aggression carried out by his friend and ally Konrad Lorenz, who declared aggression to be the "fighting instinct in beast and man which is directed *against* members of the same species" to aid in survival.[44] By 1966, the bodily basis of aggression had become neurology's holy grail and the object of considerable government funding.[45] *The Territorial Imperative* argued that in the face of evolutionary pressure, hominids respond with intelligence based on instinctual aggression and territorial defense. Violence and war had forced early humans to defend themselves, to become capable of handling weaponry; they were even necessary for consciousness itself. Individuals became tied to one another in order to outmaneuver threats.

Again, Ardrey presented the argument as universal, and again he made a political case—now explicitly in support of apartheid. He warned at length of "the chaotic future of the new black states" and then celebrated "the pariah state South Africa [for] attaining peaks of affluence, order, security, and internal solidarity rivaled by few long-established nations."[46]

AT THE APEX OF DECOLONIZATION, THE OUT OF AFRICA THEORY WAS anything but a celebration of Africa, not even a colonial romance in the style of Karen Blixen's memoir or Sydney Pollack's film that carry the same name. Violence was plastered across paperbacks, magazines, school posters. Paleoartists depicted "killer ape" scenes with gusto, including in Eastern Europe where the African origin of humanity had ostensibly taken some time to become accepted (see Figure 17.4 and 17.5). Kubrick's *2001: A Space Odyssey* was released during a particularly tense moment in 1968—the very day Martin Luther King Jr. declared, in what would be his last speech, "I have been to the Mountaintop . . . and I have seen the Promised Land." Viewers who saw the film in the cinema the next night knew that King had been assassinated, and that the Apollo 6 had burned up in the atmosphere upon its launch. The world was not exactly advancing. Kubrick's *2001* normalized Ardrey's scene, including his claim that it was not "that man had fathered the weapon. The weapon, instead, had fathered man."[47] But the film had little to say about Africa itself—the tapirs it depicted, for example, are not native to Africa. Rather, where Clarke had focused on alien civilizations, Kubrick concentrated on the link between technology and violence. The original script spoke of nuclear destruction raining down from spaceships dancing above earth. If humans were evolved killer apes, the ultimate threat to humanity was the ultimate autonomous human creation: the supercomputer HAL-9000. HAL's perfection required a human to commit an almost absurd act of violence—by deactivating

Figure 17.5. East German teaching tool that featured australopithecines as killer apes (late 1960s). Note the bludgeons, the violence against apes, and the acacias used to denote "Africa."

it in a scene imitating slow human death. Only through that battle could humanity transcend, Teilhard-de-Chardin-style, its earthly origins, and emerge reborn as a guardian of the universe. In the famed Stargate sequence toward the end of Kubrick's film, in the deathbed scene that follows, and in the climax with the noosphere-sized fetus that confronts the earth, the film aestheticized ideas about the emergence of a superior humanity.

In subsequent years, the narrative shifted. Ardrey could not control it. Most of his contemporaries cared little for him or for his politics, and they simply ignored it. Some public figures, like Ashley Montagu, were loud in their rejection of him.[48] A growing pile of scientific data, including many new discoveries of osteological remains, made Out of Africa the predominant theory of origin, and it remains so to this day. When Jacob Bronowski filmed the TV series *The Ascent of Man* for the BBC in 1973, he opened in Africa. Rather than emphasize violence, he noted instead how Dart had ostensibly recognized that the Taung Child's "teeth were *not* the great, fighting canines that the apes have." The Taung Child had been a forager and a tool-maker, not a killer.[49] So Bronowski celebrated a "human commitment to a new integration of life."[50] This was his way of returning from Ardrey's violent vision of humanity to a vision committed to human progress and self-improvement.

Since the 1970s, we have had three accounts of Out of Africa to choose from, generally speaking. The first is Ardrey's, which has been in decline but has had its own afterlives in theories of "primitive warfare," to which we will return. This study of aggression had everything to do with geopolitical concerns, as the scientific section of NATO recognized when it devoted a conference to the biology of aggression in 1980.[51] The second account is a far less violent, humanist scenario that was supported by the Leakeys, Bronowski, David Pilbeam, Jane Goodall, and those who have followed in their footsteps with a vaguely humanitarian and sympathetic outlook toward the earliest humans. The third is Afrocentrism, which has been pursued by mostly Black intellectuals since the 1980s. How do you retell Africa's story in terms not framed by the colonial states?[52] At the Metropolitan Museum in New York, the exhibition *The African Origin of Civilization* (2022–24) is explicitly grounded on Cheikh Anta Diop's work. In the Marvel movie *Black Panther* (2018), the nation of Wakanda becomes the popular fluorescent triumph of the third account: a techno-paradise in Africa guiding from time immemorial the improvement of humanity.

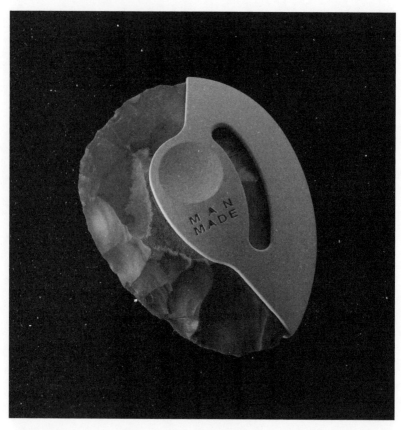

Figure 18.0. Ami Drach and Dov Ganchrow, *MAN MADE; handaxe #5*. Knapped flint and 3D printed polymer, 2014.

Chapter 18

STONE-AGE COMPUTERS

In the first half of the twentieth century, prehistorians cared only so much for paleolithic tools. Since Boucher de Perthes a century earlier, they had celebrated prehistoric "industry," peppering their books with drawings of flints and according them a role in helping think through the stages of the deep past. There were offhand comments, such as Friedrich Engels's line that in the transmission from ape to human "the hand is not only the organ of labor, but also the product of labor."[1] But no one consistently defined humanity through tools, their use, and their fabrication.[2] Until, that is, the publication of a thin 1949 book by a researcher at the British Museum, Kenneth Oakley: *Man the Tool-Maker.* In subsequent decades, scientists across the globe created the quintessentially Computer Age idea of human origins: the idea that humanity was forged in its construction of tools. What ultimately sidelined Raymond Dart's gruesome theory was not a siren call about peace and humanism, but one about tools. It claimed that the fabrication of tools, and then of tools *for making other tools*, had generated an ongoing feedback loop. Tools had built the human species.

The notion began with Oakley, who dethroned brains and skulls and instead looked at teeth, hands, and pelvises. He did not waste his reader's time, opening *Man the Tool-Maker* with a clear statement of his thesis: "Man is a social animal, distinguished by 'culture': by the ability to make tools and communicate ideas."[3] Even ideas presuppose tools, he continued, so tools were the "chief biological characteristic" of humans.[4] Where chimpanzees made tools merely in response to pressing environmental puzzles, humans make tools to anticipate future events.[5]

Oakley's originality is illustrated by his distance from the tradition he was schooled in, which had believed in the big human brain as the driver of evolution. Oakley wrote his dissertation under Wilfrid Le Gros Clark, the physical anthropologist whom we encountered objecting to UNESCO's Statement on Race. Le Gros Clark had himself followed his own mentor, the diffusionist guru Grafton Elliot Smith.

In their time, Elliot Smith and Arthur Keith had been uncontroversial when they declared the growing skull size and brain as the motor that powered the development of hominids.[6] Piltdown Man, on whom they had worked, sported a large braincase—therefore large brains had preceded toolmaking (see Figure 18.1 but also 14.1). Le Gros Clark, for his part, thought tools offered a way out of scientific racism, with its endless pondering about "innate characters." But he supported the "large brain" hypothesis, mistrusted discoveries of small-brained australopithecines in Africa, and doubted that tools "made" humanity, as Oakley thought.[7]

In 1953, Oakley announced that he and Le Gros Clark had felled the big-brained Piltdown Man. Oakley's pioneering fluoride and potassium-argon techniques for dating fossils had proved it was a forgery. He concluded further that the human origins problem now involved discovering human-made stone tools. The practical value of this approach was obvious. Stone can survive environmental condi-

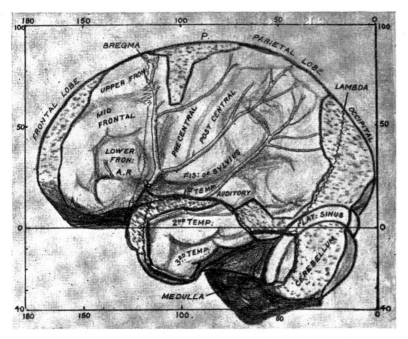

Figure 18.1. Arthur Keith, cast of the Piltdown brain. Keith thought the brain drove human evolution: Keith compared this cast with one taken from an Aboriginal Australian man and with a Neanderthal cranium. The intellect of the Piltdown man, he insisted, was highly developed.

tions that rapidly ruin skeletons. One should dig for human implements and then date them, not pray for the rare luck of discovering some questionable fossil.[8] The dating techniques he had designed helped to cleanse the fossil record of fakes and mistakes, and also to properly date the tool "industries" (as they were known) at particular locations.

Tools could also help scientists past various impasses in the study of the body. The "large brains drove evolution" model had failed to explain how bones and posture could have changed so dramatically. Its vision of the body did not allow human physique to be plastic enough.[9] Particularly troublesome were the teeth, the pelvic bone, and bipedalism, that is, the shift to walking on two feet rather than four (see Figures 18.2 and 18.0).

Figure 18.2. In 1963, Jacquetta Hawkes wrote the early chapters of UNESCO's history of humanity and presented the pelvis as a key development in human evolution.

FOLLOWING DART, OAKLEY TREATED BIPEDALISM AS A DISTINCT *dis*advantage. Standing on two feet makes you slower, less agile in turning and climbing, and more visible—hence easier prey.[10] For Dart, this weakness had demanded stealth, violence, and smarts. Oakley's image instead recalled the British experience of World War II: a stand-your-ground, day-by-day struggle of defense and protection. Oakley resolved that weapons were not intended to tear enemies limb from limb but were fundamentally defensive.[11] As hominids evolved, basic weapons and tools replaced and improved on canine teeth. Over time, they no longer needed ferocious canines and jaw muscles, now that they could cut without them. Tools had influenced physiognomy: the whole face changed. Hominids slowly transformed from "occasional" or "casual" tool users to "systematic tool makers."[12] Tools allowed them to establish traditions (for remaking tools) and then to develop burial practices, art, and religion.[13]

"Tools Makyth Man," Oakley once called his theory, paraphras-
ing *Manners Maketh Man*. His aim was to alter the strict evolutionism of his predecessors. Tools were not merely human creations, he
insisted: they interceded somewhere *between* nature and culture. This
was new. Darwin's and E. B. Tylor's successors had chafed at the bit
to narrate the rise of human culture. In the 1950s they still debated
whether human evolution was facilitated by nature or by culture—
and they now had to do without the concept of race.[14] The brain was
giving way too. Communist archaeologists had explained the material and economic basis of early "civilization," attending especially to
the agricultural revolution.[15] But was Marx's dialectical materialism
enough to explain the rise of the social order? For Oakley, no: a better materialism was now available. Hominids were "only able to survive in the face of rigorous natural selection by developing a system
of communication . . . which enabled cultural tradition to take the
place of heredity."[16] In the last sentence of *Man the Tool-Maker*, he
even stretched the argument to say that "modern" machines offered
little more than complex versions of "the simple equipment in the
tool-bag of Stone Age man: percussion, cutting, scraping, shearing,
& moulding."[17] This doesn't quite convince, given that his contemporaries were theorizing cybernetics and information theory. What
Oakley meant was that those simple actions meant that "culture/
tradition" was driving evolution and society "instead of heredity."
Tools had built communication: they had created society, all the
while changing human nature.

IN 1953, THE KENYAN-BORN ENGLISH PALEOANTHROPOLOGIST LOUIS
Leakey revised his 1934 book *Adam's Ancestors* for a new edition.
He now agreed with Oakley: the only satisfactory distinction of
human from animal was the one described in Oakley's *Tool-Maker*.[18]
    Six years later, Mary Leakey discovered a skull in Olduvai Gorge
in what is now Tanzania. The skull lay in a bed filled with tools and

Figure 18.3. Louis and Mary Leakey with a mandible, from the period when Mary discovered *Zinjanthropus* (*Paranthropus boisei*).

bones, and Louis, her husband and a master self-promoter, imme-diately announced that this "Olduvai skull," this "true man," rep-resented "the type of 'man' who made the Oldowan" tool culture. They named it *Zinjanthropus boisei*, and declared it to be a new missing link: "the oldest yet discovered maker of stone tools." It became critical to Oakley's model.[19] Louis warded off Dart: *Zinjan-thropus* was no victim of "a cannibalistic feast by some hypothetical more advanced type of man."[20] A year later, the Leakeys' expedi-tion discovered more remains at almost the same spot, and called them *Homo habilis*. These remains were much "more" human and dated to roughly the same time. Leakey now insisted on a side-by-side, competitive evolution between the two species.[21] The real tool-maker had been *habilis*: the handy, dexterous man. *Life* magazine's special volume *Early Man* (1965) staged the scene (Figure 18.4): the more advanced-looking and tool-wielding *habilis* on the left stalk a *Zinjanthropus* group that flees while its leading males defend it with outsize gestures and heavier, cruder rocks.

By 1964, *Zinjanthropus* had been relegated to the *Australopithe-cus* genus (and later to the *Paranthropus*). But even in its brief life

Figure 18.4. Jay H. Matternes, Confrontation between *Paranthropus boisei*
(*Zinjanthropus*) and *Homo habilis* (1964). Note the difference between the tools used by
habilis and the cruder stones of *Zinjanthropus*, as well as the retreat of the females and
young behind the defensive postures of the adult males.

as an exciting new genus, *Zinjanthropus* contributed to the acceptance of the Tool-Maker argument. *Zinjanthropus* allowed Leakey
to pile scorn on Ardrey's claims about primitive weaponry. Bones, he
thought, were too weak to actually slice into flesh, as Dart imagined.
Moreover, "it was the very process of making and using such tools
and weapons that brought about the big human brain," and *not* the
other way around.[22] Even in the relegation of *Zinjanthropus*, implements were the key explanatory device. The superior tool-maker
habilis had survived and led toward *Homo erectus*. *Zinjanthropus*
had made some tools, but had died out. Technical culture, not biology, showed evolution's path. Tools were not a mere drollery of evolution; they were media for life. They underwrote the slow formation

of human physique—they became an anatomical element, an "extension of Man" as Marshall McLuhan called them.[23] No species, moreover, was a direct "antecedent of Man."[24] Technology was written in stone as much as it carved new bodies, pointing little by little toward anatomically modern humans.

Zinjanthropus helped proponents of the Tool-Maker thesis overthrow theories concerning technology and culture that had been fundamentally Eurocentric, modeled around tools found especially in French dig sites. Just as Piltdown scientists had downplayed implements in favor of intelligence, so too had they disparaged non-European tool cultures as minor or incidental.[25] The assumption that tools were only used by advanced hominids was simply based on two ideas—that only advanced hominids made them, and that these advanced hominids were Europeans, because this is where stone tools were studied the longest and most carefully. *Sinanthropus* had

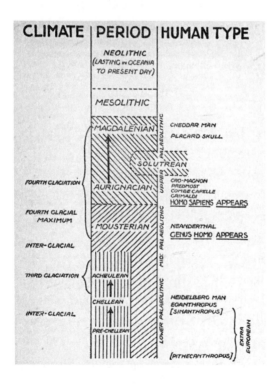

Figure 18.5.
"Development of the Science of Man," in Alfred Cort Haddon and Julian Huxley, *We Europeans: A Survey of "Racial" Problems* (1936). Note the resolute Eurocentrism of this history of technology.

DISTRIBUTION AND SUGGESTED RELATIONSHIPS OF
THE CULTURAL TRADITIONS OF EARLY MAN

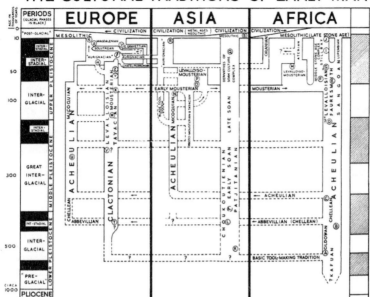

Figure 18.6. "Distribution and Suggested Relationships of the Cultural Traditions of Early Man," in Kenneth Oakley, *Man the Tool-Maker* (1949).

been an exception, but it too was thought to use relatively crude tools. Other non-European discoveries, for example, at Sterkfontein near Johannesburg, had not yet had much of an effect. Now *Zinjanthropus* sent pebble-tool culture at least a million years back in time, and provincialized Europe in models of cultural advance (see the difference between Figure 18.5, from 1936, and 18.6, from 1949). Thanks to the growing public interest in technology, *Zinjanthropus* and *habilis* were a quick success.

IN 1960, THE AMERICAN ANTHROPOLOGIST SHERWOOD WASHBURN published an article in *Scientific American* to describe the development of his thinking over the past decade. "The structure of modern man," he now concluded, "must be the result of the change in the terms of natural selection that came with the tool-using way of life."[26] Tools had diverted the course of natural selection.

Washburn had first taught at Chicago and then moved to Berkeley in 1958.[27] His earlier article "The New Physical Anthropology" (1951) used the Modern Synthesis of genetics and Darwinian evolution to reformulate the development of the human body. Each new adaptation, he argued, disturbs homeostasis, the stability of the organism. And no adaptation takes place alone: it relies on and compels a broader set of changes. Bipedalism, for example, follows a shortening of the trunk that enabled reactive "changes [in] the ilium and in the gluteal muscles in a bone-muscle complex, which makes a different way of life possible."[28] Washburn called these groupings "adaptive complexes." Over very long periods of time, they restabilized the body in new "ways of life," and they encouraged it toward newly possible goals.

Tools and culture contributed to adaptive complexes, sometimes altering the skeleton, sometimes the entire world of hormones, muscles, and nerves that fades and vanishes once the body degrades. Invisible in the fossil skeleton, these adaptive complexes had to be found elsewhere. Moreover, migrations, drift, and new environments had enabled new "ways of life," too.[29]

In short, evolution did not require a big brain: "The great increase in the size of the brain and decrease in the face was *after* the use of tools."[30] Washburn had little compunction about blaming geopolitics for big brain theories, and he endorsed a controversial evolutionary model proposed by Franz Weidenreich.[31] It was not a classic tree model but a matrix where species and races had been separated by often paper-thin differences and had consistently intermixed, pushing the body in myriad directions. Compare, for example, Arthur Keith's classic tree model (in Figure 18.7), with Weidenreich's own models that assumed much greater crossover between groups and even species that were gradually differentiating (in Figure 18.8).

Weidenreich had led the excavations on *Sinanthropus* in the 1930s

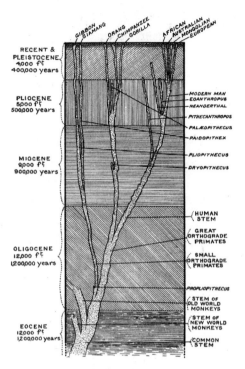

Figure 18.7 (*left*). Frontispiece of Arthur Keith, *The Antiquity of Man* (1915). Noteworthy is the place Keith grants to Piltdown Man ("Eoanthropus") and also the way his design implies continuity between species and race.

Figure 18.8 (*below*). Franz Weidenreich disapproved of traditional trees (like Keith's, Figure 18.7), which he thought presumed sexual segregation. Instead, he argued that races and species are not easily fixed, and that scholars needed to accept considerable genetic crossovers. He offered this multiregional matrix model of human evolution, with a very gradual, tentative separation from other species. Franz Weidenreich, *Apes, Giants, and Man* (1946).

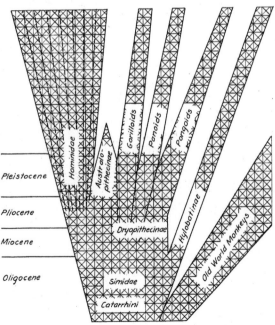

after the death of Davidson Black. In his 1946 book *Apes, Giants and Man*, he disparaged Dart's violent primal drama as circumstantial: "strange habits for an ape."[32] Weidenreich had also rejected Piltdown Man as a chimera (well before it was revealed as a forgery), and mocked the emphasis on brain size as "craniomancy."[33] He remains controversial for his "polycentric" hypothesis that humans had not evolved in one location but had mated and adapted across very different groups, even across species. By 1946, Weidenreich had also become convinced that brain size was meaningless.[34] What matters about the brain, he noted, is not its size, but the folding, creasing, and fissurization of the surface of the cortex, which reflect its thickness. These were qualities long lost in fossils. At most, skull size and shape had something to do with upright posture, but even so, "the evolution of the locomotor system preceded that of the head."[35]

Washburn cited just that passage when arguing that the hand specialization and tool use preceded the brain's growth.[36] Why did Washburn care? He understood Weidenreich to mean that biology, the brain, and race did not matter. Where would he look instead? Cybernetics—the new field that was spreading widely after the 1948 publication of Norbert Wiener's book *Cybernetics: Control and Communication in the Animal and the Machine*, as well as with the Macy Conferences between 1946 and 1953, which brought together scholars invested in modeling the human mind. The crucial issue for cyberneticians was the management of information and communication—how a machine, a mind, or an organism respond and adapt to their environments, usually through feedback mechanisms. Now Washburn did not cite Wiener or the other major figures in the movement, but its influence on him was clear. In *Cybernetics*, Wiener had described evolution as a feedback process by which random variations build on one another and enable a dynamic bodily transformation.[37] Rather than look at specific organs or body parts, Wiener went on, we must study the *systems* in which body parts acted. Washburn was theorizing a human body that resembled a

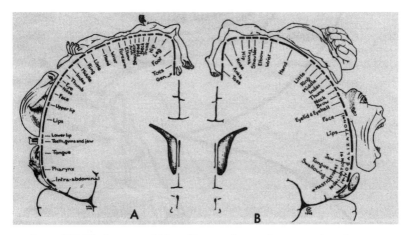

Figure 18.9. *The Sensory and Motor Homunculus* correlates areas along the central sulcus of the brain with particular body parts. Published in Wilder Penfield and Theodore Rasmussen's *The Cerebral Cortex of Man* (1953), it was rapidly picked up and widely reprinted.

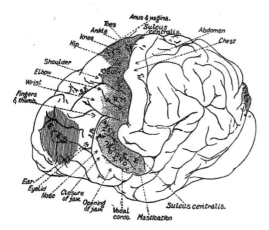

Figure 18.10. Charles Sherrington prepared this "Brain of a Chimpanzee" diagram in 1906 to show how motor functions are localized in the brain. Prehistorians contrasted it to Penfield and Rasmussen's 1953 diagram (Figure 18.9) so as to emphasize how areas representing the hand and the mouth developed for humans along with the use of tools and language.

complex computer dynamically responsive to its environment. His "adaptive complexes" took up the idea that feedback generates purposeful behavior.[38]

Washburn was also inspired by the contemporary neurobiological work that sought to localize bodily functions in the cerebral cortex. In their famous, grotesque cortical homunculi (Figure 18.9), as well as in their book *The Cerebral Cortex of Man* (1950),[39] the neuroscientists Wilder Penfield and Theodore Rasmussen pinpointed which brain sites were responsible for the sensory and motor function of particular body parts. The cortex, they showed, granted outsized importance to hands, fingers, tongues, and mouths. What was more, this was a very different picture from the one drawn of a chimpanzee brain by Charles Sherrington in 1906 (Figure 18.10), where none of these anatomical elements were pronounced. In his 1960 article in *Scientific American*, Washburn reproduced the homunculi to accentuate the significance of the face and the hand in human evolution. "Selection for more skillful tool-using resulted in changes in the proportions of the hand and of the parts of the brain controlling the hand," he wrote.[40] The cortex, he wanted to show, expanded in response to "new selection pressures associated with the evolution of complex social systems."[41] In other words, the brain recorded the capaciousness and subtlety of the human physique as it reacted to social changes. This made the brain into "an extraordinarily efficient storing computer," Washburn told his students in 1964.[42] The brain was not the seat of mental activity: it was an archive of technical and hence physiological evolution. The brain was a side-effect.

Washburn completed Oakley's and Leakey's reversal of the entire traditional order: it was not that the growing brain had led to bipedalism, with intelligence helping free the hands from locomotion. Rather, tools helped free the hands, in turn contributing to pelvic changes and then bipedalism, in turn allowing the brain to grow in particular ways. This happened at the social level, too: tools complicated hunting traditions, helped with planning, and generated more manual skills (including a mother carrying her baby). By facilitating the preparation of food that enabled teeth to become smaller, it replaced the large canines of apes, shortening the face, putting pressure on the tongue, and facilitating language. Washburn arrived at his utopia: tools "tamed" "ancient men," eliminating "the most irascible individuals . . . who could not fit into the evolving social order." "In a very real sense," he concluded, "tools created *Homo sapiens*."[43]

Washburn thought that the Leakeys' discoveries merely confirmed his ideas. To Oakley he reported: "Leakey has been lecturing
in the US and urges that all theories of human evolution be changed.
Aside from differences in terminology, he seems to be advocating
precisely what we have been teaching for years—that is, that a creature with a brain much smaller than Java Man made pebble tools
and was the first biped."[44] He celebrated *Zinjanthropus* in *Scientific
American*, contributed to its myth, and suggested it showed that the
manufacture of tools was far more important than the killing: "Man-
apes could kill only *the smallest animals*."[45]

THE PINNACLE OF THE TOOL-MAKER TRADITION, HOWEVER, CAME
in the work of the French prehistorian André Leroi-Gourhan. Largely
unknown in English, Leroi-Gourhan's work exudes an ambition worthy of Darwin, which might be what he was: the Charles Darwin of
the twentieth century. Leroi-Gourhan had a different path, though.
He had begun as an anthropologist of the Inuit and the ("vanishing") Aïnu people of Northern Japan, then shifted in the late 1930s
to research what his mentor Marcel Mauss called "techniques of the
body." Aged just twenty-five, Leroi-Gourhan contributed detailed
chapters—over a hundred pages—on cultures in Europe, the Arctic
Circle, Japan, China, and Southeast Asia to the 1937 *Encyclopédie
française* volume on *The Human Species*.[46] In his two-volume *Milieu
and Techniques* (1943–45), he tracked the use of similar tools across
vast areas to show how they influenced the diversification of cultures.
Austere in his personal and work life, he published more pages per
year than there were days; during the 1950s he brought out books
in archaeology, technology, racial theory, craniometrics, and East
Asian ethnography. He began to track evolution in the highly technical geometries of the body: angles of dentition, jugal hinges, and head
equilibrium (see Figures 18.11A and B). Then, in 1964–66, he published his magnum opus, *Gesture and Speech*, presenting a philoso-

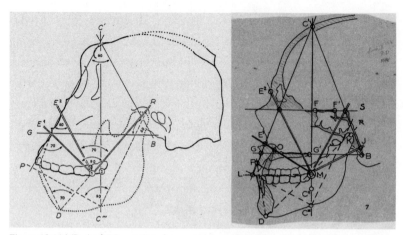

Figure 18.11A/B. André Leroi-Gourhan, craniometric drawings of skulls of *Zinjanthropus* and *Homo sapiens*. Leroi-Gourhan sought to demonstrate that tool use led to a rebalancing of the upper body, a transformation of the skull, and a shortening of the face.

phy of the evolution of gestures, rhythm, technology, and humanity. In subsequent years, his output remained prodigious, especially on excavations and paleolithic art in Europe.

Starting in the mid-1950s, Leroi-Gourhan developed a series of concepts in his lecture courses on the body as a technology. Three of these concepts played a major role in his theory of human evolution: the *liberation of the hand*, the *operational sequence*, and *operative memory*.[47]

Do body parts get "liberated?" Leroi-Gourhan remarked that, originally, the head and the hand had been stuck in thoroughly bound positions on the body of the fish and lizard (see Figure 18.12). They merely extended and assisted the body. Gradually, the head was untied from the body; then, more slowly, the hand. In hominids, the hand's "freeing" from its status as mere foot paralleled the development of tools; it enabled other changes across the body.[48] These "liberations" were not specific to humans: they had begun far earlier. Apes and humans belonged to this movement, rather than it to them.[49]

Rather than think of body parts or tools as individual bits, Leroi-

Figure 18.12. André Leroi-Gourhan, "The Liberation of the Hand" (1956).

Gourhan linked tools, fabrication, memory, teaching, and usage together into the *operational sequence*. No tool exists without the sequence of gestures involved in its use. Nor does any tool work without the corporeal and social memory needed to make and use more of its kind.[50] As tools make new gestures possible and enable new kinds of life, this operative memory grows in turn, developing new traditions, new tools, new ways of life.

What made human operative memory new and noteworthy is that it was mostly exterior to the body. Apes do not have traditions or other forms of social memory—they can use a stick but they can't develop a tradition of shaping sticks. Hominid society absorbed bodily memory. "The whole of our evolution has been directed toward placing outside ourselves what in the rest of the animal world is achieved inside by species adaptation."[51] Though outside the body, tools were nonetheless part of the anatomy. Operative memory joined anatomy and society in the gestures needed for toolmaking. And in this way, implements became parts of "the human economy." Society and technique followed a cybernetic feedback loop, a "two-way flow springing initially from the world of matter."[52] What Leroi-Gourhan called the "human economy" warped and wrapped the body itself into a tool. Hominids built and wielded tools, turning themselves into tools too. Take dentition, for example: the "facial massif" evolved in direct relation to the hands. The teeth got smaller as they reacted to the "liberated" hand and its implements.[53] Which meant that the bipedal human's face, braincase, and spine all had to morph so that now they would sustain the head equilibrium and the handling of tools. Even more than Washburn, Leroi-Gourhan thought that the body was a cybernetic device, affected by "the feedback effect of reflection." The brain recorded and redeployed these transformations.[54]

Once more, nothing was special about humans except the feedback that their technology generated between the animal self and society. Language, symbols, thought, religion—all the scientists' old props—were contingent, not necessary. *Zinjanthropus* lacked language, but it was nevertheless hard-earned proof, according to Leroi-Gourhan, that evolution for all hominids was tied to operational sequences. *Zinjanthropus* already needed tools for its anatomy to be complete.[55] Despite its "truly disconcerting" appearance and small brain, Leroi-Gourhan went on, it was "definitely human."[56] Humans had to stop seeing themselves as the sole pinnacle of evolution. What mattered instead, in Leroi-Gourhan's definition, was that "human" was every being—every *species*—that could walk upright, possess a free hand during locomotion, have a relatively voluminous brain, and use portable tools.[57]

This was nothing short of a new evolutionary theory. It did not look at domestication or sexual selection, as Darwin had, but at the use of exteriorized memory—tools, language, rhythms—that altered human bodies and constructed society. After tools came language, then cities, which also "exteriorized" evolution. Just as tools were becoming immobilized in ever-bigger operational sequences, so too did humans become ever more specialized, smaller, and immobilized parts of technical evolution—of the human economy. *Gesture and Speech* expanded this approach into a theory of humanity, its past, its future.

Like Washburn, who riffed on environmental danger, Leroi-Gourhan did not mince words as to the consequences of this technical human economy.[58] Its triumph was not some advance toward God, as Teilhard de Chardin had hoped, but rather toward self-destruction and ecological devastation.[59] Society had become "the chief consumer of humans, through violence or through work," and things only got worse. Humanity's growing dominance over the natural world meant a gluttonous human economy that ruined humanity too. The future would bring its perverse victory, "with the last small oil deposit being emptied for the purpose of cooking the last handful of grass to accompany the last rat."[60] And human beings had no control over the transformation of the human economy or over their own enslavement in it: "an anodontic human race living in a prone position and using such forelimbs as it still possesses to push buttons is not completely inconceivable."[61]

THE IDEA THAT TOOLS HAD MADE HUMANITY MADE A BROAD impression in the mid-1960s, due especially to mass science publications like the 1965 *Life* magazine special *Early Man*, which was edited by F. Clark Howell, Washburn's former student. The Tool-Maker was quickly adopted by philosophers, feminists, media theorists, and scientists as an idea that humanity is not static and biologically determined but dynamic and designed across its tools. Marshall McLuhan's conception of media not as external objects but as "extensions of man," his description of the "global village," and his slogan "the medium is the message" played on Leroi-Gourhan and Washburn's sense of the social world as a technical mediation of the body. For similar reasons, Washburn's adaptive complexes directly influenced Gregory Bateson's counterculture classic *Steps toward an Ecology of Mind* (1973).[62] Anthropologist Clifford Geertz defined humanity in cybernetic terms that echoed both thinkers: "man is precisely the animal most desperately dependent upon such extragenetic, outside-the-skin control mechanisms, such cultural programs, for ordering his behavior."[63] The goal for Geertz as for Bateson and McLuhan was to tie together the social domain with the physical body: now they could argue that technical objects fused the two. French philosophers were even more explicit. When Jacques Derrida described his project of deconstruction as a "liberation of writing," he did so while discussing Leroi-Gourhan's "liberation of the hand."[64] Gilles Deleuze and Félix Guattari, in their manifesto *Anti-Oedipus* (1972), relied on Leroi-Gourhan to avoid being swamped by the idea of consciousness and to declare that the body is made out of "desiring machines" responding to socio-technological pressures.[65] The new conception of the body felt like a serious escape from Darwinism and from philosophical tradition too. Cyberpunk—in science fiction literature, animation, and film—followed this trend, replacing body parts with machines and worrying about a humanity that had organized itself to the point of submission. Above all, there was Donna Haraway, who built the feminist cyborg in her famous "Cyborg Manifesto" (1985). Haraway had written very critically of other strands of Washburn's work.[66] Yet she relied on this image of the body as assembled, of human technology as working out a constant transformation. The examination of desire and critique of society from the 1970s on regularly relied on ideas of the mutating, dynamic, technically mediated being. Meanwhile, the human computer motif came to be cheekily used far and wide: Apple Computer called its 1982 annual report "Man Is Still the Most Extraordinary Computer of All."

WHAT IS A TOOL? IT IS WHAT THROWS OFF HUMANITY, LOCKS IT into place, threatens constantly to overcome it. It's something inhuman that supersedes the human. A tool in one's hand becomes not a medium or a part of an action, but what reshapes the body itself into a tool. The body is constantly rebuilt, subdivided, and reconstructed through adaptive complexes, operational sequences, or assemblages. These ideas were no facile postmodernism, but a detailed attempt to regraft the body through the hand, face, stone object, and matter itself. Put another way, they were a way to think about the body and its parts across technology in an age despairing over the rise of computers, the "end of work," and the decline of the traditional working class.[67] Machine simulation of human labor became the operating principle that fed the increasing despair over the loss of traditional labor. Computers and other machines could replicate manual labor and thereby replace the connection between worker and tool, which had been described as the very foundation of humanity. The debate over the "end of work" has returned with a vengeance in the last fifteen years.[68] At the same time, in rendering everything—stones and tongues and pelvises—into tools, the Tool-Maker theorists inadvertently fed into the fear that technology, as it began to escape manual- and steam-powered human industry, heralded ecological catastrophe. The tool mastered nature and dissected "Man."

Figure 19.0. Valerie Walker, "The Semi-Liberated Woman," *The Feminist Voice* (1971).

Chapter 19

THE BIRTHS AND ENDS
OF PATRIARCHY

W hat to make of that word, *Man*? I have deliberately
delayed engaging it, but it appears everywhere in the
story of human origins. It is, after all, perhaps the most
common concept, the Ur-concept, in the studies of the deep past.
Human origins have, until relatively recently, been framed as a story
of capital-M *Man*, gendered neutral yet obviously not. And has not
Man left *his* imprint all over *his* origins? We are now a full half-
century since the start of second-wave feminism, but it is still more
common to hear references to Paleolithic Man or Java Man than to
clumsy alternatives like "Neanderthal Humanity." Confused, mis-
gendered grammar often prevails, for example in titles like *The Real
Eve: Modern Man's Journey Out of Africa* and phrases like "When
we discovered Mousterian Man, one of our group had the idea of
baptizing her Augustine."[1] In France, one no longer really uses *les
hommes* (men) to signal "humans," except in two cases: *l'hommes
politiques* (politicians), and *les hommes préhistoriques*. Prehistory
has been, perhaps still is, about men and largely written and theorized by men.

How does one undo this kind of problem? The masculinized

concept of the human, long taken for obvious, seeps into so many others. Was not *Man* reflected everywhere, in concepts of the state, dreams of the future, ideals of equality and power, social systems? Not to mention the weapons with which "He" hunted, the violence "He" inflicted (including on women), the languages He spoke, the migratory conquests He undertook, the industry He built (see again Figure 4.1). Even the concept of matrilineality, as we saw in Chapter 6, was mainly a counterpoint: it explained why throughout recorded history women did not consistently hold positions of power. In the 1970s, feminist anthropologists began to cut up self-satisfied bodies of knowledge and pursue alternatives. How far back did the patriarchy go? they asked. To the beginnings of sexual dimorphism among australopithecines?—and with it, a physical and sexual imbalance favoring males?[2] To the use of tools and their contribution to social hierarchies? To the beginnings of some language somewhere, or the moment when concepts of male and female became clearly distinguished? To the imagined end of matrilineality, as Lewis Henry Morgan and Friedrich Engels had it? Feminist anthropologists and archaeologists worked on three registers: the deepest past, to redraw its image; modern science, to reveal its traps; and the present moment, for this was the time to force change.

DESMOND MORRIS'S *THE NAKED APE* WAS PUBLISHED IN 1967 TO MUCH fanfare and no less controversy. Among its notorious speculations: that evolution in hominids influenced erogenous power, which was essential to the emergence, looks, and social ethics of humanity. Natural selection enlarged penises, shaped and augmented breasts, and facilitated the disappearance of body hair. In perhaps the last grand masculinist scene in the invention of prehistory, Morris, a painter and sociobiologist, wrote with a pen that leaked erotic fantasy. He biologized social behavior, and he joyfully established the heterosexual couple as the end of evolutionary history.

Five years later, the BBC playwright Elaine Morgan responded to Morris in *The Descent of Woman*. Her reproaches were blunt. Morris, like Robert Ardrey and several others, were "Tarzanists": they had no imagination of the past outside of primitivistic jungle strongmen. At first they forget about early woman, they then "drag her onstage rather suddenly for the obligatory chapter on Sex and Reproduction, and they say: 'All right, love, you can go now,' while they get on with the real meaty stuff about the Mighty Hunter with his lovely new weapons and his lovely new straight legs racing across the Pleistocene plains."[3] Morgan was hardly the first to roll her eyes at Morris's *Naked Ape* with its reverie of sex creating Man. Nor was she exactly impressed by the implication, common among Morris and others, that woman had remained more primitive and had mostly developed in response to masculine evolution. "Any modifications in her morphology are taken to be imitations of the Hunter's evolution or else derived solely for his delectation."[4]

Morgan instead offered a theory based on the independence of Woman—as both herself and mother, shirking potential captors, twisting the hunter-gatherer image by using tools on fish and other creatures in the sea. She took advantage of the old idea that women were more associated with water than earth or fire, and she popularized the Aquatic Ape, a formerly obscure theory mentioned and discarded by Morris.[5] The theory postulated that female hominins back in the Miocene (that is, between twenty-three and five million years ago) had used not the dangerous savannah but the shallow sea to develop bipedalism, baby nurturing, and strategies of self-protection. Feline predators could not reach a mother cradling her baby in the water, and the sea had plenty of food.

From Morris to Morgan—is there a movement that more crisply reflects the passage from 1967 to 1972, from the sexual revolution to second-wave feminism? Morgan's grand theory promised the liberation of prehistoric woman. Which was direly needed: it is hard to recognize today quite how intense the masculine singular, faux-neutered "Man" was for defining humanity. Consider the titles of a few of the most frequently republished and best-known books: *The Antiquity of Man* by Arthur Keith (1911, second edition 1925); *Prehistoric Man, the Great Adventurer* by the paleoartist Charles R. Knight (1949); *Man Makes Himself* by V. Gordon Childe (1937, new editions in 1951 and 1955); *Adam's Ancestors* by Louis Leakey (1934, new editions in 1953, 1960); *The Epic of Man* (*Time-Life Books*, 1961); *Early Man* by F. Clark Howell (*Time-Life Books*, 1965); *The Emergence of Man* (again *Time-Life Books*, 1973); and again *The Emergence of Man* by John E. Pfeiffer (1969). Consider again the "Dawn of Man" battle in Kubrick's *2001: A Space Odyssey*, a film with no significant female characters. Consider the many, and supposedly varied, paradigms, from Indo-European conquerors, murderous apes, wife-capturing men, primally murdered fathers, the ostensibly neutral fossils named "Peking Man" and "Neanderthal Man" to Kenneth Oakley's *Man the Tool-Maker* (published 1949, second edition 1975), the BBC's documentary *The Ascent of Man* (Jacob Bronowski, 1973), and *Man the Hunter* (edited by Richard Lee and Irven DeVore, 1968). According to this last theory, the male was the hunter, the female the hearth-keeper and gatherer. By 1970, prehistoric females were somehow more the "second" sex than they had ever been: appendices and appendages to Man, gatherers to the dominant male hunters, swooning bare-breasted observers near the genius shaman-artists (Figure 16.5), or else huddled-up baby-carriers at the edges of social life. Consider again Figures 8.12, 8.15, 8.16, 8.17, 16.5, 16.9, and 18.4, but also the simple absence of female hominids from most images in the preceding chapters. Primitive communism had indeed featured matrilineal societies and an imputed matriarchal past among Native Americans. And artists had imagined a peaceful paradise in Knossos, which some treated as an original matriarchy. More ambiguously, "Venus" figurines (like the famous Lespugue Venus, Willendorf Venus, and the Kostienki Venuses of Figure 6.1) were widely celebrated as ancient fertility figures or

Mother-Earth forces. Nevertheless, by 1970, both primitive communism and Minoan matriarchy were considered dated ideas, and for each such depiction of women, there were a dozen prehistory tales that featured their supposed weakness and easy entrapment, and just as many images of prehistoric rape and wife-capture. As Simone de Beauvoir had noted in 1949, prehistoric Woman was shackled to her biological sex and to nature. The prevailing accounts never let their audiences forget this.

A FIRST SHIFT CAME WITH THE ENTRY OF WOMEN INTO FIELDWORK. It is not that they were altogether absent before 1970—Margaret Mead is the most famous example of an early ethnographic fieldworker, as were others among Franz Boas's students, for example Zora Neale Hurston (author of *Their Eyes Were Watching God*), Gladys Reichard, and Esther Schiff. Others had worked with better-known men, or married them and operated as couples. Mary Boyle, known as Henri Breuil's secretary, was responsible for some of the identifications of rock art that came out under his name, and she managed much of his schedule and pathways into public acclaim. Jacquetta Hawkes, née Hopkins, studied archaeology at Cambridge, worked at digs in several countries, and became a well-known BBC-based popularizer of archaeology. She wrote the bulk of volume 1 of UNESCO's massive *History of Mankind*, as well as patriotic books on Britain's antiquity.[6] At UNESCO she had to fend off the condescension of male delegates from different countries, and eventually she became a major advocate of the thesis that ancient Knossos had been a matriarchy.[7] Mary Leakey may have lacked her husband Louis's flair for self-promotion, but she was at least as formidable a paleoanthropologist and fossil hunter. Annette Laming-Emperaire, as we've seen in chapter 16, was by far the most original and systematic interpreter of cave art and a pioneer in South American paleoarchaeology. She began studying gender in cave art and tool economies in the 1950s and was at the peak of her influence in the early 1970s. Sally Binford, among the founders of processual archaeology in the later 1960s, carried out two decades of fieldwork and interpretive analysis and came to be called a "founding mother" of second-wave feminism.[8]

Still, these women were exceptions—some visible, some less so—in a man's world, and their careers were enabled, hindered, and funded by men and organizations that recognized them as just that: "exceptional" women. Sexism around fieldwork was rampant, and it was recognized as rampant. Starting in the late 1960s, this began to change. In much of the Western world, undergraduate and graduate degrees underwent institutional reforms as the student population exploded, and women entered the field (and did archaeological and anthropological fieldwork) in far larger numbers and more consistently than before. The professoriate, too, changed slowly.

IN THE 1960S, ANDROCENTRIC MODELS BECAME INCREASINGLY PREValent and much harsher, particularly because of *Man the Hunter.* Sherwood Washburn had moved on from tools, and, now working with another anthropologist, Charles S. Lancaster, described their vision like this: "When males hunt and females gather, the results are shared and given to the young, and the habitual sharing between a male, a female, and their offspring becomes the basis for the human family." They even invoked the so-called nuclear family of the postwar era: originally, they proclaimed, family hierarchy involved the male hunter as the "socially responsible provider . . . for a female and her young."[9]

By comparison, earlier approaches had foregrounded the men but treated food gathering quite differently. V. Gordon Childe had looked at the end of the last ice age: as the glaciers withdrew, he argued, many of the big herds of deer and bison moved from their former grounds. What followed was not a golden age of vegetation and hunting but a desperate jury-rigging of hunting, fishing, and gathering. This, Childe and others had argued, was a really tricky life for early humans: the environment could not be trusted to provide regular sustenance, which meant that social systems and hierarchies were short-lived and unstable.[10] The concept of the "hunter-gatherer," which referred as much to contemporary Native peoples living at subsistence levels as to peoples living before the invention of agriculture, achieved popularity in the later 1960s, around the time the hunter's masculinity came to be seen as the foundation of social organization.

The 1970s saw far more emphatic changes. Elaine Morgan's account of human origins imagined a figure that would predate Morris's "naked man" by a few million years. It cast the female hominin as the heroine of human evolution. A defender rather than an attacker, someone who reads her environment and knows to retreat, a caregiver to her child, a tool-maker. Everything we regard as human would have evolved before Man took over.

Morgan did try to explain various physiological oddities that her contemporaries usually left aside and she later moved toward establishing a more explicitly scientific basis for the aquatic ape.[11] But her work was proudly speculative, precisely because she wanted to expose the speculative nature of the claims of the Tarzanists. We need better origin myths, Morgan argued, ones that don't let men daydream ideas that legitimize the patriarchy. Even if someone like Morris got all wound up about Morgan's claims, he had no basis really for objecting. His origin point was simply much more recent than Morgan's ostensibly more primal era.

After Morgan, feminist variations on the notion of a gendered origin of humanity spread, and nonscientists joined in (not that a lack of credentials had ever stopped men). Most significant was the account offered by the American journalist Susan Brownmiller, who historicized rape all the way back to hominin apes descending from trees. Brownmiller had organized the first Speakout on Rape with the New York Radical Feminists in 1971 and was deeply involved in the attempt to rethink rape not as an occasional act but as structurally embedded in male power. Her 1975 book *Against Our Will: Men, Women, and Rape* started out with a scene of primitive rape. "When men discovered that they could rape, they proceeded to do it. . . . In the violent landscape inhabited by primitive woman and man, some woman somewhere had a prescient vision of her right to her own physical integrity, and in my mind's eye I can picture her fighting like hell to preserve it."[12] Physical integrity was the first site of self-consciousness. Brownmiller's "primitive woman" fights back and is raped. Both she and the rapist are conscious of the vicious brutality of the act. "One of the earliest forms of male bonding must have been the gang rape of one woman by a band of marauding men. This accomplished, rape became not only a male prerogative, but man's basic weapon of force against woman . . . the vehicle of his victorious conquest over her being, the ultimate test of his superior strength, the triumph of his manhood." Female hominids gravitated to protector males, which confirmed male domination, whether through rape or through protection from it. Masculinist liberal paternalism would now reign. Fear of rape became *the* cause and harbinger of social organization.[13] Brownmiller effectively fused Elaine Morgan's image of the defensive female with Robert Ardrey's scene of violent apes: all history would follow from that pattern. Her key contribution was a primal fiction that would not be judged by crude verification. How different was it, after all, from Freud's own account, with the brothers sharing the spoils of the primal murder?

Morgan's and Brownmiller's accounts were controversial—including among second-wave feminists, some of whom criticized them for simplifying and essentializing womanhood. Their impact was nonetheless immense, above all as inspirations. By the UN "International Women's Year" of 1975 (see Figure 19.1), just three years after *The Descent of Woman*, the big picture had changed dramatically. American sculptor Merlin Stone published *When God Was a Woman* in 1976, in which she pursued the original "Ancient Goddess," who had been pasted over numerous times by Judeo-Christian patriarchy. Prehistoric female rituals had been destroyed and forgotten. Stone echoed an argument pursued by the celebrated Bronze Age specialist, Marija Gimbutas, in her *The Goddesses and Gods of Old Europe* (1974). There, as in several other books, Gimbutas presented pre-Indo-European Europe as a peaceful, Goddess-loving matriarchal culture that was eventually destroyed by the "Kurgan" patriarchy of the Indo-European conquerors.[14] (Kurgan refers to the mound-like tombs that Gimbutas argued were the main physical leftovers of the Indo-European invasions, and which she thought had helped end the original matriarchy.) An all-female spirituality was gaining traction, and Gimbutas's reference to this late neolithic world

Figure 19.1. "She soars to conquer." The UNESCO Courier announces the International Women's Year, in *UNESCO Courier* (1975).

became de rigueur. So did Stone's criticism of Abrahamic monotheism for ruining the prehistoric adoration of an Earth Goddess, and the theory of Minoan matriarchy in Crete that was now being promoted by, among others, Jacquetta Hawkes.[15]

SEVERAL OTHER FEMALE ARCHAEOLOGISTS, MANY OF THEM WITH extensive field experience, also began to present detailed, complex accounts of early society. As the anthropologist Adrienne Zihlman put it in 1978, scientists usually asked "What were the women and children doing while the males were out hunting?" when, in fact, archeological digs showed a wealth of detail about what was happening "at home." The improvement of fieldwork at these digs granted female archaeologists an aura of accuracy and the force of empiricism at the very moment they were entering the field in sizeable numbers. Dismantling "Man the Hunter" became the key to undoing the gendered assumptions of decades of human origins research. As Zihlman continued, "I ask instead, 'How did human males evolve so as to complement the female role?'"[16] Some, like Zihlman's coauthor and colleague Nancy Makepeace Tanner, worked to establish a "Woman the Gatherer" scenario in which the female role came to be the test of social organization and continuity.[17] This new kind of research was rapidly recognized. Kenyan anthropologist Richard Leakey (son of Mary and Louis), in his popular BBC series *The Making of Mankind* (1978), acknowledged that "With the strength of the women's movement growing, the role of the male in Man the Hunter is being replaced by a picture of cooperative hunting-and-gathering groups in which females play a leading role."[18]

In 1984, Margaret W. Conkey and Janet D. Spector delivered a harsher verdict. All archaeologists, male and female, self-proclaimed feminists included, had claimed to be careful about not imposing their ideas on the past. They had all failed. In matters of kinship, gender, and social life, archaeology's study of gender simply lacked rigor.[19] The problem with Man the Hunter was not the obvious gendering. That was bad enough. Nor even was it Washburn's blunt projection of the nuclear family to the deep past. The real issue was that both Man the Hunter *and its rejection* postulated fixed roles for men and women, congealing and essentializing gender. "A division of labor between males and females should not be assumed but rather be considered a problem or a feature of social structure to be explained."[20] The critics of Man the Hunter also lacked "an explicit theory of human social life." Archaeology needed to develop "a specific paradigm for the study of gender."[21] This, we have seen, was the basis of Conkey's criticism of André Leroi-Gourhan and Annette Laming-Emperaire's theories of cave painting. To Conkey what mattered was the study of complex human social arrangements—not the paintings on the cave walls, and not the speculative, overarch-

Figure 19.2. Representation of "Lucy" (*Australopithecus afarensis*) by Maurice Wilson (late 1970s). The sexualization of Lucy suggests that Wilson was influenced by Desmond Morris, though Wilson, unusual among paleoartists, did not shy away from depicting genitals. This image also contributed to the longstanding debate about whether Lucy was a climber or a walker.

ing stories of hunters and of gatherers. After a decade of fits and starts, feminist archaeology could begin in earnest. Rather than locate objects found in a dig back into a social organization where gender roles were already known, the archaeologist had to visualize over and over the place and meaning of objects that would be forever ambiguous.

TOWERING OVER ALL OF THIS WORK, IN THE 1970S AND LATER, WAS Lucy. A partial skeleton discovered in 1974 in the Afar Triangle in Ethiopia, she made front-page news around the world as the oldest hominin yet found—an *Australopithecus afarensis,* as the species was named, dating back to over three million years ago. That the skeleton was female proved fortuitous. She was baptized after the Beatles song "Lucy in the Sky with Diamonds," implying that she was so perfect you almost needed to be on drugs to realize she was real. She was nicknamed the Grandmother of Humanity, pushing aside the male figures who had played the symbolic role of Adam.[22] (See, in this context, Maurice Wilson's contemporary illustration of Lucy, included here as Figure 19.2, though Wilson was clearly influenced by Morris). Other female fossils, for example Cinderella and Twiggy (both found at Olduvai), had never won celebrity. Nor had Augustine, Leroi-Gourhan's female "Mousterian Man." Now, even other fossils were named in Lucy's honor. One partial skeleton, discovered by Laming-Emperaire and her team later that year in Brazil and eventually identified as a Paleo-Indian, was named Luzia.

The "Grandmother of Humanity" idea gained further kind of scientific credential about a decade after Lucy's discovery, when the "Mitochondrial Eve" thesis was proposed and publicized in the mid-1980s. The Mitochondrial Eve relied on logical inference. Because a part of our DNA, the mitochondrial genome, is passed from generation to generation only through the mother, it was possible to track genetic variation and mutations backward, and thus infer one point in the past to which all intermediate generations could be traced.[23] In the public imagination, this was translated into the image of one single mother for all living humanity. It even became possible, by 1988, for *Newsweek* to publish a female-led version of Zallinger's March of Progress design: it skipped the men and instead included both Lucy and Eve (see also Figure 19.3).

Images of prehistoric femininity were double-edged. They represented an overdue correction. But this correction wasn't without its own pitfalls—note, for example, how often the figure of the mother appears in the previous pages. Archaeologists like Conkey were arguing that gendered roles had to be destabilized. Could the matriarchal figurines of the Gravettian period, the Lespugue, Kostienki, and Willendorf "Venuses" and so many others alongside them, be freed of the long interpretive tradition that presented them as Mother Earth–style goddesses? Should (or even *could*) Lucy be spared this kind of overdrawn, overbearing maternal symbolism?[24]

EARLIER FEMALE-CENTERED ACCOUNTS OF HUMAN ORIGINS CAME in for revision by the 1970s, too. We saw in Chapter 6 how Simone de Beauvoir undercut Engels's approach to matrilineality and primi-

Figure 19.3. Female hominins lead evolution. This illustration by Ib Ohlsson for *Newsweek* magazine (1988) features Lucy and the Mitochondrial Eve, and it replaces males with skulls. Most female hominins are accompanied by their young. Not so the modern woman who, mirroring Zallinger's "Modern Man" (Figure 8.1), confidently strides out of the timeline to the right.

tive communism by emphasizing how static his idea of femininity was, how distant from the actual situations women constantly face. By 1970, from Kate Millett to Shulamith Firestone to various radical groups like *The Feminist Voice*, feminists treated Engels's *The Origin of the Family* as a text that had once been foundational to socialist feminism, all the while criticizing it severely for its pretenses and failures. (Valerie Walker abbreviated these criticisms into a cartoon, the figure with which this chapter opened, which appeared in *The Feminist Voice*). Many simply denied the primitive-matriarchy notion altogether. But others used it to imagine an intellectual or spiritual principle in which woman remained the force of order in antiquity, whether as mother or as prehistoric goddess. The promise of a post-patriarchy often came to be phrased as a radical proposal for a restoration of matriarchy, for example by feminist groups like Cell 16 in the US. In the words of one (male) supporter of Cell 16, "it is the male principle in human beings that has brought us historically to the verge of extinction; if we are to survive it will be because the female principle, once omnipotent in pre-history, is returned to power so that our warped existence can be set right again after being awry for ten thousand years."[25]

Even more dramatic were the attacks on Freud, who was routinely vilified by feminists as the paragon of the patriarchy. His Oedipus complex, and especially his discussion of it in *Totem and Taboo*, were the great proofs of his alleged über-sexism. Some, like Firestone, argued that the Oedipus complex was the product of capitalism, others that it was just wrong. The English feminist Juliet Mitchell addressed both Engels and Freud in her *Psychoanalysis and Feminism* (1974). Regarding Engels, Mitchell effectively echoed Beauvoir: his study concentrated on work and technology, not on the realities of gender oppression.[26] He had recognized neither the symbolic nor the psychic ways in which masculinism functions. Her view of Freud was more complex: his pessimism regarding the condition of women stemmed not from his supposed "reactionary spirit" but from his observations of the society around him.[27] His work was "not a recommendation for a patriarchal society but an analysis of one."

Mitchell criticized other feminists on the principle that feminist analysis should be relentless and not celebrate some naturist utopia of the Engels, Stone, or Gimbutas varieties. (Fifteen years after Mitchell, Donna Haraway would agree: better woman the cyborg than woman the natural being.) A post-patriarchal society worth fighting for would not simply restore a primitive matriarchy, the "reign of nurturing, emotionality and non-repression."[28] This was not to deny socialism, which Mitchell identified with her feminism. But she turned the hope of primitive communism and its future restoration on its head. "Long before a situation is analysed, people wish for its overthrow," she complained, urging instead: "know the devil you've got."[29]

But what does it mean to know the devil you've got? Mitchell's answer serves, in fact, as a major point of inspiration for the book you are now reading. She objects to the tendency to turn the rejection of the patriarchy into a question of its origins. To ask "When did it all start?" is to ignore "the position and role of women in society" in order to instead offer easy, convenient shortcuts.

> It seems to me that "why did it happen" and "historically when?" are both false questions. The questions that should, I think, be asked in place of these, are: how does it happen and when does it take place in our society?[30]

As for Conkey, so too for Mitchell: myth had to be hunted, analysis should never end. And despite the great shifts effected by feminist archaeology and anthropology in the 1970s and 1980s, there is little reason to lower Mitchell's flag and proclaim the analysis complete.

Figure 20.0. Emile Bayard, "The First Organized Battles in the Stone Age, or the Entrenched Fort of Furfooz," in Louis Figuier, *L'Homme primitif* (1870). The engraving invokes Horace Vernet's paintings *The Capture of Constantine* (1837) and *Defense of Mazagran* (1840), which depict French battles in the conquest and imperial rule of Algeria.

Chapter 20

IS VIOLENCE INGRAINED, AND HOW?

Throughout my research for this book, I have been surprised at how some sentences—not only concepts and expressions, but indeed entire sentences—get quoted over and over. Few have had as winding an afterlife as Darwin's two sentences about uniting "all nations and races," which we saw already in Chapter 5, on the politics of evolution, and in Chapter 15, on UNESCO's attempt to overcome racism:

> As man advances in civilization, and small tribes are united into larger communities, the simplest reason would tell each individual that he ought to extend his social instincts and sympathies to all the members of the same nation, though personally unknown to him. This point being once reached, there is only an artificial barrier to prevent his sympathies extending to the men of all nations and races.[1]

In 1871, the passage was a near perfect representation of the ideology of liberal imperialism: civilization stands opposed to war. *We* extend our sympathies to the men of all races. (They do not extend their own sympathies, and some of them we conquer for their own good so they can learn to extend their sympathies.) Once they are taught this imperative, they'll become good individuals; they will become "Man." Other empires need to be taught it, too. We are moral, and the others will become so as well.

Eighty years after Darwin, Ashley Montagu fought to include the passage in full in UNESCO's first Statement on Race. Now it meant something slightly different. Cooperation, Montagu thought, was natural, and tribalism had caused racist hierarchies, World War II, and the destruction of Indigenous peoples.[2] For Montagu as for many at UNESCO, Darwin's comment on universality spoke of the dangers of modern warfare. *We* who fought in the World Wars have fought against tribalism and nationalist racism. *We* have also been tribal and we must extend our sympathy to all. The Irish biologist, communist sympathizer, and public intellectual J. D. Bernal agreed. In *The Freedom of Necessity* (1949), he paraphrased Darwin's passage to mean *not* that "tribes were always at war" but that war was basically a modern invention. Back in prehistory, war "had not been invented, because the need for it on any serious scale did not exist."[3] Tribalism was today's, not yesteryear's, danger.

Fast-forward another sixty years. Today, the passage plays the opposite role. It is routinely used to mean that things keep getting better and better. That "primitives" were always at war, that war belongs with them, not us. Modern liberals love the phrase. Steven Pinker invoked the passage as an epigraph to his triumphalist *The Better Angels of Our Nature: Why Violence Has Declined* (2011).[4] In his own *The World Until Yesterday*, Jared Diamond (of *Guns, Germs, and Steel* fame) detailed violence in Papua New Guinea as evidence that "traditional societies" engage constantly in intensive warfare.[5] The sociobiologist E. O. Wilson also paraphrased Darwin: prehistoric genetic adaptations mean that "we are hampered by the Paleolithic Curse . . . We are addicted to tribal conflict, which is . . . deadly when expressed as real-world ethnic, religious, and ideological struggles. People find it hard to care about other people beyond their tribe or country."[6] Yuval Noah Harari, author of the mega-bestseller *Sapiens: A Brief History of Humankind* (2011) and reigning prophet of prehistory's future, agreed. In *Homo Deus* (2015) he adapted Darwin's maxim for his own history of life and information when he declared that all through human history and roughly until the start of the current millennium, war was one of three things people worried about every day. Today, he says, they don't—surely not those people who live just a few miles east of Harari's own home.

Only recently, the very same words spoke against, and not for, the narcissistic adulation of our own society. Time and again, Darwin's famous passage has been dressed up as a great scientist's grand conclusion. But it is pure ideology, and has been from the start. So, what instead is the exact relationship between war, human origins, and modernity? Over the past half-century, one of two answers has been given. One group of scholars has argued that war is essential to human nature at its most primitive—that war pervades life in archaic societies and that the advance of civilization is the same thing as the overcoming of war. A second group has argued contrarily that war is a consequence of the establishment of hierarchies that came with civilization. War is the opposite of civilization, or war is civilization. Each answer is about something more than the place of war in human life: it is about humanity itself.

THE EUROPEAN CLAIM ABOUT INDIGENOUS SAVAGERY GOES ALL THE way back to Hernán Cortés and the siege of Tenochtitlan in 1520, if not to Columbus. Cortés famously pronounced, of the Tlaxcalan allies who fought with him against the Aztecs, that "no race, however savage, has ever practiced such fierce and unnatural cruelty."[7] But whereas Europeans routinely attributed savagery to Native peoples, the early modern theorists of sovereignty—Bodin, Grotius, Pufendorf, Vattel—plainly declared that war involves sovereign states. It was not a matter of tribes, nor of disorganized, indiscriminate killing. Generalized and uncontrolled violence did exist. But it wasn't war.

The establishment of prehistory as a research field in the nineteenth century dredged the problem back up. John Lubbock, in *Pre-Historic Times* (1865), distinguished between more and less violent Indigenous nations, but his text nonetheless relished in describing the "whetted appetite" of some of their gods who (he claimed) drove them to violence, cannibalism, and warfare. Lubbock described the Fijians, Sioux, and Māori in much the same way as settler-colonial states did when they projected violence onto the peoples they had colonized.[8]

In the twentieth century, and just as Native Americans were becoming set-piece villains in Hollywood films, scholars came to doubt whether any relationship could be drawn between violence and Indigenous peoples. In a 1936 profile of Franz Boas, *Time* magazine presented a debate between the famed German-Jewish anthropologist and "Britain's Anthropologist," Arthur Keith, on war and evolution. "Keith declared that Nature keeps her human orchard healthy by pruning, that war is her pruning hook. Dr. Boas took vigorous exception: 'War eliminates the physically strong; war increases all the devastating scourges of mankind such as tuberculosis and genital diseases; war weakens the growing generation.'"[9] Boas's antiracism required that human affairs, violence included, be decoupled from nature. Five years later, Bronisław Malinowski criticized both those who discovered in the past a "golden age of perpetual peace" and those who promoted primitive war as "an essential heritage of man . . . from which man never will be able to free himself."[10]

Malinowski wrote this in the altogether brief period when the "civilized peoples of the earth" caused seventy-plus million deaths and untold numbers of other casualties. Early definitions of "genocide" described it as a distinctly modern crime.[11] Hiroshima made clear that powerful states now had the means of eliminating life altogether. Would nuclear war return humanity to the Stone Age—to the caves after atomic destruction, as Dr. Strangelove proposed? The question became palpable soon after the bombings of Japan. John Hersey's report on Hiroshima, published by *The New Yorker* in 1946 and immediately republished as a book, answered yes. In its pages, human beings desiccated by radiation wandered aimlessly in a devastated, apocalyptic landscape. The Polish-born mathematician Jacob Bronowski, who would become an influential public intellectual—he wrote and narrated the BBC's *The Ascent of Man* (1973)—presented his visit to Nagasaki as the event that made him commit to a humanism rooted in science and opposed to all war. The question remained: if ferocious war is central to human existence, on whom to pin the blame? Ancients or moderns?

THE 1945 UNITED NATIONS CHARTER RESOLVED IN ITS PREAMBLE "TO save succeeding generations from the scourge of war, which twice in our lifetime has brought untold sorrow to mankind." And UNESCO's constitution blared out that wars begin in the minds of men. War is something impersonal, caused by humans, but a force that escapes their grasp. Ending it demands incessant efforts to maintain the peace.

Immediately after the Second World War, interest in Indigenous war was limited. Public debates concerned instead the terror of atomic power, militarism, the chance of a third world war. Anthropologists like Margaret Mead, Ruth Benedict, and Harry Holbert Turney-High played down the importance of the raiding practiced by some Indigenous peoples. Benedict "tried to talk of warfare to the Mission Indians of California, but it was impossible. Their misunderstanding of warfare was abysmal."[12] Turney-High even wrote an entire book to distinguish "primitive" from "modern" war. Other social scientists agreed.[13]

In the 1960s, the confrontation with modern war led again to questions about the beginnings of war. The sheer scale of recent violence, real and imagined, led some to ask: Was war ingrained? J. Robert Oppenheimer presented his experience of the first atomic explosion in 1945 as a primal event, and he famously reported thinking of Vishnu in the *Bhagavad Gita*, as if a primal, divine otherness now had reality: "I am become death, the shatterer of worlds."[14] The ancient epic, in Sanskrit (which is still often treated as the primal Indo-European language), seemed appropriate for absolute destruction. Meanwhile, Freud's primal murder—by this point something of an embarrassment to most psychoanalysts—made a public comeback.

Robert Ardrey's naturalization of brutality and territoriality begins to make sense against this backdrop too. His allies, animal ethologists Konrad Lorenz and Nicolaas Tinbergen, accentuated, as Lorenz put it, the "fighting instinct in beast and man." Freud (like many others) had argued that civilization suppressed and deformed innate aggression. Tinbergen agreed: civilization, he declared, caused a disequilibrium of the instincts.[15] Aggression riveted the reading public: the books of Lorenz, Ardrey, and Desmond Morris (Tinbergen's doctoral advisee) became major bestsellers. Few objected openly to their claims—Hannah Arendt was a notable exception. For all her fascination with the idea, Arendt condescended that animal aggression was ultimately irrelevant. "In order to know that people will fight for their homeland we hardly had to discover instincts of 'group territorialism' in ants, rats, and apes . . . I am surprised and often delighted to see that some animals behave like men; I cannot see how this could either justify or condemn human behavior."[16] Most intellectuals thought that ethology offered evidence that our prehuman past was alive within us, that war is in our genes, hidden under the thin veneer. The idea of an innate aggression instinct was occasionally bolstered by sometimes unexpected scientific findings, most significantly by Jane Goodall's late 1970s observation of systematic killings among chimpanzees.[17]

The observation that war may be ingrained in human nature was no abstract matter. "Primitive warfare" was easily conflated with anticolonial struggles from Algeria to East Africa to Vietnam and Afghanistan, which were regularly in the news. Jean-Paul Sartre famously explained anticolonial violence as follows: "The native cures himself of colonial neurosis by thrusting out the settler through force of arms. When his rage boils over, he rediscovers his lost innocence and he comes to know himself."[18] Sartre did not mean that the colonized are violent, rather he was agreeing with Frantz Fanon's diagnosis that colonial oppression was overwhelming: revolution was bound to be violent. The point was perfectly legitimate, but it further unsettled those who acted as if Europe was violently besieged. If violence traveled South-to-North, Europeans and Americans had all the more reason to embrace Gandhi's and Martin Luther King Jr.'s nonviolence as civilized. Violence was unethical, strategically counterproductive, savage. Any opposition to an idealized, universal Man was easily declared an inhuman, animal brutality. Civilization meant extending one's "social instincts and sympathies to . . . the men of all nations and races."

Postwar economic development projects also strengthened the claim that Native peoples should be studied for their violence. To jumpstart development, economists like W. W. Rostow looked to economic markers like technological advance, productivity, population density, and agricultural efficiency. The intended contrast between Europe and America on the one hand, and impoverished, supposedly stagnant preagricultural societies on the other, could not be more striking. When photographs and films from Native battles and hunts—for example among the !Kung or the Yanomami—appeared in magazines or arthouse cinemas, it was partly to highlight their visually striking difference from "us."[19] When Robert Gardner's ethnographic film *Dead Birds* was released in 1963, public responses focused on a battle between Hubula villages in New Guinea (Gardner referred to the Hubula by the exonym "Dani"). As the film made clear, the battle had an almost ritualized quality: it was an organized but limited war in which each side mobilized large numbers so as to inflict a single, spiritually demoralizing fatality on their enemy.[20] The goal of war was not killing per se, but a balance of power and a ceremonial exchange in a complex, quite egalitarian, and tense society. Yet the scene of dozens of near-naked men fighting on each side had different, and quite primitivizing consequences (see Figure 20.1). The most striking example of primitivization came through Kubrick's clear reuse of some of Gardner's stylistic choices in the battle of the apes in "The Dawn of Man" sequence in *2001: A Space Odyssey*. Just as in *Dead Birds*, Kubrick set up a stage with war parties facing off, with one main combatant from each side stepping forward into a pitched man-on-man fight, reaching closer to strike while their side yells out its support. Kubrick replicated the soundscape of war cries, the awkward, sharp movements of the men fighting, and above all, the highly compressed depth of field through which Gardner had heightened the drama of the battle. *2001* completed the transformation from the Hubula's struggle for a balance of spiritual power to killer apes inaugurating humanity by fighting to kill.

"Until recently," Sherwood Washburn and Charles S. Lancaster wrote, "war was viewed in much the same way as hunting. Other human beings were simply the most dangerous game."[21] For Washburn and Lancaster, war extended the ubiquity and pleasurable aggression in ancient hunting. War, like hunting, directed the use and improvement of tools; it enabled social differentiation. In the late 1960s, the connection between hunting and war was supported by studies of animal extinction by the Soviet climatologist Mikhail I. Budyko and the American geologist Paul S. Martin. The "Over-

Figure 20.1. Still from the battle scene in *Dead Birds* (Robert Gardner, 1963). For the Hubula people, warfare participates in balancing power, and Gardner focused on its intricate, limited, ceremonial character. Gardner rejected Western scorn of the Hubula as primitive and violent, but in his direction of the film he was at best an inconsistent advocate. *Dead Birds* influenced depictions of prehistoric life—see for example Kubrick's *2001* and Figure 18.4.

kill Hypothesis," as Martin's work on the subject came to be called, postulated that whenever humans had arrived at new geographic locations, they had rapidly decimated large mammal populations.[22]

Is hunting necessarily a matter of violent and warlike behavior? Predictably, Robert Ardrey swooped in to affirm as much in *The Hunting Hypothesis* (1976). The illustrator Zdeněk Burian also drew image after image of the entrapment, killing, and flaying of mammoths, rhinos, and bison by Native-American-looking Cro-Magnons (see Figure 20.2), often in warpaint, sometimes fighting over the kill.[23] By 1972, when the anthropologist Napoleon Chagnon published *Yanomamö: The Fierce People*, the image of ingrained violence was quite standard. Chagnon posited that the Yanomami (whom he

Figure 20.2. In the 1970s, Zdeněk Burian painted a series of scenes about the capture and killing of mammoths; like contemporary scholars, he sometimes linked hunting to warfare.

referred to as "Yanomamö") of Brazil lived a "chronic state of warfare." Theirs was a perpetual state of nature, with man as wolf to man.[24] Indigenous warfare, he went on, was constant, unending. Scholars who celebrated a peaceful humanity and ignored this "reality" were hallucinating.

SUCH WAS THE BEGINNING OF THE "FIRST" OF THE TWO SCHOOLS OF thought on war and civilization—the one that argued that aggression is innate, that Indigenous peoples are fundamentally violent, that civilization is fundamentally about the taming of violence. War and violence arose with humanity, and only thanks to modern and (Western) humanism could it be overcome.

The question of when war arose is mostly conceptual. In fact, in many cases the same evidence could be interpreted in opposite ways. Arendt wondered: Did long-range weapons (arrows and spears) release the "aggressive instinct" (as Tinbergen argued) or did they show that such an instinct plays only a minimal role in war (as Otto Klineberg claimed)?[25] But this is not simply a matter of instincts or spears: if you conceive of hunting and war-making as coextensive, as Ardrey, Tinbergen, Burian, and (to some extent) Chagnon did, you're more likely to think of early humans as natural warmongers. But if you think that war proper is a different matter than hunting or than relatively contained violence, you're more likely to also think that organized militaries or states are necessary for hierarchy and war. André Leroi-Gourhan, for example, argued that aggression and violence had been part and parcel of "the human economy" ever since australopithecines, but their effects and social place depended on social and technical evolution. "Aggression in primitive peoples takes very different forms from those of war as rendered possible by the existence of large sedentary units."[26]

So, as of the late 1960s, a second school of opinion arose. Some scholars and activists radicalized the positions of Mead, Benedict, and Turney-High and argued that war is more modern and that, in any case, it is a fundamentally different matter than occasional violence in Indigenous societies. The latter did not rise to war because such fights could never be as devastating as organized state violence. Claude Lévi-Strauss had offered an early version of this claim. In *Tristes tropiques* (1955)—perhaps the most widely-read work of anthropology ever written—he presented the Nambikwara of Brazil as deeply nonviolent and their expressions of hostility as silly, even pathetic. They had no kings, no real chiefs. Against them stood the massive power of "our" hegemonic modernity that throws "our shit in the face of humanity."[27] Levi-Strauss extended his argument from his 1952 UNESCO pamphlet and contrasted Man (the all-powerful, singular force of Western humanism) with humanity as defined by diversity. Summoning "our master, our brother" Jean-Jacques Rousseau, he insisted that the Nambikwara represented nonstate peoples: they may not be living in primitive communism but they were a people without writing, without capital, without authority, without real violence in them.[28] It was an excessive stance: Lévi-Strauss had taken modernity full circle to Rousseau, reinventing the pure first human.[29] The Nambikwara came to represent the poorest, the guiltless and powerless, the least "civilized" Indigenous peoples.

That same decade, a series of influential psychiatrists interested in the mind of non-Western peoples worked to refute the idea that war is a human instinct. Yes, some psychoanalysts still postulated that tribal aggression had such a long history as to be effectively innate. To them, modern limitations—"the stress of self-imposed inhibition of manifest aggressivity"—caused "psychological injury."[30] But most thought it best to set aside the speculation in Freud's *Totem and Taboo*.[31] A new field of "transcultural psychiatry" took to studying Indigenous peoples in the context of colonial violence: some transcultural psychiatrists were simply old-style colonial psychiatrists in new clothes. But the more original ones emphasized colonialism's *creation* of aggression. The most famous student of the psychic traumas of colonialism was Frantz Fanon, psychiatrist by day and anticolonial revolutionary by night (and day).[32] Others, such as Henri Ellenberger, Georges Devereux, and Fritz Morgenthaler, became supremely interested in the different varieties of aggression, and they too attributed much of it to colonialism. Devereux studied war and aggression among the Mohave in Arizona and California, and later he offered a detailed account of their rationales for war, which he considered recent developments and not perennial behaviors.[33] For this generation of ethnographer-psychoanalysts, war and aggression were prompted by state violence rather than by an innate or primitive character. Indigenous peoples were not primitive: they were colonized, their societies damaged, their beliefs distorted, their aggression dramatized.

During the same period, the concept of the rebel targeted both colonialism and capitalism, effectively fusing the two. By the mid-

1960s, the old Leninist model of Marxist revolution seemed ever less workable. New libidinal, natural, anticapitalist heroes took center stage, from the anticolonial fighter to the urban guerilla, from the sexual revolutionary to the Native depicted as close to nature. In the months after the general strike and student revolution in Paris in May 1968, Jean-Paul Sartre's magazine *Les temps modernes* fashioned itself into a revolutionary headquarters. That October, it published the breakthrough essay by the American anthropologist Marshall Sahlins, "The Original Affluent Society."[34] Sahlins argued that Indigenous peoples who appear to us poor are in fact anything but. Hunter-gatherers—a new term in the 1960s–1970s—simply had a standard of living that was "easily satisfied." It is our own time that is "the era of hunger unprecedented." Civilization doesn't eradicate starvation: it creates it. Few would contest the starting point of this argument, namely that poverty and starvation in modernity were morally shocking.[35] Sahlins took that fact to a logical conclusion. "Poverty is not a certain small amount of goods. . . . Poverty is a social status."[36] Indigenous peoples were the last remaining true rebels, actively refusing the logic of capitalism. In *Stone Age Economics*, the book where Sahlins expanded on the essay, he further proposed that "primitive society" is not a warmongering one, but is "at war with War."[37] Indigenous societies are not just uninterested in wealth, but profoundly invested in rejecting a descent into generalized warfare. Warfare is waged only to avoid social dissolution and worse violence, not out of fierceness or cruelty. Sahlins's opposition to *Yanomamö: The Fierce People* was absolute. And indeed, Sahlins never stopped decrying Chagnon and his influence as invidious and contemptible.

An early adherent of Sahlins's was Pierre Clastres, a French-born ethnographer of the Tupi in Brazil and the Aché in Paraguay. Writing in a 1970s climate suspicious of the state's tendency to become omnipotent and tyrannical, Clastres radicalized Sahlins's anarchism to propose that Indigenous societies are more than hostile to wealth and hierarchy, and that they are in fact passionate about stopping the creep of the state. Like Lévi-Strauss, Clastres emphasized that a chief's authority was fragile and temporary: a warrior-king was accepted only for the duration of hostilities and out of utter need.[38] Afterward he may be rewarded, but his authority waned. The powerful state that had obsessed philosophers like Hegel and Marx was for Clastres the greatest of enemies.[39] Not least because it invaded the refusenik Indigenous societies from without and reduced them to savagery and slavery.

Such approaches to the question of war were clearly linked to current events: student revolutions, the embrace of nonviolence, disappointment with the Soviet promise grinding down to a gulag halt, and opposition to the Vietnam War, which many saw as an instance of overwhelming American power failing against a weak, small, almost guerilla state. Exactly contemporary to Clastres's argument was anthropologist James C. Scott's *The Moral Economy of the Peasant*. Scott focused on peasant rebellions in early twentieth-century Vietnam, which he considered techniques of resistance to state authority. The book sent Scott down a path of writing about the force of state violence, and especially about the reaction of those threatened by it.

Even if Indigenous peoples were violent, such violence could not really compete with that of great powers that could "bomb them back to the Stone Age." The rejection of the native-as-warmonger also fed off of (and fed into) a slowly growing public perception of Native Americans as peaceful, intimate with nature, dominated by white supremacy, and distorted in their public representations by outsiders. But perhaps nothing mattered more than a disgust increasing during the 1970s with the neocolonial effects of international economic development—the ongoing expropriation and continuing impoverishment of Africa, South and South-East Asia, and Latin America. Resistance to the state, as Clastres described it, was resistance to a capitalism that used the state to triumph.

BY THE 1990S, EACH OF THE TWO POSITIONS ON THE ORIGINS OF WAR was loudly decrying the other for supposedly achieving institutional dominance and refusing reality. Genocidal civil wars in Yugoslavia and Rwanda, along with broader anxieties about power and international development, brought into sharp relief the question of who is responsible for violence and war. Is it the state? And who polices a violent state? In what ways are individuals and groups responsible for mass violence? In Europe especially, war was perceived as part of an earlier era that had been overcome—except the implosion of Yugoslavia was dragging the continent back to an era of war. That sense of time—that civilization meant the overcoming of war, that those left behind needed help to develop into a peaceful and advanced society—was consistent with the approach that treated Indigenous peoples as living amid violence. Meanwhile, Sahlins and Clastres had long ago become standard references in anarchist circles and the academic Left: they were especially useful for a critique of the colonial state.

Within the academy, a third anthropological position developed, which focused on colonial and anticolonial violence in the "tribal zone." R. Brian Ferguson and Neil L. Whitehead, the editors of the book *War in the Tribal Zone: Expanding States and Indigenous Warfare*, focused on the borders of expansionist states, especially the area where state intrusion created havoc among existing non-state communities. Violence was a borderlands issue, they argued, it concerned the zone where the state deploys violence against non-state peoples living both within and outside of it, and by confronting them, rouses them too to violence.[40] Ferguson and Whitehead's tribal-zone idea was influential within the academy but did not break the dominant opposition between the two schools.

The competition continued, albeit on new grounds. In 1996, in *War Before Civilization*, Lawrence H. Keeley denounced "neo-

Rousseauians" who had created a "pacified past" where "war was rare, ritualized, abnormal, and foreign to human psychology." Archaeological evidence dating to the early Neolithic indicated fortifications around settlements. This meant that it was necessary to think of the Neolithic as a period full of defensive and aggressive warfare. Keeley emphasized the "intensity, dangerousness, and effectiveness of primitive war . . . the casualty rates, the destruction, and the gains or losses of territory and other vital possessions."[41] Two years later, Christopher Boehm countered that egalitarianism had prevailed in every hunter-gatherer society: the purpose of equality, Boehm claimed, was to shut down violent aggression. Then, in 2000, Raymond Kelly inverted the conventional terms and wrote not of intensity of warfare but of intensity of *peace*.[42] He did not reject Keeley's challenge based on the archaeological record, but instead he focused on how societies avoid expanding localized violence into war. He theorized that across history, war had developed in three stages: first, a long era of "coalitionary" killing aimed mostly at gaining local advantage; later, a period of "intrinsic defensive advantage" that established greater capacities for trade and exchange rather than war; and finally, as of about 14,000 years ago, the era of proper war, waged in forms we would recognize today.[43] Compared to the "hundreds of thousands of years" of the first two eras, war proper was a recent invention. It differed considerably by region, and it resulted in the emergence of both warlike and warless societies.

The particular date when war began in earnest has remained a matter of ongoing debate. In 2015, for example, remains of a formerly unknown species from 330–230,000 years ago, now named *Homo naledi*, were found deposited in a chamber of a complex cave system in South Africa. Signs of possible violence raised the question anew: Do war and murder go back to the beginning of humanity? (And did *sapiens* exterminate Neanderthals, *naledi*, *Homo floresiensis* in Indonesia or others in a genocidal fury?) Other signs indicated that the bones had been deposited in the cave—which replayed the debate about Neanderthal burials. Other recent archaeological discoveries, for example, of artifacts and bones dating back 15,000 years or so have raised similar questions about violence, selfhood, and community.

Darwin's expression—that only an artificial barrier prevents us from extending our sympathies to all of humanity—is today mostly of value to those who take a position that modernity is peaceful and the future will be better still. To read Stephen Pinker is to be bathed in the warmth of *pax nostra*, to be appalled anew at the gruesome bloodbath of the Neolithic and to celebrate how far from violence we have come. To read Chagnon on the warlike character of the Yanomami—or to watch films like *Cannibal Holocaust* (Ruggero Deodato, 1980), which impute casual rape, murder, and cannibalism to that same nation—is to become inured to the violence inflicted on them all the way to the present.[44] No complexity and no hope survives this theory of unmitigated progress.

But sometimes the second school also seems to have rather too much clarity in its vision of the deep past. James C. Scott, after a career spent studying state power and sources of resistance to it, argued in *Against the Grain: A Deep History of the Earliest States* (2017) that the rise of agriculture entailed the invention of storage (and theft) and hence state formations that in turn produced war, hierarchy, and slavery.[45] And the recent, remarkable doorstopper *The Dawn of Everything* by David Graeber and David Wengrow depicts early humanity as creative, anarchic, making-it-up-as-you-go-along. Although these are not the only voices arguing that war began relatively late, with the rise of the state, it matters that a direct line of intellectual descent takes us from Sahlins and Clastres through Scott to Graeber: Scott was inspired by Clastres and painted a similar picture of the state's intrusion into nomadic societies; Graeber was Scott's colleague at Yale, and he coauthored a book on kings with Sahlins. A certain kind of scholar needs the state to have been brutal and repressive from the start, and those who were not within its purview—nomads, hunter-gatherers—to have been limber anarchists who offered a superior vision for life. "Anarcho-primitivist" authors have offered variations on this argument, proposing that hunter-gatherer life offers a guide for our collective future, or even that it was the only setting in which true equality ever existed.[46]

Is this all an academic spat? After all, who really cares if there was more war or less war among some early humans? Some were probably more, others less, violent. Yet today the obsession with "primitive warfare" stands sentry before any access to the deep past. It takes us right back to the state of nature; it is revealing of scientific culture, and Western culture more broadly. The ongoing belief in a state-of-nature as a time and place of brutality and suffering—or an Eden-like paradise—continues to shape how Westerners treat much of the rest of the world.

Meanwhile, very real violence against Indigenous peoples continues. Voluntarily isolated people have continued to suffer attacks and be destroyed. In the 1980s and 1990s, some state authorities continued to present some of their "Natives" as dangerous, even as descendants of cannibals (if not cannibals themselves). In Brazil, Papua New Guinea, and elsewhere, such claims attracted tourists and supported ongoing expropriation.[47] Jared Diamond, in *The World Before Yesterday* (2012), speaks of the radical transformation since "yesterday's" traditional societies, and criticizes Robert Gardner's film *Dead Birds* and the work of Gardner's collaborator Karl Heider. Gardner, as we have seen, had presented war-making among the Hubula as having a restorative aim—to pacify the dead. Heider had gone further, and had dubbed the Hubula *Peaceful Warriors*.[48] Diamond instead thunders that they had effectively misrepresented what he regarded as an extremely brutal situation. But he offers this argument exactly at a time when natural gas and mining projects in Papua New Guinea involve the seizure of native land and heighten tensions between communities. Most famous of all advocates of the "savage Natives" position is Jair Bolsonaro, who as a Brazilian congressman in 1998 voiced his wish that Brazil had destroyed its Indigenous peoples. As president from 2019–22 he actively encouraged gold mining in the Amazon at the expense of Yanomami and other peoples and their habitats, with malaria, malnutrition, and mercury poisoning as the direct effects.[49]

To both schools, the past performs tasks for the present. Neither position has really exited the eighteenth century. One targets the violence of the weak; the alternative sees a ready-made target, the state, and its aide, capitalism. Readers become the real winners in this ongoing debate: we get to select the image of humanity that most appeals to us, regardless of its cost.

Figure 21.0. Walton Ford, *The Flaming Fields* (2020), watercolor, gouache, and ink on paper. 83 1/2 × 60 1/4 in.

A Storm Blowing from Paradise

What are you saying? We agonized in vain for the sake of
a cloud?

—EURIPIDES, *Helen*

D
o we have a better understanding of human origins today
than scientists and the public did in the eighteenth or
nineteenth centuries?

When it comes to the details, the answer is surely yes. As I was fin-
ishing this book, Svante Pääbo was awarded the 2023 Nobel Prize for
his leadership in mapping the Neanderthal genome, an achievement
with medical consequences for everything from immune responses
and blood clotting to depression and susceptibility to COVID-19.
Thanks to paleoanthropologists, we know a great deal about the
trading networks, religious rituals, food intake, and social organization of "prehistoric" societies. Genetics has begun to produce a map
of the deep past unlike any other—with the promise that once we
have enough antique specimens, we will be able to use genetic variation to plot the course and movements of humanity.

Yet the epic myths on which this book has focused remain very
much alive. Many have been recycled under new names. We still talk

of thin veneers, disappearing or "uncontacted" peoples, of the Nean-
derthal/*sapiens* difference, of "Man." Stadial theory—the *civilized/
barbarian/savage* argument—was used in the post-1945 division of
first/second/third worlds, which euphemistically became a distinction of developed/developing/underdeveloped economies. Killer apes
rearticulated the violent state of nature and the motif of the "savage beneath the thin veneer"—and they eventually fed into theories
of primitive warfare. Many other ideas have a habit of returning
after being debunked. For a century now, for example, scientists have
been rejecting the theory that humans colonized South America by
crossing the South Pacific on boats, rather than by traversing a glacier bridge on the Bering Strait, only for other scientists to bring it
back. Shamanism—more than half a century old and long abandoned as an explanation for cave paintings—is back and it is exciting.[1] One of the rioters who stormed the US Capitol on January 6,
2021, famously donned faux-Visigoth furs and a helmet, stood at
the Senate rostrum, and later called himself the QAnon Shaman.
Right-wing populist regimes regularly describe refugees as flooding "Asiatic" or "African" hordes. The dream of a peaceful primal
era, whether matriarchal or not, is a regular feature in critiques of
state violence, especially violence inflicted on Indigenous peoples.
University professors and high-school teachers downplay the faults
of scientists—notably Darwin—because it's hard to teach evolution or science without heroes. Many decontextualize genetic tests
on ancient hominids and perceive the results as hard evidence that
describes in detail whole ways of life.

The persistence of old ideas is common even in debates about
quite technical matters. For example: How old is *Homo sapiens*?
100,000 years old or 500,000? For almost a century, the term "Cro-
Magnon Man" described early forms of *Homo sapiens* by referencing the site in France where that earlier specimen was first discovered.
Then, in the 1970s, the description "early *sapiens*" swallowed up
"Cro-Magnon Man" into a large group of early fossils, including,

for example, *Homo heidelbergensis*, which was at times considered an intermediate form that had departed from Africa some 400,000 years ago. "Early *sapiens*" was now described awkwardly as dating back half a million years. Over time, this theory dissipated: scientists focused again on the relatively more recent movements of *Homo sapiens*: leaving Africa about 70–100,000 years ago, arriving in Europe about 60–50,000 years ago. In the last few years, since *sapiens* bones dating some 315,000 years ago were found in Morocco and elsewhere, far from the usual "cradles" of humanity, the entire picture is being remade.[2] The older, slower development theory going back to half a million years has returned. Which is fine—except it has returned as an innovation. The demand for scientific innovation and for the rapid popularization of new discoveries creates a cycle of constant (re)discovery.

Genuinely new theories have emerged, especially out of genetics and neuroscience. But because novelty is so valued and because some scientists (like the public at large) tend to teleport rather too quickly into the deep past, these theories enjoy a very public life even when they are misguided, incomplete, or pernicious.

THE MITOCHONDRIAL EVE THEORY, MENTIONED IN CHAPTER 19, IS perhaps the most popular of the recent genetic ideas. The Mitochondrial Eve follows from the idea that mitochondrial RNA passes on to a fetus from the mother alone. Ever since it was first floated in the late 1980s and 1990s, Eve has regularly been extrapolated to mean that all humans can trace their descent to a single female. The theory has been a subject of scientific controversy from the start. Its reference to the Biblical Eve has had its own share of implications, and its usefulness for the Out of Africa approach to human origins has been overwhelming. Scientists in opposing camps have accused each other of racism: some supporting Out of Africa treat the advocates of the competing multiregional hypothesis as camouflaged polygenists, while their critics note that Out of Africa sometimes sounds rather like "Thankfully We Escaped Africa."[3] Often, the idea of a biological Eve is just too irresistible. In the BBC's *History of the World* (2012), Andrew Marr presents a large Black family crossing a natural bridge between two mountains. Among them is a humble Eve, the single mother of all now-living humans. The scene of the movement of ancient populations out of Africa is full of pathos. High above the abyss, Eve's child is terrified. But they make it across, and thanks to their bravery, humanity's movement from Africa begins. (In other words, the BBC did not notice it was supporting a "Thankfully We Escaped Africa" story.) Then there's Yuval Noah Harari, who forges an Über-Eve: "Just 6 million years ago, a single female ape had two daughters. One became the ancestor of all chimpanzees, the other is our own grandmother." The point is cute and catchy and completely absurd. (New species form not of a single act of procreation, but thanks to several generations of mutations across a population, and often thanks to the mixture of several populations, and so on.) Why do we need a single Eve, a single grandmother, a single point of origin for all?[4] No more serious but at least more fun is Luc Besson's film *Lucy*, with Scarlett Johannson in the lead as a woman who becomes a cyborg and achieves mastery of her "full brain capacity." At the climax of the film, Lucy travels across human history with a couple of swipe-lefts, eventually to commune, like God touching Adam in the Sistine Chapel, with Lucy the *Australopithecus afarensis*.

Neuroscientific references, for their part, often feel stitched together without care for the actual research they claim to rest on. The most popular of these concepts is the "reptilian deep brain," a psychologist's answer to the psychoanalyst's troublesome Unconscious. Paul D. MacLean first developed it in his "triune theory of the brain" in the 1950s, and by the 1990s the expression was in wide circulation. MacLean distinguished between a "reptilian" system in the basal ganglia (which determined the behavior of lower animals), a "paleomammalian or limbic" system, and a "neomammalian" one.

They are layered atop one another—much as Herbert Spencer had proposed that the brain adds layers of complexity over the course of evolution—and the neomammalian brain (the neocortex) applies only to humans. As in Spencer's theory, and as in the Killer Ape thesis, at the bottom of the Lizard Brain thesis we find a subhuman force. One purpose of the reptilian brain was to replace the Freudian Unconscious with a properly neurobiological explanation of our ancient animal selves.

The triune theory was excluded relatively quickly by other brain scientists, but this hasn't stopped commentators from resurrecting the lizard brain and from citing neuroscience as its foundation.[5] The political commentator Joe Klein declared Donald Trump to be dominated by his lizard brain in February 2016. Klein took the lizard brain to be scientifically current and accurate—"an actual part of our brain, the amygdala—the most primitive part."[6] Months later, just before the 2016 election, guru Deepak Chopra went on Conan O'Brien and authoritatively diagnosed Trump with a "reptilian brain on overdrive."[7] The lizard just sticks well (and to Trump especially) as a liberal battle cry—we the rational, they the reptiles.

Other ostensibly neuroscientific theories rely on concepts straight out of marketing and self-help. In the 1990s, the archaeologist Stephen Mithen theorized a "creative explosion" due to human "cognitive fluidity," which he argued began some 60–40,000 years ago.[8] Mithen asks—as many today do—why *sapiens*, that is, anatomically modern humans, outlived other hominids. He argues that it was due to "a different type of mentality" that was a result, he firmly believes, of language and creativity.[9] But how do you measure "cognitive fluidity" or "a different type of mentality?" Mithen relies on parallel archaeological efforts as well as contemporary philosophy of mind, which he fuses in order to declare that *Homo sapiens* 50,000 years ago suddenly became far more creative, compared to the "remarkable monotony" of millions of years before and the genus *Homo* in general (see Figure 21.1).[10] The Achilles' heel of this argument is its circular logic—really its reliance on "creativity," an idea whose legs Mithen and his colleagues took for granted throughout, even as they debated its exact meaning.[11] Put another way, archaeologists now rely on genetics, anthropology, cognitive science, gender studies, and economics as they do on archaeology. They use evermore complex microscopes, gene mapping approaches, and calculations. They hope to be able to make this past relevant—but can't help but use common concepts that make sense to us.

By the time that Yuval Noah Harari wrote *Sapiens*, the "Creative Explosion" theory had lost its luster. Harari rehashed it and blended it with Richard Wrangham's 1999 theory that cooking made us human.[12] As is his style, Harari upped the ante. Some 300,000 years ago, the beginnings of cooking enabled fundamental mutations in human biology: we became dramatically smarter in a "Cognitive Revolution," he opines. So it was cooking that stormed the brain's Winter Palace. Harari posits that what is singular about humans is the ability to "create" a shared "imagined reality" that guides a community toward the future. Who doesn't love a theory that humanity resides in "our" creativity? "Our" creativity 80,000 years ago, but also your creativity and mine. As others have pointed out, this is the perfect theory for hedge fund managers, gig-economy creators, and virtual reality gamers. Our future relies on the "imagined reality" that we can construct—that's what makes it "human." Just as the first cognitive revolutionaries way back in time were rendered obsolete by their creation, "we" too shall be transfigured by our newly "imagined reality."

The still-younger discipline of neuroaesthetics has offered similarly kitschy conclusions, for example, that cave art is the most ancient confirmation that our brains are "primed" by what they perceive. In *The Tell-Tale Brain* (2010), influential neuroscientist V. S. Ramachandran first proposes that Lascaux's artists "must have used some of the same aesthetic laws used by modern artists."[13] No sooner does he make this claim about eternal aesthetic laws than he races to

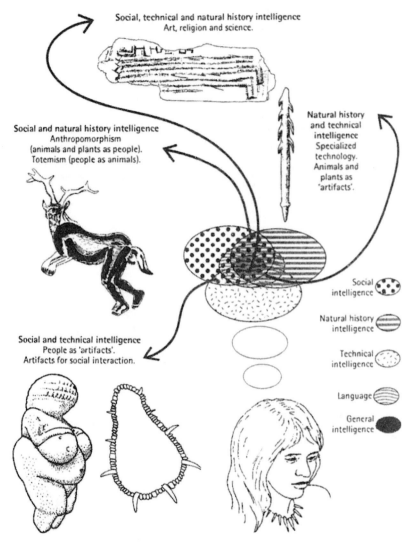

Figure 21.1. "The 'cultural explosion' as a consequence of cognitive fluidity." Steven Mithen, *The Prehistory of the Mind* (1996).

tell us about a Harvard psychologist's ideas, then he pivots to Shakespeare, and then pivots again, this time to talk about an imaginary gene mutation that might have allowed us to "imagine orgasms." A mere page after first mentioning Lascaux, Ramachandran returns to it but has forgotten all about eternal laws. He introduces the exact opposite idea: our ancestors lacked mental imagery altogether and cave paintings enabled them to acquire it. Before a hunt,

> it was easier to engage in realistic rehearsal if they had actual props, and perhaps these props are what we today call cave art. They may have used these painted scenes in much the way that a child enacts imaginary fights between his toy soldiers, as a form of play to educate his internal imagery. Cave art could also have been used for teaching hunting skills to novices. Over several millennia these skills would become assimilated into culture and acquired religious significance. Art, in short, may be nature's own virtual reality.[14]

Except we have no access to nervous systems of beings that died 400 centuries ago. So *perhaps* it was easier to rehearse with cave paintings, *perhaps* they were props, *perhaps* painting skills turned art into "nature's virtual reality." *May have* and *should have* and *may be.* Or perhaps this is lovely speculation appropriate to the age of PowerPoint presentations. Like the dreamy mutation that might allow us to imagine orgasms, it is enchanting and also silly.

THE BETTER QUESTION IS: TWO AND A HALF CENTURIES IN, CAN WE call prehistory a success? So many of the debates are the same as they were in the beginning. The compulsion endures. The compulsion, that is, to offer a complete theory of humanity by referencing the deep past, to get rid of recorded history, to forget how little we know and answer all the questions. We still quibble about whether the first humans were peaceful or violent, whether early societies were patriarchal, how we should understand the value of states and science and even the destruction of Indigenous worlds. We telescope at will back in time to draw meaning from the deep past about ourselves and the world we want. We look for our beach, humanity's beach, in some distant place-time.

One result is that good, reliable science becomes harder to iden-tify. The past loses its elasticity, its complexity, its electricity. And at a time when we are regularly plied with disinformation and lies, most theories of prehistory only add to the pile. Worse, we end up with an idealized humanity. When policymakers, industry leaders, and the powerful identify humanity with the ancient hominid frolicking in the savannah, creating tools, priming its brain, being creative, and so on, they spread an image of the human that is fundamentally deceit-ful. (And they do: Barack Obama and Bill Gates have waxed enthu-siastic about Harari's *Sapiens*.) It allows them to misidentify and make excuses for the real humanity that burns forests and oil and cares little for the poverty right outside our door or on the other side of the planet. And most of the rest of us, too, commit to fantastical, distorted futures based on a perfected universal story. We dissociate from the world that we ourselves have hurt.

So, whatever the various advances in technology or research methods, the original illusions endure. We continue to imagine that our relation to origins is direct and pure and meaningful. We over-look the biological determinisms and ignore the colonial baggage. We subscribe to the slogans. We pretend to own humanity across the empire of time. And prehistory continues to be all about "us" rather than about the reality, so far as it can be known, of early social life. Or, better yet, about the inconceivable qualities of a profoundly different world.

THE MOST OBVIOUS AND GREATEST COST OF THE 250-YEAR OBSES-sion with human origins research has been borne by the Indige-nous peoples whose destruction was rationalized because they were "primitives" who were "vanishing" anyway; by Jews, Roma, homo-sexuals, and others deemed subhuman by Nazism; by all those who were racialized by ideas about prehistoric humanity; and by refugees, still disdained today as a watery mass and a horde. Humanity still bleeds because of our obsession with defining some group of our fellows according to a supposedly savage past.

There is perhaps no clearer sign today of the folly of prehistory than this: every scholar with a fragment of an old bone or a theory about the workings of Neanderthal vocal cords is heard, publicized, photographed like Hamlet peering into Yorick's skull, and celebrated as making a paramount contribution to The History of Humanity. But any nation, tribe, ethnic group that lays claim to objects that have been wrenched from it and placed in museums has to show standing in court. Indigenous peoples often lack such standing. They need funding and perhaps state support, which is rarely forthcoming, then they have to establish their claims in public; to compete using existing laws, which tend to be hostile; to lobby for new laws while waiting out years of foot-dragging by museums and by the governments that rule over them; to somehow avoid being tarred, even in a hush-hush manner, as primitives; and so on. Whoever seeks to have their past restored needs to use the language of archaeology and the networks and mechanisms of international cooperation. Meanwhile, scholars can easily play a double game: they claim the grand truth of the History of Humanity while nonetheless allowing that others have traditions "to be respected." Museums profess their sensitivity, but they still own the objects. Few are willing to accept that their epic of human grandeur robs and injures. Objects may house spirits or be plainly alive to the claimants seeking their return in order to restore alternative ways of life. Yet their worlds have been sundered, and the law treats their leftovers as superstitions.

I do not report this reality because others have not, but because it is banal. In it, "our" humanism is always slated to win. And these "debates" feed right back into our fetishization of humanity. At this point, the concept of the human is the emptiest concept of all. The entire language in which prehistory arose, sustained itself, and continues today—a language that, as the scientists of our origins never tire of telling us, teaches us what it means to be human—is a language of real power disavowed. If it has on occasion helped break down some divisions between humans, it enables others to live on.

The problem lies less with science or museums rather than with the humanist impulse that accompanies them. The story told in this book is in part a story of scientific horrors. But it is not a story, for the most part, of evil philosophers or scientists, nor at all a story in which science is the enemy. It is a story of the lengths that we will go to to convince ourselves that we share something more than (most of our) DNA with hominids from tens and hundreds of thousands of years ago: that what we share with them is meaningful, and that it is our "human nature." When early humanity is presented as violent or weak, we pronounce ourselves triumphant. When it is presented as strong or complex, we empathize with it. Prehistory has been a mirror like the one Italo Calvino described in *Invisible Cities*: "at times the mirror increases a thing's value, at times denies it. Not everything that seems valuable . . . maintains its force when mirrored."[15] It is time for our existence, our politics, and our work for a more equal, more just, and plainly fairer world not to be refracted through that mirror.

THE CONCEPT OF HUMANITY THAT FOLLOWS FROM PREHISTORY rests on hierarchies and exclusions, on ideals of purity and power. Even today many imagine humans in a state of nature, in an Indigenous community, on untrodden land, building tools and living in a dangerous and beautiful world over which they would gradually come to reign. The fantasy allows us to forget that in reality, humans have almost nothing in common with our paleolithic forefathers. We live in the world we have created. We bask in a techno-society made of wires and concrete, surveillance and staggering exploitation, and we cover everything with this humanist illusion. For some three centuries now, the image of the original humans has been key to ideas of human perfection, and yet its most important effect has been dehumanization: of the rest of the world by the West, and of the West by itself. Even after World War II, when international organizations, governments, and intellectuals committed to a universalist program to teach the world about human equality and the insignificance of racial and ethnic origins—even then the project of human nature worked to politicize differences, to tie development loans and capital to a particular Western, liberal vision of modernity that was in conflict with its Soviet and anticolonial competitors.

Today, the prehistory we have invented still distorts. And this is not only because of state power or capitalism or technology. Instead, we pass on this idealized humanity through the words we use every day. These words carry a charge, an intensity that fills our thinking and shapes our values even when they have little to do with origins. It is this role for prehistory that is most interesting and most troubling. Can we not finally begin to recognize that that humanity is an invention, a pretend foundation? Why should we not define humanity in some other way than through a convenient grand narrative that begins with the first humans? To emphasize, as I do, the profound problems with the ideology of human origins is not to give up on human equality, on difference, or on the shared dreams of a fairer and better future. On the contrary. It is to recognize that we need a stronger foundation than this ersatz definition, than a past so distant we can always fill it with our inventions. Our orientation toward the deep past starts out in awe—as many writers note, in childlike awe— but it ends in prejudice and violence.

Throughout this book, I have been drawn to those thinkers who engaged with deep passion the question of what humanity is, who recognized the dangers and violence of this line of inquiry, and who did not hide their contempt for purity and certainty: André Leroi-Gourhan, Juliet Mitchell, Georges Bataille, Sigmund Freud. For them and some others, the human has been its own worst enemy, and humanism, which has always hidden violence, has run dry of hope. If we follow their lead, what do our origins look like? When we recognize that *we* have built them, and that we have built them in such a way as to justify violence? Can we still identify with the "birth" of humanity? We should not. For there to be any future for humanity, we must see that the deep past, however enchanting, isn't worthy of our love. It must become another time. And we have to come to a new definition of humanity, one in which we are no longer imagining ourselves as pure, but in which we recognize that we have always been compound beings, webs of meaning, and cyborgs. A definition in which we are violent because of what we do now, not because of what those hominids back then might have done. One in which our sense of home and intimacy—not some imagined antiquity—becomes the motivation for undoing the ecological destruction we have wrought. One in which doubt about our claims and doubt about our answers and doubt about the capital "h" we put in Humanity reaches everywhere and should again become our operating principle, a skepticism that never rests, a skepticism in the service of a better theory for tomorrow.

Acknowledgments

I started my research in 2012, long before I knew this might be a book. Several institutions supported me with grants: the American Council of Learned Societies (Charles A. Ryskamp Fellowship, 2015–16); the Partner University Fund "Crossroads in Intellectual History" from the French American Cultural Exchange (FACE) Foundation; the Centre National de la Recherche Scientifique (CNRS/UMI 3199); the Université Paris Sciences-et-Lettres/NYU Global Alliance; and NYU's Global Research Initiative. I am grateful to each of these funding bodies for all the time they made possible for research, thinking, and writing.

I have also benefited from the kindness of librarians and archivists across many institutions. In France: Fondation Teilhard de Chardin, Paris; Muséum National d'Histoire Naturelle, Paris (Paul Rivet, Henri Breuil, and Marcel Mauss archives); Collège de France, Paris (Henri Breuil library; Marcel Mauss, André Leroi-Gourhan, and Georges Dumézil archives); Bibliothèque Nationale de France, Département de Manuscrits Occidentaux, Paris (Georges Bataille and Claude Lévi-Strauss archives); Maison Archéologie-Ethnologie René Ginouvès, Université de Nanterre (André Leroi-Gourhan archive); UNESCO House, Paris (archives on the Statements on Race and early-postwar archives on UNESCO projects concerning Indigenous peoples). In the United States: Wenner-Gren Foundation, New York (archives on Teilhard de Chardin, Raymond Dart, anthropology symposia, South Africa, and Sherwood Washburn); Rockefeller Archive Center, Sleepy Hollow, NY (interwar anthropology projects, also with thanks to Freddy Foks for sharing from his research on Grafton Elliot Smith); Yale University, New Haven, CT (Roman Jakobson Archive; with thanks to Jamie Phillips); Bancroft Library, University of California-Berkeley (Sherwood Washburn files), and elsewhere. In St. Petersburg, Russia (Sankt-Peterburgskii Filial Arkhiva), Archive of Nikolai Marr (with thanks to Alexandra Poloshukina). In the United Kingdom: British Library, London (John Lubbock, Baron Avebury materials); National Archives, Kew, UK (W. J. Perry Papers, W. H. R. Rivers Papers); Cambridge University Library (Alfred Cort Haddon). And in Germany, the library of the Max Planck Institute in the History of Science and the Free University of Berlin. The pandemic made digital research essential, and I am most grateful to librarians at NYU's Bobst Library for their support.

The rights holders for most of the images reproduced in this book were absolutely wonderful. I cannot name them all so I name none here in order not to be unfair; but they have my profound appreciation.

In constructing my arguments, I have danced a fair bit around the work of scholars whom I admire. I have benefited a great deal from their work, and although it has not been easy to sidestep them, I have generally organized my argument to complement rather than retrace or argue with theirs: Pratik Chakrabarti, Emily Kern, Michael Kunichika, Erika L. Milam, Nathan Schlanger, Marianne Sommer, Nadine Weidman, and especially Maria Stavrinaki. I strongly recommend their books and articles for their rigor, force, and importance. I am also grateful to Veronika Kusumaryati and her Hubula correspondents for their guidance, which I have followed. Over the decade of this research, I have published several articles and essays from this project, and I am grateful to the editors and managing editors of the Centre Georges Pompidou catalog *Préhistoire: une énigme moderne,* and the journals *History of Science* and *RES: Anthropol-*

ogy and Aesthetics—especially *RES*'s model editor and intellectual Francesco Pellizzi. Francesco's passing in August 2023 is a source of extraordinary sadness for me: it was a privilege to think with him, and he treated me always with wonderful generosity, a mentor's seriousness and laughter, and intellectual and personal friendship. It was a pleasure to coorganize a conference at NYU-Paris with Erika Milam. I presented other chapters or ideas at the Cornell-Weill History of Psychiatry Working Group (my favorite reading group, with thanks to George Makari and Orna Ophir), Berkeley, Yale University, Columbia University, the University of Wisconsin-Madison, the Society for French Historical Studies, and the American Historical Association. Thank you to my interlocutors in these fora.

Since arriving at NYU in 2008, I've avoided reteaching courses. "The Birth of the Human," a class I taught for the Core Curriculum, is the sole exception: by teaching it several times since 2013, I learned this material and rethought it as I inched toward the book. My teaching assistants for its different iterations were heroes; they also often helped me think through a problem: Alexander Arnold, Jonas Knatz, Alex Langstaff, Emily Stewart Long, Matt Maclean, Jamie Phillips, John Raimo, Anne Schult, Alec Shea, Joshua Sooter, Devin Thomas. Many of the students in these classes threw questions at me that forced changes in my approach and my attitude. I am also thankful to Hayley Ackerman and Victoria Aranowicz for their assistance.

I usually write late at night where it really feels as if, as the song goes, προχωρούσα / μέσα στη νύχτα / χωρίς να γνωρίζω κανέναν / κι'ούτε κανένας / κι'ούτε κανένας με γνώριζε. That's a real feeling but it's false. I am intensely thankful to Rania and to Leon and Isabelle, who create daily the conditions in which it is possible to write and where we can talk about and play with ideas; to my sister Sarra, my parents Maria and Nikos, and our wonderful greater family that is intensely on my mind as I write now. And so many of my colleagues and friends walked this path with me; sang with me; tolerated my thoughts-out-loud; and sometimes held these thoughts in their hands and gave them back to me all better; they made me talk about things I knew little about and realized I had to learn; they said no when I needed to hear it. For their direct help on this project, I want to thank Robyn d'Avignon, David W. Bates, Zvi Ben-Dor Benite, Olivier Berthe, Danielle Judith Zola Carr, Stephanie DeGooyer, Dan Edelstein, Liz Ellis, Udi Greenberg, Dagmar Herzog, Lotte Houwink ten Cate, Jonas Knatz, Martti Koskenniemi, Meredith Martin, Mark Mazower, Todd Meyers, Eric Michaud, Priya Nelson, Samantha Paul, Francesco Pellizzi, Jamie Phillips, Andy Rabinbach, Camille Robcis, Gisèle Sapiro, Anne Schult, Sophie Smith, Natasha Wheatley, Larry Wolff, and Nasser Zakariya. I wish I had enough space to describe your contributions in detail. Amia Srinivasan offered careful criticisms of the penultimate version of the manuscript, and I am most grateful for them. I outlined large sections of the book *chez* my cousin Xenia Geroulanos in Paris, and I owe her and Rodrigo Rey Rosa a huge debt. Early enough in my research, Maria Stavrinaki invited me to coedit a special issue of *RES: Anthropology and Aesthetics*. That project was a crash-course and changed my horizons. We then worked together repeatedly and edited translations of each other's texts. A lot of my thinking has developed thanks to, and in conversation with, hers—Μαρία, ευχαριστώ.

Ekin Oklap, Alex Christie, Anna Haddelsey, and Jacqueline Ko at the Wylie Agency have been immensely supportive. This book would not exist without Ekin's original trust and all of their shared commitment to it. I want to add that, through their engagement, Ekin and Jackie have changed the way I think about writing. So has Dan Gerstle at Liveright, who patiently and carefully tamed a messy manuscript, saw through an often-shortsighted argument and transformed both into an actual book that has gone far beyond my expectations. It was my first time working with an editor with a real vision of the book, and I hope mine kept up. I'm extremely grateful to each and every one of you—had someone told me in advance that I would work with you, and that you'd be as extraordinary as you are, I would not have even believed any of it. More recently, Zeba Arora managed an astonishing array of details, and Rebecca Rider (with the help of Rebecca Homiski and Rebecca Munro) copyedited with very real diligence and care—my thanks to you are absolute.

Throughout this book, I have tried to offer a strong criticism of our pretensions to grandeur, our thirst for powerful stories, our belief that we grasp the whole picture and spin it into a thorough system of knowledge. It's always nice to blame others, the ideologies of the enemy; but this project is more of a criticism, as philosophical as it is historical, of *our* delusion—that we grasp the origin as much as the end, that we control the definitions, that we master knowledge. Of course, this puts me in a corner, and the historian Pierre Vidal-Naquet obliges me with a great description of my predicament. He writes: "To write history using philosophy, as a historian, means being willing to enter into the difficult but necessary game of being a philosopher when amongst philosophers, cutting up the text 'where the natural joints are', like a good butcher as Plato says in the *Phaedrus*. And then, one has to go further, comparing what has not been compared before, setting in serial order what has never been serialized, revealing the ideological fragility of what has hitherto been considered knowledge, demolishing what are claimed to be certainties, trying to turn up the cards to see what they conceal, in short playing the role which has always fallen to the historian, the role of the traitor." A butcher and a traitor, okay, albeit with apologies for all who helped me on this path.

Notes

Introduction: The Human Epic

1. Raymond V. Audette and Troy Gilchrist, *Neander-Thin: A Caveman's Guide to Nutrition* (Dallas: Paleolithic Press, 1996).
2. Duo Xu, Pavlos Pavlidis, Recep Ozgur Taskent, Nikolaos Alachiotis, Colin Flanagan, Michael DeGiorgio, Ran Blekhman, Stefan Ruhl, and Omer Gokcumen, "Archaic Hominin Introgression in Africa Contributes to Functional Salivary *MUC7* Genetic Variation," *Molecular Biology and Evolution* 34, no. 10 (2017): 2704–15, and see the reporting around that article.
3. The expression of breathing life into the past is very common; consider, for example, Jeremy DeSilva, "Did the First Americans Arrive via Land Bridge? This Geneticist Says No.," *New York Times*, February 8, 2022.
4. On prehistory and art, see the brilliant book by Maria Stavrinaki, *Saisis par la préhistoire: Enquête sur l'art et le temps des modernes* (Dijon: Presses du Réel, 2019); translated by Jane Marie Todd as *Transfixed by Prehistory: An Inquiry into Modern Art and Time* (Princeton, NJ: Zone Books, 2022).
5. Jed Z. Buchwald and Mordechai Feingold, *Newton and the Origin of Civilization* (Princeton: Princeton University Press, 2013); Anthony Grafton and Noel Swerdlow, "Technical Chronology and Astrological History in Varro, Censorinus and Others," in *Classical Quarterly* 35 (1985): 454–65; Anthony Grafton, *Joseph Scaliger: A Study in the History of Classical Scholarship. Volume II: Historical Chronology* (Oxford: Clarendon, 1994).
6. Hans Blumenberg, *Work on Myth* (1979; Cambridge, MA: MIT Press, 1988).

Chapter 1: The Infancy of Humanity

1. Jean-Jacques Rousseau, *The Confessions* (London: Penguin, 1953), 28–30.
2. Jean Starobinski, *Jean-Jacques Rousseau. La transparence et l'obstacle* (Paris: Plon, 1955); Stefanos Geroulanos, *Transparency in Postwar France* (Stanford: Stanford University Press, 2017), 1–7, 267–75.

3. Jean-Jacques Rousseau, *Emile, or On Education* (New York: Basic Books, 1979), 59.

4. Anthony Pagden, *European Encounters with the New World* (New Haven, CT: Yale University Press, 1993).

5. On Adario and the Native American man who influenced his creation, see David Bell's criticism of David Graeber and David Wengrow's *The Dawn of Everything* (2021). David A. Bell, "A Flawed History of Humanity," *Persuasion* [online community], November 19, 2021.

6. On natural religion deriving from Jesuit concepts, see Harro Höpfl, *Jesuit Political Thought* (Cambridge: Cambridge University Press, 2004) and Alan Charles Kors, *Atheism in France, 1650–1729*, vol. 1 (Princeton: Princeton University Press, 1990).

7. Giambattista Vico, *The First New Science*, ed. Leon Pompa (Cambridge: Cambridge University Press, 2002), 33–34.

8. Johann Gottfried Herder, *Another Philosophy of History* (London: Hackett, 2004), 79.

9. Jean-Jacques Rousseau, *Discourse on the Origin of Inequality*, trans. Donald A. Cress (London: Hackett, 1992), 84–85.

10. Jean-Jacques Rousseau, *Essay on the Origin of Language*, in Jean-Jacques Rousseau and Johann Gottfried Herder, *On the Origin of Language: Two Essays*, trans. John H. Moran and Alexander Gode (Chicago: University of Chicago Press, 1986), 30–31.

11. Hesiod, *Works and Days*, in *The Homeric Hymns and Homerica*, trans. Hugh G. Evelyn-White (Cambridge, MA: Harvard University Press, 1914), 110–69.

12. Dan Edelstein, *The Terror of Natural Right* (Chicago: University of Chicago Press, 2009), 12.

13. Rousseau, *Discourse on the Origin of Inequality*, 40.

14. Benjamin Straumann, *Roman Law in the State of the Nature* (Cambridge: Cambridge University Press, 2015).

15. Höpfl, *Jesuit Political Thought*, esp. 257–61.

16. The conspicuous exception had been Michel de Montaigne, but his famous essay "Of Cannibals," which was certainly an inspiration for Rousseau, was hardly a blueprint for examining the past.

17. Ioannis Evrigenis, *Images of Anarchy* (Cambridge: Cambridge University Press, 2014), 159.

18. Thomas Hobbes, *Leviathan: The English and Latin texts*, ed. Noel Malcolm (Oxford: Oxford University Press, 2012), 13:21; 2:195–96.

19. Cătălin Avramescu, *An Intellectual History of Cannibalism*, trans. Alistair Van Blyth (Princeton: Princeton University Press, 2011), 7, 9.

20. John Locke, *Second Treatise*, in *Two Treatises of Government*, ed. Peter Laslett (Cambridge: Cambridge University Press, 1967) §49, 319.

21. John Locke, *Second Treatise*, §108, 357.

22. John Locke, *Second Treatise*, §111, 360.

23. Rousseau, *Discourse on the Origin of Inequality*, 11.

24. Rousseau, *Discourse on the Origin of Inequality*, 41.

25. Rousseau, *Discourse on the Origin of Inequality*, 41. Rousseau worked

on the Abbé de Saint Pierre's project for perpetual peace and he arguably derived the childhood-as-first-age argument from the Abbé's theory of progress.

26. Rousseau, *Discourse on the Origin of Inequality*, 19.

27. Hesiod, *Works and Days*.

28. Quoted by George Boas, *The Cult of Childhood* (London: Warburg Institute, 1966), 42.

29. *Encyclopédie, ou dictionnaire raisonné des sciences, des arts et des métiers, etc.*, eds. Denis Diderot and Jean le Rond d'Alembert (Paris: Briasson, 1751–65), 5: 651–2.

30. Isaak Iselin, *Philosophische Muthmaßungen über die Geschichte der Menschheit* (1786; Hildesheim: Olms, 1976). Simone Zurbuchen, s.v. "Iselin, Isaak," in *The Cambridge History of Eighteenth-Century Philosophy*, ed. Knud Haakonssen (New York: Cambridge University Press, 2006), 2:50. Fernando Vidal, *The Sciences of the Soul* (Chicago: University of Chicago Press, 2011), 191, 196.

31. Adam Ferguson, *An Essay on the History of Civil Society*, rev. ed. (London: Millar and Cadell, 1768), 2.

32. Herder, *Another Philosophy of History*, 8.

33. Herder, *Another Philosophy of History*, 9, 12, 33. Eventually, once Sanskrit got Europeans all excited, Herder moved the Orient eastward and settled on India as humanity's geographical cradle. See Nicholas Germana, "Herder's India," in Larry Wolff and Marco Cipolloni, eds., *The Anthropology of the Enlightenment* (Stanford: Stanford University Press, 2008), 119–39. Raymond Schwab's *The Oriental Renaissance* (New York: Columbia University Press, 1984), explained in detail how the imagination of the East generated self-criticism, something generally downplayed in the large body of writing since Edward Said's indispensable *Orientalism* (1979).

34. Herder, *Another Philosophy of History*, 10.

35. Claude Henri de Saint-Simon, *The Political Thought of Saint-Simon* (Oxford: Oxford University Press, 1976), 220–21.

36. Herder, *Another Philosophy of History*, 75.

37. Robert Chambers, *Vestiges of the Natural History of Creation* (London: Churchill, 1844), 355.

38. John Lubbock, *The Origin of Civilization and the Primitive Condition of Man* (London, 1861), 4–5.

Chapter 2: Europe's "Indigenous" Noble Savages

1. Charles de Brosses, *Du culte des dieux fétiches* (Geneva, 1760), translated as "On the Worship of Fetish Gods," in *The Returns of Fetishism* eds. Rosalind C. Morris and Daniel H. Leonard (Chicago: University of Chicago Press, 2017), 44–132. See also William Pietz, *The Problem of the Fetish*, ed. Francesco Pellizzi, Stefanos Geroulanos, and Ben Kafka (Chicago: University of Chicago Press, 2022).

2. Hayden White presented the noble savage theme as an attempt of European intellectuals to assert their authenticity: White, "The Noble Savage Theme

as Fetish," in *Tropics of Discourse* (Baltimore: Johns Hopkins University Press, 1985).

3. Tacitus, *Germania*, chapter 2.

4. Patrick J. Geary, *The Myth of Nations: The Medieval Origins of Europe* (Princeton: Princeton University Press, 2001).

5. Eric Michaud, *The Barbarian Invasions* (Cambridge, MA: MIT Press, 2019), 4.

6. Christopher Krebs, *A Most Dangerous Book: Tacitus's Germania from the Roman Empire to the Third Reich* (London: Norton, 2012), 151.

7. Michel Foucault, *"Society Must Be Defended": Lectures at the Collège de France, 1975–76*, eds. Mauro Bertani and Alessandro Fontana (New York: Picador, 2003), 140–65. Michaud, *The Barbarian Invasions*, 9, 102.

8. Montesquieu, *The Spirit of the Laws*, trans. Anne M. Cohler et al. (Cambridge: Cambridge University Press, 1989), 646.

9. Montesquieu, *The Spirit of the Laws*, 647.

10. Edward Gibbon, *The History of the Decline and Fall of the Roman Empire*, vol. 1 (London: Strahan and Cadell, 1776), 222.

11. Gibbon, *History of the Decline and Fall*, 225.

12. Gibbon, *History of the Decline and Fall*, 227.

13. Gibbon, *History of the Decline and Fall*, 217.

14. Gibbon, *History of the Decline and Fall*, 231.

15. Herder, *Another Philosophy of History*, 33. Cited in Michaud, *Barbarian Invasions*, 5.

16. Tillmann Lohse, "A Collapsing Migratory Regime?," in *European History Yearbook* vol. 19: *Victimhood and Acknowledgment* (Berlin: De Gruyter, 2018), 156.

17. Georg Wilhelm Friedrich Hegel, *Lectures on the Philosophy of World History*, trans. Robert F. Brown and Peter Crafts Hodgson (Oxford: Clarendon, 2011), 460.

18. Richard Wagner, "A Communication to my Friends" (1851), in *Wagner on Music and Drama*, eds. Albert Goldman and Evert Sprinchorn (New York: Dutton, 1964), 264.

19. John Bowen, "The Historical Novel," in *A Companion to the Victorian Novel* eds. Patrick Brantlinger and William B. Thesing (London: Blackwell, 2005), 253.

20. Michaud, *Barbarian Invasions*, 12.

21. Paul Azan, ed., *Par l'épée et par la charrue: Écrits et discours de Bugeaud* (Paris, 1948), 79. Aimé Césaire quotes this passage in his *Discourse on Colonialism* (1950).

22. Jules Michelet, *The People* (London: Longman, Brown, Green, and Longmans, 1846), 13.

23. Mike Tyson, interview with Paul Holdengräber, New York Public Library, November 12, 2013.

24. *The Works of Sir William Jones*, by Lord Teignmouth [John Shore] (London: Stockdale and Walker, 1807), 34–35.

25. Friedrich von Schlegel, "Die bis jetzt so dunkle Geschichte der Urwelt." in *Über die Sprache und Weisheit der Indier* (Heidelberg: Mohr und Zimmer, 1808), iii.

26. Schlegel, "Die bis jetzt so dunkle Geschichte," 81.

27. Schlegel, "Die bis jetzt so dunkle Geschichte," 175.

28. See Rasmus Rask's influential prize essay *Undersøgelse om det gamle Nordiske eller Islandske Sprogs Oprindelse* [Investigation of the Origin of the Old Norse or Icelandic Language] (1818), trans. Niels Ege (Amsterdam: Benjamins, 2013), 78–79.

29. Jakob Grimm, "Vorrede," in *Deutsche Grammatik* (1818; Göttingen: Dieterichschen Buchhandlung, 1819), vi. In addition to the fairy tales and to his work in linguistics, Grimm later published a commentary on, who else, Tacitus.

30. Wilhelm von Humboldt, *Ueber die Verschiedenheit des menschlichen Sprachbaus . . .*, 3rd ed. (1836; Berlin: Calvary, 1883), 256.

31. Franz Bopp, *Vergleichende Grammatik* (Berlin: Dümmler, 1857), xxiv.

CHAPTER 3: THE CREATURES DEEP TIME INVENTED

1. Benoît de Maillet, *Telliamed* (Basel: Libraires associés, 1749), 1:178–79.

2. Georges Louis Leclerc, comte de Buffon, *The Epochs of Nature*, ed. and trans. Jan Zalasiewicz, Anne-Sophie Milon, and Mateusz Zalasiewicz (Chicago: University of Chicago Press, 2018); discussed in Maria Stavrinaki, *Transfixed by Prehistory: An Inquiry into Modern Art and Time*, trans. Jane Marie Todd (New York: Zone Books, 2022), 9–10, 33, 339n.2.

3. Martin J. S. Rudwick, *Bursting the Limits of Time* (Chicago: University of Chicago Press, 2005) and *Worlds Before Adam* (Chicago: University of Chicago Press, 2008); Paolo Rossi, *The Dark Abyss of Time* (Chicago: University of Chicago Press, 1984).

4. James Hutton, *Theory of the Earth* (1788), in *Contributions to the History of Geology*, vol. 5, ed. George W. White (Darien, CT: Hafner, 1970), 128.

5. See, for just one example, V. Gordon Childe, *Man Makes Himself* (London: Watts, 1936), 42–44.

6. Charles Darwin, *On the Origin of Species* (London: Murray, 1859), 464.

7. John Ruskin, *Modern Painters V* (London: George Allen, 1906), 338.

8. Ruskin, *Modern Painters V*, 338.

9. Ruskin, *Modern Painters V*, 339.

10. John Ruskin, *The Turner Gallery at Marlborough House* (London: Smith, Elder and Co, 1857).

11. Ruskin, letter to Henry Acland, May 24, 1851, in *The Works of John Ruskin*, eds. E. T. Cook and Alexander Wedderburn (London: Allen, 1909), 36:115. See also the discussion in Maria Stavrinaki, " 'We Escape Ourselves': The Invention and the Interiorization of the Age of the Earth," *RES: Anthropology and Aesthetics* (2018), 20–37.

12. Rebecca Bedell, "The History of the Earth: Darwin, Geology and Landscape Art," in *Endless Forms: Charles Darwin, Natural Science and Visual Arts* eds. Diana Donald and Jane Munro (New Haven, CT: Yale Center for British Art, 2009), 49–79. See also Stavrinaki, *Transfixed by Prehistory*, chap. 1, 2.

13. William Buckland, "On the Discovery of Coprolites," *Transactions of the*

Geological Society 2:3:1 (1829), 235; cited in Rudwick, *Worlds Before Adam*, 155.

14. Henry De la Beche, *Notes on the Present Condition of the Negroes in Jamaica* (London: Cadell, 1825).

15. For the connection of these figures to New World and enslaved populations, see Surekha Davies, *Renaissance Ethnography and the Invention of the Human* (Cambridge: Cambridge University Press, 2017).

16. See also "The Age of Reptiles," the frontispiece of George Fleming Richardson's *Geology for Beginners*, 2nd ed. (London: Longman, Brown, Green and Longmans, 1843). This replaced an earlier frontispiece that had three lizard-like dinosaurs facing off—a decidedly less intense situation than the one Richardson took from Beche. Meanwhile, Robert Farren had a large, slightly updated copy of Beche's drawing painted on large canvas, so he could illustrate his Cambridge lectures. See Zoë Lescaze, *Paleoart: Visions of the Prehistoric Past* (New York: Taschen, 2017), 38.

17. Darwin, *Origin of Species*, 406.

18. Darwin, *Origin of Species*, 62.

19. Darwin, *Origin of Species*, 58.

20. On ruins, see Julia Hell, *The Conquest of Ruins* (Chicago: University of Chicago Press, 2019); Julia Hell and Andreas Schönle, eds. *Ruins of Modernity* (Durham, NC: Duke University Press, 2010); Sabrina Ferri, *Ruins Past: Modernity in Italy, 1744–1836* (Oxford: Oxford University Press, 2015); Andrew Hui, *The Poetics of Ruins in Renaissance Literature* (Fordham University Press, 2016); Duncan Bell, *Remaking the World* (Princeton: Princeton University Press, 2016), chap. 5.

21. Johann Gottfried Herder, *Another Philosophy of History and Selected Political Writings* (London: Hackett, 2004), 79.

22. C.-F. Volney, *Les Ruines, ou Méditation sur les révolutions des empires* (Paris: Desenne, 1791), x, xi, 7, 12, and *A New Translation of Volney's Ruins* (Paris: Levrault, 1802), ix, x, 8, 13.

23. At the time, the Macellum was understood as a Temple of Serapis.

24. Charles Lyell, *Principles of Geology* (London: Murray, 1830), 1: 359–60.

25. Hugh Falconer, "Abstract of a Discourse on the Fossil Fauna of the Sewalik Hills," *Journal of the Royal Asiatic Society of Great Britain and Ireland* 8 (1846): 105–11. Discussed in Pratik Chakrabarti, *Inscriptions of Nature* (Baltimore: Johns Hopkins University Press, 2020), 1.

26. Boucher de Perthes, *De l'homme antédiluvien et de ses œuvres* (Paris: Jung-Treuttel, 1860), 8.

27. Robert Chambers, *Vestiges of the Natural History of Creation* (London: Churchill, 1844), 318, 309–10. Patrick Brantlinger, *Dark Vanishings: Discourse on the Extinction of Primitive Races, 1800–1930* (Ithaca, NY: Cornell University Press, 2003), 21.

28. Darwin, *Origin of Species*, 282.

29. Charles Darwin, *The Descent of Man* (London: Murray, 1871), 1:18–19; see also 1:30.

30. Charles Lyell, *Geological Evidences for the Antiquity of Man* (London: Murray, 1863), 465.

31. This language made it all the way to Franz Boas: see *The Mind of Primitive Man* (New York: Macmillan, 1911), 238.

32. E. B. Tylor, *Researches into the History of Mankind* (London: Murray, 1865), 2, and "Phenomena of the Higher Civilisation Traceable to a Rudimental Origin among Savage Tribes," *Anthropological Review* 5, no. 18 (1867): 303–14.

33. E. B. Tylor, *Primitive Culture* (London: John Murray, 1871), 1:15, see also 1:6.

34. John Lubbock, *The Origin of Civilisation and the Primitive Condition of Man* (1871), 1.

35. Ernst Haeckel, *Natürliche Schöpfungsgeschichte* (Berlin: Reimer, 1868), 13, and *History of Creation* (London: Henry S. King & Co., 1876), 16.

36. Friedrich Rätzel, *Anthropo-Geographie* (Stuttgart: Engelhorn, 1891).

37. Tylor, *Primitive Culture*, 2nd ed. (London: Murray, 1874), 1:61.

38. Alfred Russel Wallace, "The Origin of Human Races and the Antiquity of Man Deduced from the Theory of 'Natural Selection,'" in *Journal of the Anthropological Society of London* 2 (1864): clxix.

39. In French "race" ultimately could refer to a geographic differential, an "innate nature," a locale where a skull had been discovered, or a national origin (the prehistorian Armand de Quatrefages put his work to use by writing a controversial book on "The Prussian Race" just after France lost the Franco-Prussian War). On the discussions of race, see Jennifer Michael Hecht, *The End of the Soul* (New York: Columbia University Press, 2003) and Alice Conklin, *In the Museum of Man* (Ithaca, NY: Cornell University Press, 2013).

40. Darwin, *Origin of Species*, chap. 9.

41. E.g., Gabriel de Mortillet, *Le préhistorique: Antiquité de l'homme* (Paris: Reinwald, 1883), 411. See also François-Xavier Fauvelle-Aymar, François Bon, and Karim Sadr, "L'Ailleurs et l'avant," *L'Homme* 184 (2007): 25–45.

42. Armand de Quatrefages de Bréau, *Fossil Men and Savage Men* (1884). Saul Dubow, *Scientific Racism in Modern South Africa* (Cambridge: Cambridge University Press, 1995), 51.

43. William King, "The Reputed Fossil Man of the Neanderthal," *Quarterly Journal of Science* (1864), 88–97. Many other authors, including even anti-evolutionists, jumped on the bandwagon of comparisons, so that titles proliferated such as *Fossil Men and Their Modern Representatives: An Attempt to Illustrate the Characters and Condition of Pre-historic Men in Europe, by Those of the American Races* (1880). That particular title was by the Canadian geologist John William Dawson (an antievolutionist). Other similar ones include Abel Hovelacque, *Notre ancêtre: Recherches d'anatomie & d'ethnologie sur le précurseur de l'homme* (Paris: E. Leroux, 1878) and also *Les débuts de l'humanité: L'homme primitif contemporain* (Paris: O. Doin, 1881).

CHAPTER 4: HUMANITY, DIVIDED BY THREE

1. Adam Smith, *Lectures on Jurisprudence*, LJ (A) i.27. A. R. J. Turgot also used a four-part schema in his *Plan for Two Discourses on Universal His-*

tory, which used the same categories. Smith made the four-stage theory his standard, despite his discussion of a two-stage division between "savage nations" and "civilized and thriving nations" in *An Inquiry into the Nature and Causes of the Wealth of Nations* (London: Strahan, 1776), I:2.

2. Edmund Burke, *The Correspondence of Edmund Burke*, ed. George Herbert Guttridge (Cambridge: Cambridge University Press, 1961), 3:351. Edward Gibbon agreed: he played out in his German barbarians the Enlightenment schema for the development of humanity. J. G. A. Pocock, "Gibbon and the Shepherds," *History of European Ideas* 2 (1981): 193–202.

3. Henri de Saint-Simon, "Mémoire sur la science de l'homme" (1813), in *Science de l'homme*, by Barthélemy-Prosper Enfantin and Henri de Saint-Simon (Paris: Victor Masson, 1858).

4. Burke, for example, had simply added a fourth term to it, those "erratick manners of Tartary."

5. Georges Cuvier, *The Animal Kingdom Arranged in Conformity with Its Organization* (London: Whittaker, 1827), 1:94–98. Patrick Brantlinger, *Dark Vanishings: Discourse on the Extinction of Primitive Races, 1800–1930* (Ithaca, NY: Cornell University Press, 2003), 20.

6. Richard Wittman, "Space, Networks, and the Saint-Simonians," *Grey Room* 40 (2010): 24–49; Antoine Picon, *Les saint-simoniens* (Paris: Belin, 2002).

7. Saint-Simon, "Mémoire sur la science de l'homme," 256, 428.

8. On how "fetish" transformed from a "pidgin word" used to mark value differentials in Portuguese trades in West Africa to the basis of "fetishism" as an Enlightenment theory of primitive religion, see William Pietz, *The Problem of the Fetish* (Chicago: University of Chicago Press, 2022).

9. For Saint-Simonian texts on the triad, see "De la classe ouvrière," *Le Producteur* 3 (April–June 1826): 306–7; "Troisième lettre au Baron d'Eckstein," *Le Producteur* 4 (July–September 1826): 358; "Coup-d'œil historique: Sur le pouvoir spirituel," *Le Producteur* 5 (1926): 65. For Comte's, see Auguste Comte, "Considérations philosophiques," in *Le Producteur, journal de l'industrie, des sciences et des beaux-arts*, vol.1 (Paris: Sautelet, October–December 1825), 293–94.

10. See also Saint-Simon, "Mémoire sur la science de l'homme," 225, 268.

11. Bazard attacked Comte for misguided atheism and plagiarisms: Saint-Amand Bazard et al., *Doctrine de Saint-Simon: exposition, première année, 1828–1829* (Paris: Au Bureau de l'organisateur, 1831), 369–370, and see sessions 14 and 15.

12. Michel Chevalier, *Cours d'économie politique* (Paris: Capelle, 1842), 68–70.

13. Chevalier, *Cours d'économie politique*, 76.

14. Chevalier, *Cours d'économie politique*, 72–77.

15. Enfantin, "À Sa Majesté Napoléon III," in Enfantin and Saint-Simon, *Science de l'homme*, xiii–xiv.

16. Johannes Feichtinger, Franz L. Fillafer, and Jan Surman, eds. *The Worlds of Positivism: A Global Intellectual History* (New York: Palgrave-Macmillan, 2017).

17. John Stuart Mill, *Auguste Comte and Positivism* (London: Trubner, 1865).

18. E. B. Tylor, *Primitive Culture*, 2nd ed. (London: Murray, 1873), I:477–78.

19. Charles Lyell, *Geological Evidences for the Antiquity of Man* (London: Murray, 1863), 377.

20. Anthropologists like Adolf Bastian (*Beiträge zur vergleichenden Psychologie: Die Seele und ihre Erscheinungsweisen in der Ethnographie* [Berlin: F. Dümmler, 1868]), psychologists like Wilhelm Wundt (*Elemente der Völkerpsychologie*, [Leipzig: Kröner, 1911]), art historians like Aby Warburg (*Images from the Region of the Pueblo Indians*, translated with an interpretive essay by Michael P. Steinberg [1922; Ithaca and London: Cornell University Press, 1995]).

21. James Lorimer, *The Institutes of Law: A Treatise of the Principles of Jurisprudence* (Edinburgh: Clark, 1872), 204.

22. Lorimer, *The Institutes of Law*, 3rd ed. (London: Blackwood, 1883) 1:101.

23. François Bon, "The Division and Discord of Prehistoric Chronologies," *RES: Anthropology and Aesthetics* 69–70 (Spring–Autumn 2018), 78.

CHAPTER 5: THE CONFLICT OF THE SCIENCES

1. Gustave Flaubert, "Quidquid volueris," in *Oeuvres completes*, ed. Jean Bruneau (Paris: Suil, 1964), 102–13. See also Marie Josephine Diamond, "Flaubert's 'Quidquid Volueris': The Colonial Father and the Poetics of Hysteria," *SubStance* 27, no. 1 (1998): 71–88.

2. Charles Darwin, *On the Origin of Species* (London: Murray, 1859), 488.

3. Charles Darwin, "To Charles Lyell, 25 October [1859]," Darwin Correspondence Project, Letter 2510.

4. Darwin, *Origin of Species*, 467.

5. John Evans, "Flint Implements in the Drift," *Athenaeum* (June 25, 1859): 841; John Evans, "On the Occurrence of Flint Implements in Undisturbed Beds of Gravel, Sand, and Clay," *Archaeologia* 38 (1860): 280–307; Joseph Prestwich, "On the Occurrence of Flint-Implements, Associated with the Remains of Animals of Extinct Species in Beds of a Late Geological Period, in France at Amiens and Abbeville, and in England at Hoxne," *Philosophical Transactions of the Royal Society of London* 150 (1860): 277–317. Donald Grayson, *The Establishment of Human Antiquity* (New York: Academic Press, 1983). A. Bowdoin Van Riper, *Man among the Mammoths: Victorian Science and the Discovery of Human Prehistory* (Chicago: University of Chicago Press, 1993).

6. Jacques Boucher de Perthes, *De l'homme antédiluvien et de ses œuvres* (Paris: Jung-Treutel, 1860), 26, 4.

7. Darwin to T. H. Huxley, 26 [February 1863], Darwin Correspondence Project, Letter 4013.

8. Johannes Fabian, *Time and the Other* (New York: Cambridge University Press, 1983), 17.

9. The classic book on the subject is Maurice Olender, *The Languages of Paradise: Race, Religion, and Philology in the Nineteenth Century* (Cambridge, MA: Harvard University Press, 1992).

10. Jean-Jacques Rousseau, *Essay on the Origin of Language*, in Jean-Jacques Rousseau and Johann Gottfried Herder, *On the Origin of Language: Two*

Essays, trans. John H. Moran and Alexander Gode (Chicago: University of Chicago Press, 1986), 5.

11. August Schleicher, *Compendium der vergleichenden Grammatik der indogermanischen Sprachen* (Weimar, 1861). See also Jean-Paul Demoule, *Mais où sont passés les indo-européens? Le mythe d'origine de l'Occident* (Paris: Seuil, 2014).

12. Adolphe Pictet, *Les origines indo-européennes ou les Aryas primitifs* (Paris: Joël Cherbulier, 1859), 3, 7, 8.

13. Pictet, *Les origines indo-européennes*, 9.

14. Saussure, posthumous review of Pictet, *Les Aryas primitifs*, in *Journal de Genève* (April 17, 1878), 3. The review continues on April 19 and April 25, 1878, with even more critical comments.

15. August Schleicher, *A Compendium of the Comparative Grammar of the Indo-European, Sanskrit, Greek and Latin Languages* (London: Trübner, 1874), 6, 7.

16. Friedrich Max Müller, *Lectures on the Science of Language Delivered at the Royal Institution* (London: Longmans, Green, 1862), 14.

17. Max Müller to Darwin, "From Friedrich Max Müller 29 June 1873," (Letter 8957) and "From Friedrich Max Müller 13 October [1875]," (Letter 10194) in Darwin Correspondence Project. Other linguists also demurred: August Schleicher expressed his own objections in *Die darwinische Theorie und die Sprachwissenschaft* (Weimar: Hermann Böhlau, 1873).

18. Max Müller, *Lectures on the Science of Language*, 332–33.

19. Max Müller, *Lectures on the Science of Language*, 345–47.

20. Ernst Haeckel, for example, thought language was key to evolution. See Mario A. di Gregorio, "Reflections of a Nonpolitical Naturalist: Ernst Haeckel, Wilhelm Bleek, Friedrich Müller and the Meaning of Language," *Journal of the History of Biology* 35, no. 1 (2002): 79–109; Robert J. Richards, "The Linguistic Creation of Man: Charles Darwin, August Schleicher, Ernst Haeckel, and the Missing Link in Nineteenth-Century Evolutionary Theory," in *Experimenting in Tongues: Studies in Science and Language*, ed. Matthias Doerres (Stanford: Stanford University Press, 2002).

21. Max Müller, *Lectures*, 4, 12.

22. This was not altogether new: as Raymond Schwab showed in *La Renaissance orientale* (Paris: Payot, 1950), 321–27, biology and Indo-European linguistics had comfortably cohabited since the beginning of the century.

23. E. B. Tylor, *Researches into the Early History of Mankind and the Development of Civilization* (London: Murray, 1865), 326.

24. E. Lartet and H. Christy, "Sur des figures d'animaux graves ou sculptées et autres produits d'art et d'industrie rapportables aux temps primordiaux de la période humaine," *Revue Archéologique* 9 (1864): 233–67. Gabriel de Mortillet, "Promenades Préhistoriques à l'exposition universelle" (Paris: Reinwald, 1867).

25. The Institute subsumed the racially oriented and polygenist *Anthropological Society of London*, and the monogenist, originally religious *Ethnological Society of London* (which had evolved out of the Aboriginal Protection Soci-

ety). Some of its practitioners cared about skull measurements, and some remained polygenists for a long time. Others tried to come up with alternatives to that old racial division that was so thoroughly rejected by the Darwinians, and so politically debunked by the American Civil War.

26. E. B. Tylor, *Primitive Culture* (London: Murray, 1871), 1:6–7.

27. Tylor, *Primitive Culture* (1871), 1:257; 1:5–6. He went on to emphasize this: "This hypothetical primitive condition corresponds in a considerable degree to that of modern savage tribes, who, in spite of their difference and distance, have in common certain elements of civilization, which seem remains of an early state of the human race at large."

28. Tylor, *Primitive Culture* (1871), 1:19.

29. John Lubbock, *Prehistoric Times, as Illustrated by Ancient Remains, and the Manners and Customs of Modern Savages* (London: Williams and Norgate, 1865).

30. Lubbock, *Prehistoric Times*, 31, 268 and following.

31. Lubbock, *Prehistoric Times*, 268.

32. John Lubbock, *The Origin of Civilisation and the Primitive Condition of Man: Mental and Social Condition of Savages* (London: Longmans, Green, 1870), 1.

33. Charles Darwin, *The Descent of Man* (London: John Murray, 1871), 1: 3.

34. The *Voyage of the Beagle* was originally published as *Journal of Researches*. Charles Darwin, *Journal of Researches*, 2nd ed. (1839; London: John Murray, 1845), 504, 505. Patrick Brantlinger, *Dark Vanishings: Discourse on the Extinction of Primitive Races, 1800–1930* (Ithaca, NY: Cornell University Press, 2003), 155–56.

35. Darwin, *The Descent of Man*, 1:111–12.

36. Darwin, *The Descent of Man*, 1:132, 1:111.

37. Darwin, *The Descent of Man*, 1:137.

38. Darwin, *The Expression of Emotions in Man and Animals* (London: Murray, 1872), 235.

39. Darwin, *The Descent of Man*, 1:100–1.

Chapter 6: Mother Love: Primitive Communism

1. On its fate, see Sarah Johnson, "Farewell to *The German Ideology*," *Journal of the History of Ideas* 83 (2022): 143–70.

2. Karl Marx and Friedrich Engels, "Manifesto of the Communist Party," in *Marx-Engels Collected Works*, by Karl Marx and Friedrich Engels (henceforth *MECW*) (New York: International Publishers, 1976), 6:481.

3. Jean-Jacques Rousseau, *Discours sur l'origine et les fondements de l'inégalité parmi les hommes*, eds. Blaise Bachofen et Bruno Bernardi (1755; Paris: Flammarion, 2008), 109.

4. Henry Sumner Maine, *Ancient Property: Its Connection with the Early History of Society, and Its Relation to Modern Ideas* (London: Murray, 1878).

5. Karuna Mantena, *Alibis of Empire* (Princeton: Princeton University Press, 2010), 131.

6. Henry Sumner Maine, *The Early History of the Property of Married Women* (Manchester: A. Ireland and Co., 1873), 3.

7. Mantena, *Alibis of Empire*, 132.

8. Maine, *Early History of the Property of Married Women*, 21.

9. Emile de Laveleye, *De la propriété et de ses formes primitives* (Paris: Baillière, 1874). Gareth Stedman Jones, *Karl Marx* (Cambridge, MA: Harvard University Press, 2016), 585. Laurence Krader, ed. *The Ethnological Notebooks of Karl Marx* (Assen: Van Gorkum, 1974), 34–42.

10. August Bebel, *Die Frau und der Sozialismus* (Zürich: Volksbuchhandlung, 1879), 13.

11. John McLennan, *Primitive Marriage: An Inquiry into the Origin of the Form of Capture in Marriage Ceremonies* (Edinburgh: Black, 1865), 45–46.

12. McLennan, *Primitive Marriage*, 46.

13. Charles Darwin, *The Descent of Man, and Selection in Relation to Sex* (London: John Murray, 1871), 2: 362–63. In an 1883 letter to Karl Kautsky, Engels rejected Darwin's opinion and authority on the matter. *MECW* 46 (1992): 437–38.

14. Johann Jacob Bachofen, *Das Mutterrecht* (Stuttgart: Krais und Hoffmann, 1861).

15. Lionel Gossman, *Basel in the Age of Burckhardt* (Chicago: University of Chicago Press, 2000). Alexis Giraud-Teulon also expounded Bachofen's matriarchal cosmology into an argument specifically concerning property and kinship: Giraud-Teulon, *Les Origines de la famille* (Geneva: Cherbuliez, 1874). Alfred Espinas was more ambivalent, complementing social formations—between, for example, a "primitive society" and a horde of men—thereby breaking the reliance on a single past system: Espinas, *Des sociétés animales* (Paris: Baillière, 1878).

16. Lewis H. Morgan, *Systems of Consanguinity and Affinity in the Human Family* (Washington, DC: Smithsonian Institution, 1871).

17. Lewis H. Morgan, *Ancient Society* (New York: Holt, 1877), v.

18. Morgan, *Ancient Society*, vii, and see also p. 61.

19. Morgan, *Ancient Society*, vii.

20. Morgan, *Houses and House Life of the American Aborigines* (Washington, DC: Government Printing Office, 1881), 2.

21. Morgan, *Ancient Society*, 345.

22. Morgan, *Ancient Society*, 63, emphasis mine.

23. Morgan, *Ancient Society*, 345–6.

24. Freud would of course adopt this argument in *Totem and Taboo* and especially in *Moses and Monotheism*.

25. Engels to Karl Kautsky, February 10 and March 2, 1883, in *MECW*, 46:436–38 and 451–53. Engels to Eduard Bernstein, February 27, 1883, *MECW*, 46: 450.

26. International Institute of Social History (Amsterdam), Marx/Engels Papers 270, Inv. B162 [B146], "Lewis H. Morgan." Engels, *Origin of the Family*, *MECW*, 26 (1990):134 and following. Morgan also offered tactical advantages. If Engels published *The Origin of the Family* as a mere commentary on Morgan instead, he could even avoid Chancellor Bismarck's Anti-Socialist Law, which prohibited socialist activism. But eventually he decided against:

"either good but bound to be prohibited; or allowed but lousy. The latter I cannot do." Engels to Karl Kautsky, April 26, 1884.

27. Engels, *Origin of the Family, MECW,* 26:131.
28. Engels, *Origin of the Family, MECW,* 26:131.
29. Engels, *Origin of the Family, MECW,* 26:148. In passing, he mocks Wagner—declaring that shortly before his death, Marx had objected to the "complete misrepresentation of primitive times in Wagner's *Ring of the Nibelungen*": Engels, *Origin of the Family, MECW,* 26:147n.
30. Engels, *Origin of the Family, MECW,* 26:148.
31. Engels, *Origin of the Family, MECW,* 26:157.
32. Engels, *Origin of the Family, MECW,* 26:158.
33. Engels, *Origin of the Family, MECW,* 26:164, translation amended.
34. Engels did not notice that Morgan had cited Bachofen repeatedly—he proposed they reached the same conclusions independently, which meant the conclusions were right. Morgan, *Ancient Society,* 349–50. Engels, *Origin of the Family, MECW,* 26:165. This was not quite accurate, but the main issue is that Engels flattened the two.
35. Engels, *Origin of the Family, MECW,* 26:165.
36. Morgan, *Ancient Society,* 552.
37. Engels, *Origin of the Family, MECW,* 26:132.
38. Engels, *Origin of the Family, MECW,* 26:173.
39. Engels, *Origin of the Family, MECW,* 26:165, translation amended.
40. Paul Lafargue, Marx's son-in-law, used Engels and primitive communism in his *The Evolution of Property from Savagery to Civilization* (London: Swan Sonnenschein and Co., 1890).
41. Rosa Luxemburg, *The Accumulation of Capital,* trans. Anges Schwartzschild (1913; London: Routledge and Kegan Paul, 1951), 377–78.
42. Cathy Porter, *Alexandra Kollontai: A Biography* (London: Virago, 1980), 68.
43. Alexandra Kollontai, "Communism and the Family" (1920), in *Selected Writings,* trans. Alix Holt (Westport, CT: Lawrence Hill and Co., 1980), 250. But see also her "Theses on Communist Morality in the Sphere of Marital Relations" in that same volume.
44. Stalin, "Dialectical and Historical Materialism," in *History of the Communist Party of the Soviet Union (Bolsheviks): Short Course,* Central Committee of the CPSU (Moscow: Foreign Languages Publishing, 1939).
45. Bronisław Malinowski, *Crime and Custom in Savage Society* (London: Kegan Paul, Trench, Trubner, 1926).
46. W. H. R. Rivers, "An Inquiry into Socialism and Human Nature," in *Psychology and Politics* (London: Kegan Paul, 1923).
47. Alain Testart, *Le Communisme primitif,* vol. 1 (Paris: Maison des Sciences de l'Homme, 1985).
48. Simone de Beauvoir, *The Second Sex,* trans. H. M. Parshley (London: Jonathan Cape, 1953), 80.
49. Beauvoir, *Second Sex,* 83.
50. Beauvoir, *Second Sex,* 6, 16–17.
51. See the brilliant book by Cathy Gere, *Knossos and the Prophets of Modernism* (Chicago: University of Chicago Press, 2009), 149.

52. Gere, *Knossos and the Prophets of Modernism*, 149.
53. On the forgeries, see Kenneth Lapatin, *Mysteries of the Snake Goddess: Art, Desire, and the Forging of History* (New York: Houghton Mifflin, 2002); Gere, *Knossos and the Prophets of Modernism*; and Kevin Butcher and David Gill, "The Director, the Dealer, the Goddess, and Her Champions: The Acquisition of the Fitzwilliam Goddess," *American Journal of Archaeology* 97 (1993): 383–410.
54. Morgan, *Ancient Society*, 358.

CHAPTER 7: THE DISAPPEARING NATIVE

1. *Information Respecting the Aborigines in the British Colonies: Extracts from the Report Presented to the House of Commons, by the Select Committee on that Subject* (London: Darton, Harvey, 1838), 16.
2. *Information Respecting the Aborigines*, vi.
3. Philip J. Deloria, *Indians in Unexpected Places* (Lawrence, KS: University Press of Kansas, 2004).
4. Zvi Ben-Dor Benite, *The Ten Lost Tribes* (Oxford: Oxford University Press, 2009).
5. Danilyn Rutherford, *Living in the Stone Age: Reflections on the Origins of a Colonial Fantasy* (Chicago: University of Chicago Press, 2018), x.
6. Lucas Bessire, *Behold the Black Caiman* (Chicago: University of Chicago Press, 2014).
7. Patrick Brantlinger, *Dark Vanishings: Discourse on the Extinction of Primitive Races, 1800–1930* (Ithaca, NY: Cornell University Press, 2003).
8. Charles Darwin, *Journal and Remarks, 1832–1836*, vol. 3 of *Narrative of the Surveying Voyages of His Majesty's Ships Adventure and Beagle* (London: Colburn, 1839), 520. (Reprinted as *Voyage of the Beagle*).
9. Darwin, *Journal and Remarks*, 533–34.
10. Charles Darwin, *On the Origin of Species* (London: Murray, 1859), 382.
11. Charles Darwin, *The Descent of Man* (London: Murray, 1871), 1:238. Hobbes's phrase is from Thomas Hobbes, *Leviathan: The English and Latin texts*, ed. Noel Malcolm (Oxford: Oxford University Press, 2012), 1:13, 193–94.
12. George W. Stocking, Jr., *Victorian Anthropology* (New York: Free Press, 1987), 279–80.
13. Darwin, *The Descent of Man*, 2nd ed. (London: Murray, 1874), 210–11, 214.
14. Stocking, *Victorian Anthropology*, 283.
15. Alfred Russel Wallace, "The Origin of Human Races and the Antiquity of Man Deduced from the Theory of 'Natural Selection,'" in *Journal of the Anthropological Society of London* 2 (1864): clxix, italics mine.
16. Wallace, "The Origin of Human Races," clxxxii.
17. Armand de Quatrefages, *Hommes fossiles et hommes sauvages* (Paris: Baillière, 1884), 292–400.
18. Armand de Quatrefages, *The Pygmies* (1887; London: Macmillan, 1895), 169–70.
19. Quatrefages, *The Pygmies*, 87–88.

20. Thomas H. Huxley, *Man's Place in Nature* (New York: Appleton, 1896), 216; quoting Desmoulins, *Histoire naturelle des races humaines* (1826). This passage does not appear in the 1863 original edition.

21. Alison Bashford and Joyce Chaplin, *The New Worlds of Thomas Robert Malthus* (Princeton: Princeton University Press, 2016).

22. Joseph Hornung, "Civilisés et barbares," *Revue de droit international et de législation comparée* 17 (1885): 5–7.

23. Charles Salomon, *L'occupation des territoires sans maître* (Paris: Giard, 1889), 68. Quoted in Martti Koskenniemi, *The Gentle Civilizer of Nations* (Cambridge: Cambridge University Press, 2001), 106.

24. See Andrew Fitzmaurice, *Sovereignty, Property and Empire, 1500–2000* (Cambridge: Cambridge University Press, 2014), 268.

25. Lewis Morgan, *Ancient Society* (New York: Henry Holt, 1877), 3.

26. Morgan, *Ancient Society*, vii–viii.

27. Aby Warburg, *Images from the Region of the Pueblo Indians of North America* (Ithaca, NY: Cornell University Press 1995), 2.

28. Warburg, *Images*, 54.

29. Victor Segalen, *Essay on Exoticism*, trans. Yaël Rachel Schlick (Durham, NC: Duke University Press, 2002), 57.

30. Franz Boas, *The Mind of Primitive Man* (New York: Macmillan, 1911), 263–64. On Boas and his Kwakwaka'wakw interlocutors, see Aaron Glass, *Writing the Hamat'sa: Ethnography, Colonialism, and the Cannibal Dance* (Vancouver: UBC Press, 2021), and Isaiah Lorado Wilner, "Body Knowledge," *Journal of the History of Ideas* 83, no. 1 (2022): 111–42, and 83, no. 2 (2022): 229–55.

31. W. H. R. Rivers, *History of Melanesian Society* (Cambridge: Cambridge University Press, 1914), 1:vii. See also Rivers's "The Disappearance of Useful Arts" in *Festskrift Tillägnad Edvard Westermarck*, eds. Ola Castrén, Yrjö Hirn, Rolf Lagerborg, and Axel Wallensköld (Helsingfors: Simelius, 1912) and "The Ethnological Analysis of Culture" *Science* 34, no. 874 (September 29, 1911): 385–97.

32. W. H. R. Rivers, ed., *Essays on the Depopulation of Melanesia* (Cambridge: Cambridge University Press, 1922), 84, 87.

33. Mabel Cook Cole and Fay-Cooper Cole, *The Story of Primitive Man: His Earliest Appearance and Development* (Chicago: University of Knowledge Incorporated, 1941).

34. Bronisław Malinowski, *Argonauts of the Western Pacific* (1922; London: Routledge, 1932), xv.

35. Malinowski, "Ethnology and the Study of Society," in *Economica*, no. 6 (1922): 208–19, 208.

36. We can even see the logic of the concept replaying itself as anthropologists, one after another, climbed on the backs of their precursors only to criticize them for failing to think through the consequences of "disappearance." M. F. Ashley Montagu wrote in the mid-1930s that "wherever the white man has penetrated," the story was the same, involving "ruthless cruelty, injustice, dispossession, and wholesale murder of native populations." In Australia, he tracked the annihilation of entire populations: from some

two thousand individuals in 1896, "to-day, owing chiefly to the destructive effects of missionary and other white influences, the population of the Arunta [today: Arrernte] is reduced to some 350 souls, whilst their culture is practically destroyed." M. F. Ashley-Montagu, *Coming into Being among the Australian Aborigines* (London: Routledge, 1937), 15 and 15n1. His book reworked Malinowski's own dissertation topic, and was published with a foreword by Malinowski. So Ashley Montagu was quietly targeting him too when issuing his "clarion call to the hearts and heads, to the humanity, of white Australia." Ashley-Montagu, "The Future of the Australian Aborigines," *Oceania* 8 (1938): 343.

37. Leo Frobenius, *The Childhood of Man* (Philadelphia: Lippincott, 1909), 134. Smuts knew Frobenius's work.

38. Saul Dubow, *Scientific Racism in Modern South Africa* (Cambridge, UK: Cambridge University Press, 1995), 51; Peder Anker, *Imperial Ecology: Environmental Order in the British Empire, 1895–1945* (Cambridge, MA: Harvard University Press, 2002), 163.

39. Mabel Cook Cole and Fay-Cooper Cole, *The Story of Primitive Man: His Earliest Appearance and Development* (Chicago: Field Museum, 1941).

40. Claude Lévi-Strauss to Marcel Mauss, November 10, 1935, in BnF. NAF 28150 (181). I take this reference from Ian Merkel, "Terms of Exchange: Brazilian Intellectuals and the Remaking of the French Social Sciences" (PhD Dissertation, New York University, 2018), 152.

41. Claude Lévi-Strauss, *Tristes Tropiques*, translated by John Weightman and Doreen Weightman (1955; New York: Penguin, 1992), 414, 375.

42. Claude Lévi-Strauss, "The Work of the Bureau of American Ethnology" (1965), in *Structural Anthropology Two* (Chicago: University of Chicago Press, 1976), 59.

CHAPTER 8: NEANDERTHALS, "OUR DOUBLES"

1. Charles Q. Choi, "Neandertal–Human Trysts May Be Linked to Modern Depression, Heart Disease," *Scientific American* (LiveScience), February 12, 2016.

2. Rudolf Virchow, "Untersuchung des Neanderthal-Schädels," *Verhandlungen der Berliner Gesellschaft für Anthropologie, Ethnologie und Urgeschichte* 4 (1872): 157–65. Translation from Andrew Zimmermann, *Anthropology and Antihumanism in Imperial Germany* (Chicago: University of Chicago Press, 2001), 70.

3. Hermann Schaaffhausen, "Zur Kenntnis der ältesten Rasseschädel," in *Archiv für Anatomie, Physiologie, und wissenschaftliche Medizin* (1858), 465, 470–71; George Busk, trans., "On the Crania of the Most Ancient Races of Man," in *Natural History Review* 2 (April 1861): 163–64, 167.

4. Charles Lyell, *The Geological Evidences for the Antiquity of Man* (London: Murray, 1863), 89.

5. Lyell, *Geological Evidences for the Antiquity of Man*, 82.

6. Lyell, *Geological Evidences for the Antiquity of Man*, 89, 84.

7. Lyell, *Geological Evidences for the Antiquity of Man*, 101.

8. Armand de Quatrefages, *Hommes fossiles et hommes sauvages: Études d'anthropologie* (Paris: Baillière, 1884), 61, 64. Translated as *The Human Species* (New York: Appleton, 1879). Note also Leroi-Gourhan's comments in *Gesture and Speech*, 12.

9. William King, "The Reputed Fossil Man of the Neanderthal," *Quarterly Journal of Science* (1864): 88–97.

10. Marianne Sommer, "Mirror, Mirror on the Wall: Neanderthal as Image and 'Distortion' in Early 20th-Century French Science and Press," *Social Studies of Science* 36, no. 2 (2006): 207–40.

11. Gabriel von Max's painting *Pithecanthropus Alalus* (1894) emphasized speechlessness, weakness, lonely fear.

12. Marcellin Boule, *Les hommes fossiles: Éléments de paléontologie humaine* (Paris: Masson, 1920), 187.

13. Rachel Caspari and Sang-Hee Lee, "Older Age Becomes Common Late in Human Evolution," *Proceedings of the National Academy of Sciences* 101, no. 30 (July 27, 2004): 10895–900.

14. Marcellin Boule, "L'homme fossile de La Chapelle-aux-Saints," *Années de paléontologie* (1911): 91–107, also 129–45.

15. Boule, *Les hommes fossiles* and "L'homme fossile de La Chapelle-aux-Saints."

16. Boule, *Les hommes fossiles*, 189. To present this finding as uncontroversial, Boule systematically tabulated in his book all then-recent discussions of paleolithic burials.

17. Boule, *Les hommes fossiles*, 247, 457. See the excellent discussion, also of František Kupka's image, by Sommer, "Mirror, Mirror," 207–40; Sommer overstates Boule's emphasis on Neanderthal's distance from *sapiens*.

18. Boule, *Les hommes fossiles*, 487.

19. Arthur Keith, *The Antiquity of Man* (London: Williams and Norgate, 1916), 124.

20. Boule, *Les hommes fossiles*, 247, 407, 476.

21. G. J. Sawyer, Viktor Deak, Esteban Sarmiento, and Richard Milne, *The Last Human: A Guide to Twenty-Two Species of Extinct Humans* (New Haven: Yale University Press, 2007), 243.

22. Boule also supervised a far less brutish reconstruction of the mosaic of the Neanderthal's facial muscles. Discussed in Boule, *Les hommes fossiles*, 205–8 and elsewhere.

23. G. Elliot Smith, "Neanderthal Man *Not* Our Ancestor," *Scientific American* 139, no. 2 (August 1928): 112–15.

24. Another drawing in G. Elliot Smith—*Human History* (London: Jonathan Cape, second edition, 1934)—clearly derives from H. G. Wells's *The Outline of History* (New York: Macmillan, 1921), despite its different, somewhat ridiculous nose.

25. Sommer, in "Mirror, Mirror," notes that when the La-Chappelle-aux-Saints Neanderthal was brought out, the papers depicted him as either animal or human, while Boule positioned him in a gradation.

26. Wells, *Outline of History*, 65, 69–70.

27. Harry Johnston, *Views and Reviews: From the Outlook of an Anthropologist* (London: Williams and Norgate, 1912), 201–2.

28. Johnston, *Views and Reviews*, 200, 201–2, 203. On Johnston and his *The Negro in the New World*, see Susan D. Pennybacker, "The Universal Races Congress, London Political Culture, and Imperial Dissent, 1900–1939," *Radical History Review* 92 (2005): 103–17.

29. G. P. Wells, H. G. Wells, and Julian Huxley, *The Science of Life* (London: Waverley, 1929). Julian Huxley and Alfred Cort Haddon, *We Europeans* (New York: Harper, 1936).

30. Wells, Wells, and Huxley, *Science of Life*, 105, 252, 859. Huxley and Haddon, *We Europeans*, 37–38, 107.

31. Francisco J. Ayala and Camilo J. Cela-Conde, *Processes in Human Evolution: The Journey from Early Hominins to Neanderthals and Modern Humans*, 2nd ed. (Oxford: Oxford University Press, 2017), 398.

32. Carleton S. Coon, *The Races of Europe* (New York: Macmillan, 1939), 24, 25–28.

33. André Leroi-Gourhan, *Prehistoric Man* (New York: Philosophical Library, 1957), viii.

34. Georges Bataille, "Lecture, January 18, 1955," in *The Cradle of Humanity* (New York: Zone Books, 2009), 88. Bataille relies on Johannes Maringer's research.

35. Georges Bataille, "The Passage from Animal to Man and the Birth of Art," in *Cradle of Humanity*, 73.

36. Leroi-Gourhan, *Prehistoric Man*, viii.

37. J. E. Weckler, "Neanderthal Man," *Scientific American* 197, no.6 (December 1957): 93, 96.

38. Wilfrid E. Le Gros Clark, *The Antecedents of Man* (1959; London: Harper and Row, 1963) and *The Fossil Evidence for Human Evolution* (Chicago: University of Chicago Press, 1955), 57–60.

39. Le Gros Clark, *Fossil Evidence*, 57, emphasis mine.

40. Dragutin Gorjanović-Kramberger, *Der diluviale Mensch von Krapina in Kroatien* (Wiesbaden: C.W. Kreidel, 1906).

41. Consider the contrast in, e.g., Erik Trinkaus, "Cannibalism and Burial at Krapina," *Journal of Human Evolution* 14, no. 2 (1985): 203–16.

42. F. Clark Howell and the Editors of *Life Magazine*, *Early Man* (New York: Time-Life Books, 1965), 45.

43. Sylvia Massey Czerkas, *Dinosaurs, Mammoths, and Cavemen: The Art of Charles R. Knight* (New York: E. P. Dutton, 1982), fig. 156.

44. Giles Oakley, "My Neanderthal Heritage: Memories of Maurice Wilson (1914–87)" Library and Archives [blog], London's Natural History Museum. July 6, 2012.

45. Dimitra Papagianni and Michael A. Morse, *The Neanderthals Rediscovered: How Modern Science Is Rewriting Their Story* (London: Thames and Hudson, 2013).

46. Papagianni and Morse, *Neanderthals Rediscovered*; Svante Pääbo, *Neanderthal Man: In Search of Lost Genomes* (New York: Basic Books, 2015); Rebecca Wragg Sykes, *Kindred: Neanderthal Life, Love, Death and Art* (London: Bloomsbury Sigma, 2021).

47. Fred H. Smith, "The Fate of the Neandertals," *Journal of Anthropological*

Research 69, no. 2 (2013): 191. Responding to Carleton Coon's 1939 cartoon, Smith also declares, "I am certain we would spot the Neandertal in a second if he walked down the street in my Tennessee home town." It's hard not to hear a dog whistle given the history of this expression.

48. Smith, "Fate of the Neandertals," 190.

49. Philip Cohen, "Who's the Daddy?" *The New Scientist*, March 24, 2001; see also Michael Marshall, "Neanderthal-Human Sex Bred Light Skins and Infertility," *The New Scientist*, January 29, 2014.

50. It too meets with the critique that "the last Neanderthals made no decisive contribution to the gene pool of [non-African] humans alive today." Friedemann Schrenk and Stephanie Müller, in collaboration with Christine Hemm, *The Neanderthals*, trans. Phyllis G. Jestice (2005; London: Routledge, 2009), 100.

CHAPTER 9: THE THIN VENEER

1. Adam Sedgwick to Charles Darwin, December 24, 1859, *The Life and Letters of Charles Darwin*, ed. Francis Darwin (London: Murray, 1887), 2:249.

2. Arthur Schopenhauer, "On Human Nature," in Paul Wood, Leon Wainwright, and Charles Harrison, *Art in Theory: The West in the World* (Chichester, UK: Wiley, 2021), 248.

3. Charles Darwin, *The Descent of Man* (London: Murray, 1871), 1:104.

4. Darwin, *Descent of Man*, 1:166–67.

5. Darwin, *Expression of Emotions in Man and Animals* (London: Murray, 1872), 235.

6. Darwin, *Descent of Man*, 1:46.

7. Thomas H. Huxley, *Evolution and Ethics* (London: Macmillan, 1894), 81.

8. H. Rider Haggard, *Allan Quatermain* (1887; London: Longmans, Green, 1888), 6, 5.

9. Jack London, "The Somnambulists" (1906) in *Revolution and Other Essays* (New York: Macmillan, 1910), 41.

10. London, *Call of the Wild* (New York: Macmillan, 1903), 43, 166–67, 198–99. See also the discussion of London's relation with Spencer in Lawrence I. Berkove, "Jack London and Evolution," in *American Literary Realism 36*, no. 3 (2004): 243–55, and Marianne Sommer, "How Cultural Is Heritage?," in *A Cultural History of Heredity* eds. Staffan Müller-Wille and Hans-Jörg Rheinberger (Berlin: Max-Planck-Institut für Wissenschaftsgeschichte Pre-Prints, 2005), 3:233–53.

11. Edward Rice Burroughs, *The Return of Tarzan* (1913; London: Methuen, 1920), 12, 25.

12. Rounsevelle Wildman, *Tales of the Malayan Coast, from Penang to the Philippines* (Boston: Lothrop, 1899).

13. For the pronouncement that civilization is not a veneer and *must* reach colonial subjects, consider Lord Russell of Killowen, "International Law," *Law Quarterly Review* 48 (1896): 335, discussed in Martti Koskenniemi, *Gentle Civilizer of Nations: The Rise and Fall of International Law 1870–1960* (Cambridge: Cambridge University Press, 2009), 108.

14. James Frazer, *The Golden Bough* (London: Macmillan, 1900), 3:74.
15. Kathleen Frederickson, *The Ploy of Instinct: Victorian Sciences of Nature and Sexuality* (New York: Fordham University Press, 2014), 4.
16. On Spencer, see Robert Richards, *Darwin and the Emergence of Evolutionary Theories of Mind and Behavior* (Chicago: University of Chicago Press, 1987), and Robert M. Young, *Mind, Brain and Adaptation in the Nineteenth Century* (Oxford: Oxford University Press, 1990), chaps. 5–6.
17. Versions of Spencer's logic echoed far and wide—from the psychologist of brain injury John Hughlings Jackson (who used Spencer's argument on "dissolution") and the German founder of experimental psychology Wilhelm Wundt to the American philosopher and psychologist William James and the founder of behaviorism John B. Watson. See J. Hughlings Jackson, "The Croonian Lectures on Evolution and Dissolution of the Nervous System," *British Medical Journal* 1, no. 1213 (1884): 591–93 and William James, *The Moral Equivalent of War, International Conciliation* 27 (February 1910).
18. Gustave Le Bon, *De Moscou aux monts Tatras: Étude sur la formation actuelle d'une race* (Paris: Delagrave, 1881). See Stefanos Geroulanos, "The Plastic Self and the Prescription of Psychology, 1890–1920," in *Republics of Letters* 3, no. 2 (2014).
19. Le Bon, *La Civilisation des Arabes* (Paris: Imprimeurs de l'Institut, 1884), 652. On the broader theme of the survival of "organic memories," see Laura Otis, *Organic Memory* (Lincoln: University of Nebraska Press, 1994).
20. Le Bon, *The Crowd* (1895; New York: Macmillan, 1896), 7–8, 41–42, 48–50.
21. Sigmund Freud, *Totem and Taboo*, in *The Standard Edition of the Complete Psychological Works of Sigmund Freud*, vol. 13 ed. by James Strachey (London: Hogarth, 1966), 1.
22. Freud, *Totem and Taboo*, 66.
23. Nicholas Murray Butler, *A World in Ferment* (New York: Scribner, 1917), 89. Butler explicitly claims that this isn't about war "with untamed barbarians or wild Indians" but a blight on civilization.
24. Nellie McClung, *In Times like These* (Toronto: McLeod, 1917), 19.
25. R. A. Reiss, *Report upon the Atrocities Committed by the Austro-Hungarian Army during the First Invasion of Serbia* (London: Simpkin, Marshall, 1916), 185.
26. LeRoy Eltinge, *Psychology of War* (Fort Leavenworth, KS: Army Service Schools Press, 1915), 3–4.
27. George Washington Crile, *A Mechanistic View of War and Peace* (New York: Macmillan, 1915), 37.
28. Walter B. Cannon, *Bodily Changes in Pain, Hunger, Fear and Rage* (New York: Appleton, 1915), 289.
29. Walter B. Cannon, "Stimulation of Adrenal Secretion by Emotional Excitement," *Proceedings of the American Philosophical Society* 1 (1911): 226–27.
30. On Cannon, see Stefanos Geroulanos and Todd Meyers, *The Human Body in the Age of Catastrophe* (Chicago: University of Chicago Press, 2018).
31. F. Clark Howell and the Editors of Time-Life Books, "How the Savage Lives on in Man" in *Early Man* (Young Readers Nature Library, Adapted from the Life Nature Library), (Alexandria, VA: Time-Life Books, 1979), 113–15.

32. Sigmund Freud, "Thoughts for the Times on War and Death" (1915), in *The Standard Edition of the Complete Psychological Works of Sigmund Freud*, vol. 14, ed. by James Strachey (London: Hogarth, 1957), 299.

33. Henri Bergson, *The Two Sources of Morality and Religion* (1932; London: MacMillan, 1935), 248.

34. Ernst Jünger, *Der Kampf als inneres Erlebnis* (Berlin: E.S. Mittler, 1922), 7. See also, for a similar formulation, Ernst Jünger, *Sturm* (Stuttgart: Klett-Cotta, 1979), 25.

35. John W. Dower, *War without Mercy: Race and Power in the Pacific War* (New York: Pantheon Books, 1986), 53.

Chapter 10: On the Antiquity of the Psyche

1. Sigmund Freud, "Infantile Sexuality," in *Three Essays on the Theory of Sexuality*, in *The Standard Edition of the Complete Psychological Works of Sigmund Freud*, vol. 7, 1901–1905, *A Case of Hysteria, Three Essays on Sexuality and Other Works* (London: The Hogarth Press, 1953), 173–206.

2. Sigmund Freud, *The Origins of Psycho-Analysis* (London: Imago, 1954), letter 127.

3. Sigmund Freud, "From the History of an Infantile Neurosis" (1918) in *The Standard Edition of the Complete Psychological Works of Sigmund Freud*, vol. 17, ed. by James Strachey (London: The Hogarth Press, 1955), 18, 10.

4. On the editing of *The Interpretation of Dreams* see Lydia Marinelli and Andreas Mayer, *Dreaming by the Book* (New York: Other Press, 2003).

5. Jeffrey Masson, ed. *Complete Letters of Sigmund Freud to Wilhelm Fliess* (Cambridge, MA: Harvard University Press, 1985), 363.

6. Sigmund Freud, *Totem and Taboo*, in *The Standard Edition of the Complete Psychological Works of Sigmund Freud*, vol. 13, ed. James Strachey (London: Hogarth, 1955), 141.

7. Ribot presented psychiatric pathologies as atavisms in *Heredity* (London: King, 1875), 7, 34. My thanks to Giuseppe Bianco for this reference.

8. Charles Blondel, *La Conscience morbide* (Paris, 1914), 247, 250, and "La croyance à l'extase selon M. Pierre Janet," *Revue de métaphysique et de morale* 3, no.1 (1928): 111.

9. Wilhelm Wundt, *Elemente der Völkerpsychologie* (Leipzig: Kröner, 1912).

10. Freud, *Totem and Taboo*, 1; see also 3, 22, 95, 123, 177.

11. Freud, *Totem and Taboo*, 125, citing Darwin, *The Descent of Man* (London: Murray, 1871) 2:362–63.

12. Freud, *Totem and Taboo*, 143.

13. Sigmund Freud, *Group Psychology and the Analysis of the Ego*, in *The Standard Edition of the Complete Psychological Works of Sigmund Freud*, vol. 18, ed. by James Strachey (London: Hogarth, 1955), 121.

14. "The primal horde may arise once more out of any random crowd." Freud, *Group Psychology*, 123.

15. Freud, *Group Psychology*, 127.

16. Wilhelm Reich, *The Mass Psychology of Fascism* (1933; New York: Farrar, Straus and Giroux, 2000).

17. Carl G. Jung, "On the Psychology of the Unconscious," in *Two Essays on Analytical Psychology*, trans. R. F. C. Hull (Princeton: Princeton University Press/Bollingen Press, 1966), 69.

18. See, for example, Carl G. Jung, *Archetypes and the Collective Unconscious* (London: Routledge, 1968), 43.

19. Carl G. Jung, "Therapeutic Principles of Psychoanalysis" (1912), in *Jung Contra Freud* (Princeton: Princeton University Press, 2012), 121.

20. Jung, "Archetypes of the Collective Unconscious" (1954), in *Archetypes and the Collective Unconscious*, 22; Jung, "The Role of the Unconscious" (1918) in *Civilization in Transition* (New York: Pantheon, 1964), 10–11. See also Carl G. Jung, "The Concept of the Collective Unconscious" (1936), in *Archetypes and the Collective Unconscious*.

21. Carl G. Jung, "Archetypes of the Collective Unconscious" (1934), in *The Integration of the Personality* (New York: Farrar and Rinehart, 1939), 52–53. The revised 1954 edition of this essay starkly avoids the "Western" focus of this point.

22. Jung, "Archetypes of the Collective Unconscious" (1934) in *Integration of the Personality*, 92; also Jung, *Archetypes and the Collective Unconscious*, 23.

23. Carl G. Jung, "The State of Psychotherapy Today" (1934) in *The Collected Works of C.G. Jung* (Princeton: Princeton University Press, 2014) 10:165.

24. Joseph Campbell, *The Hero with a Thousand Faces* (1949; Novato, CA: New World Library, 2008), 14.

25. Jordan Peterson, *Maps of Meaning: The Architecture of Belief* (New York: Routledge, 1990), xviii.

CHAPTER 11: THE HORDES AND THE FLOOD

Epigraph: Alexander Kluge, *Difference and Orientation: An Alexander Kluge Reader*, ed. Richard Langston (Ithaca, NY: Cornell University Press, 2019), 112.

1. "Viktor Orban's Speech against Brussels, Migrants and Globalism" and "Viktor Orban—Globalist NWO Agenda—'Pre-Planned' Europe Invasion Threatens 'Christian Traditions," YouTube, March 15, 2016. Also Paul Hanebrink, *A Specter Haunting Europe* (Cambridge, MA: Belknap, 2018), 275.

2. Klaus Theweleit, *Male Fantasies, I: Women, Floods, Bodies, History* (Minneapolis: University of Minnesota Press, 1986), 230.

3. Patrick Brantlinger, *Dark Vanishings: Discourse on the Extinction of Primitive Races* (Ithaca, NY: Cornell University Press, 2003), 104.

4. G. W. F. Hegel, *Lectures on the Philosophy of World History*, trans. H. B. Nisbet (Cambridge: Cambridge University Press, 1975), 175–76, 194. G. W. F. Hegel, *Lectures on the Philosophy of World History*, trans. Robert F. Brown and Peter Crafts Hodgson (Oxford: Clarendon, 2011), 197.

5. H. Rider Haggard, *Cetywayo and His White Neighbours* (1888; London: Kegan Paul, 1896), 3.

6. Edwin Arnold, "The Duty and Destiny of England in India," *North Ameri-*

can Review 154/423 (1892): 170. Discussed in Duncan Bell, *Reordering the World* (Princeton: Princeton University Press, 2016), 60.

7. M. P. Shiel, *The Yellow Danger, Or, what Might Happen in the Division of the Chinese Empire Should Estrange all European Countries* (London: Richards, 1898), 288. Discussed in Sven Lindqvist, *A History of Bombing*, trans. Linda Haverty Rugg (1999; London: Granta Books, 2002), §60.

8. Shiel, *Yellow Danger*, 271.

9. Shiel, *Yellow Danger*, 290.

10. Alison Bashford, *Global Population: History, Geopolitics, and Life on Earth* (New York: Columbia University Press, 2014), 109.

11. Manfred Görtemaker, *Deutschland im 19. Jahrhundert. Entwicklungslinien: Schriftenreihe der Bundeszentrale für politische Bildung*, vol. 274 (Wiesbaden: VS Verlag für Sozialwissenschaften, 1996), 357.

12. Lothar von Trotha to Heinrich von Schlieffen, October 4, 1904, German Federal Archive, Berlin Reichskolonialamt, 1001, 2089, quoted in René Lemarchand, *Forgotten Genocides Oblivion, Denial, and Memory* (Philadelphia: University of Pennsylvania Press, 2013), 166n34. The "over there/ back here" motif has hardly disappeared, but I do not have the room to cover it here. See, for just one well-known example, George W. Bush, "President Bush Addresses the 89th Annual National Convention of the American Legion": "Every day we work to protect the American people. Our strategy is this: We will fight them over there so we do not have to face them in the United States of America." Office of the Press Secretary, White House Archives, August 28, 2007.

13. Ibn Khaldūn, *The Muqaddimah*, trans. Franz Rosenthal (Princeton: Princeton University Press, 1969). Aziz Al-Azmeh, *Ibn Khaldūn in Modern Scholarship* (London: Third World Center for Research and Publishing, 1981); Marwa Elshakry, "The Invention of the Muslim Golden Age," in *Power and Time* eds. Dan Edelstein, Stefanos Geroulanos, and Natasha Wheatley (Chicago: University of Chicago Press, 2020), 80–102.

14. Gustave Le Bon, *La civilisation des arabes* (Paris: Firmin-Didot, 1884), 362–64.

15. Le Bon, *La civilisation des arabes*, 33.

16. Gustave Le Bon, *Les premières civilisations* (Paris: Marpon et Flammarion, 1889), 19, 124, 143.

17. Houston Stewart Chamberlain, *Foundations of the Nineteenth Century*, trans. John Lees (1889; London: Lane, 1911), 361–63.

18. Chamberlain, *Foundations of the Nineteenth Century*, 256.

19. Le Bon, *Les premières civilisations*, 615.

20. H. G. Wells, *The Outline of History*, 3rd ed. (New York: Macmillan, 1921), 1097.

21. In a later variation, Hannah Arendt presented the purpose of totalitarian movements to be the reduction of a people to a "horde." To make the point that "rootlessness is a characteristic of all race organizations," she compared the Boers to African "tribes who had also roamed the Dark Continent for centuries—feeling at home wherever the horde happened to be, and fleeing like death every attempt at definite settlement." Hannah

Arendt, *The Origins of Totalitarianism* (1949; New York: Harcourt Brace, 1973), 196.

22. Thomas Nail, *The Figure of the Migrant* (Stanford: Stanford University Press, 2015), chap. 9.

23. Hegel, *Lectures on the Philosophy of World History*, 460.

24. The Cambridge anthropologist Alfred Cort Haddon subscribed to the general English distaste of the Germans—so he presented a very simple map that carried no dates or natural obstacles whatsoever: Huns and "Germanic" peoples were simply undifferentiated vectors on the map. Alfred Cort Haddon, *The Wanderings of Peoples* (1911; Cambridge: Cambridge University Press, 1927), Map II.

25. See Karl Hermann Jacob-Friesen's map "The Paths of the Main German Races at the Time of the Migrations of Peoples" in Albert E. Brinckmann, *Geist der Nationen: Italiener, Franzosen, Deutsche* (Hamburg: Hoffman und Campe, 1938, 1948), reproduced in Eric Michaud, *The Barbarian Invasions* (Cambridge, MA: MIT Press, 2019), 177–78.

26. Harry Johnston, *Views and Reviews* (London: Williams and Norgate, 1912), 160.

27. J. B. Bury, "Appendix," in *Decline and Fall of the Roman Empire*, by Edward Gibbon, ed. J. B. Bury (New York: Macmillan, 1914), 343.

28. Peter Gatrell, *A Whole Empire Walking: Refugees in Russia during World War I* (Indianapolis: Indiana University Press, 2005), 200.

29. Anne Schult, "Numbers and Norms: Robert René Kuczynski and the Development of Demography in Interwar Britain," *History of European Ideas* 46 (2020): 715–29; also Schult, "Interwar Statistics, Colonial Demography, and the Making of the Twentieth-Century Refugee," *Journal of Global History*, 18, no. 1 (2023): 131–51.

30. Hanebrink, *A Specter Haunting Europe*, 56, 55.

31. Cicely Hamilton, *Theodore Savage* (London: Leonard Parsons, 1922). Discussed in Lindqvist, *A History of Bombing*, §109.

32. *L'Action Française*, February 18, 1939. Cited, alongside *Ouest-Éclair* (February 8, 1939) and *Le Journal des débats* (February 10, 1939), in *Outcast Europe: Refugees and Relief Workers in an Era of Total War 1936–48* by Sharif Gemie, Laure Humbert, and Fiona Reid (London: Bloomsbury, 2012), 54–55.

33. Theweleit, *Male Fantasies*, I:229.

34. Theweleit, *Male Fantasies*, I:232.

35. Carl Schmitt, *The Crisis of Parliamentary Democracy* (Cambridge, MA: MIT Press, 1988), 23. Later, he sprinkled some Slavophobia into his arguments. Schmitt, "The Age of Neutralizations and Depoliticizations" (1929) in *The Concept of the Political*, trans. G. Schwab (1932; Chicago: University of Chicago Press, 2007), 80–96.

36. Heinrich Mann, *Der Haß* (Amsterdam: Querido, 1933), 71.

37. Joseph Goebbels, "Communism with the Mask Off," in *The Third Reich Sourcebook*, eds. Anson Rabinbach and Sander L. Gilman, trans. Lilian M. Friedberg (Berkeley: University of California Press, 2013), 134. See my essay "The Temporal Assemblage of the Nazi New Man," *Power and*

Time: Temporalities in Conflict and the Making of History, eds. Edelstein, Geroulanos, and Wheatley (Chicago: University of Chicago Press, 2020), 173–200.

38. Hanebrink, *A Specter Haunting Europe*, 146.

39. Richard J. Evans, *The Third Reich in Power, 1933–1939: How the Nazis Won Over the Hearts and Minds of a Nation* (New York: Penguin Group, 2005), 359.

40. Hitler, "Secret Memorandum on the Four-Year Plan," in Rabinbach and Gilman, *Third Reich Sourcebook*, 652.

41. Adam Tooze, *The Wages of Destruction* (London: Allen Lane, 2006), 323.

42. Mark Mazower, *Hitler's Empire: How the Nazis Ruled Europe* (New York: Penguin Books, 2009), 142. Walter von Reichenau, "Orders for Conduct in the East," and also "Secret Report of the Security Service of the Reichsführer SS: The Image of Russia in the Populace," in Rabinbach and Gilman, *Third Reich Sourcebook*, 740, 825–26.

43. Unsigned memorandum ("Bormann memo"), July 16, 1941, in *Documents on German Foreign Policy, 1918–1945*, series D (1937–1945), (Washington, DC: US Government Printing Office, 1964), 13:151. Hitler had already said as much about the Urals in 1936: Max Domarus, ed., *Hitler: Reden und Proklamationen* (Munich: Süddeutscher Verlag, 1965) 1.2.642.

44. Adolf Hitler, *Hitler's Table Talk 1941–44*, ed. Hugh Trevor-Roper (New York: Enigma, 2000): "9–10 October, 1941," 40–43; "5 November 1941," 89; "23rd April 1942," 329; "7 July 1942," 427; "26 August 1942," 500. Richard J. Evans, *The Third Reich at War* (New York: Penguin, 2009), 111, 317.

45. Hitler, *Hitler's Table Talk 1941–44*, "23rd April 1942," 329.

46. Hitler, *Hitler's Table Talk 1941–44*, "5th–6th January 1942," 140. "Memorandum by the Director of the News Service and Press Department (Schmidt), Berlin Nov. 30, 1941" in *Documents on German Foreign Policy*, D:13, doc. 525, p.908.

47. Also Franz Alfred Six, "The Shape of Twentieth-Century Europe, March 3, 1943," in *Documents on the History of European Integration*, eds. Walter Lipgens and Wilfried Loth, series D, (Berlin: De Gruyter, 1989–2020), 13:121.

48. Felix Kersten, *The Kersten Memoirs* (London: Hutchinson, 1956), 132–37, cited in Mazower, *Hitler's Empire*, 205.

49. Georges Bataille, "À propos de récits d'habitants d'Hiroshima," in *Critique* 8–9 (1947), translated as "Concerning the Accounts Given by the Residents of Hiroshima," in *American Imago* 48 (1991): 498.

50. Frantz Fanon, "On Violence," in *The Wretched of the Earth* (New York: Grove Press, 2004), 8.

51. Fanon, "On Violence," 7, emphasis added.

52. Jordanna Bailkin, *The Afterlife of Empire* (Berkeley: University of California Press, 2012), 31, quoting Elspeth Huxley, *Back Street, New Worlds: A Look at Immigrants in Britain* (New York: W. Morrow, 1964), 87.

53. For the published version, see J. Enoch Powell, *Freedom and Reality* (Kingswood, UK: Elliot Right Way Books, 1969), 289.

54. Henri Baudet, *Paradise on Earth: Some Thoughts on European Images of Non-European Man* (1959; Westport, CT: Greenwood, 1976), 4.

55. Georges Dumézil writing for the *Nouvelle Revue Française* in 1941, quoted in Jean-Paul Demoule, *Mais où sont passés les Indo-Européens?* (Paris: Seuil, 2014), 156.

56. Georges Dumézil, *Mythe et épopée* (Paris: Gallimard, 1971), 1, back cover.

CHAPTER 12: NAZIS

1. Adolf Hitler, *Mein Kampf*, trans. Ralph Manheim (New York: Mariner, 2002), 225, 227–28, 235–36, 329, 646, 667.

2. Stefanos Geroulanos, "The Temporal Assemblage of the Nazi New Man," in *Power and Time: Temporalities in Conflict and the Making of History*, eds. Dan Edelstein, Stefanos Geroulanos, and Natasha Wheatley (Chicago: University of Chicago Press, 2020), 173–200.

3. Albert Speer, *Inside the Third Reich* (New York: Macmillan, 1970), 56. Joachim Fest, *Albert Speer* (Malden, MA: Polity, 2007), 67–68. Speer's recollections on the subject are likely invented.

4. Alfred Rosenberg, *Der Mythus des 20. Jahrhunderts* (1930; Munich: Hoheneichen Verlag, 1939), includes his rejection of criticisms directed at his "Wotanism."

5. Hitler, *Mein Kampf*, 290.

6. Hermann Rauschning, *Hitler Speaks* (London: Eyre and Spottiswoode, 1939) 87. It is accepted that Rauschning's text is not documentary. Either way, Rauschning's passage speaks just as well to the availability (for Rauschning) of the idea of barbarians rejuvenating decadent civilizations, an idea that, as we have seen, worked well with conservative fears of nomads/barbarians. Nevertheless, in his 1943–44 course on German history, Ernst Kantorowicz cited this very passage as an accurate illustration of what he called Nazism's "rebarbarization" of Germany, its "tribal ideal" and its "prehistorism"—an obsession with the extra-historical purity of the race and return to the times outside of recorded history. See Leo Baeck Institute, Ernst Kantorowicz Collection, Box III, AR7216.8/2, Loc 549/5, p.1288 (170).

7. On the "hyperborean" path of culture and Nazi legends, see Dan Edelstein, "Hyperborean Atlantis: Jean-Sylvain Bailly, Madame Blavatsky, and the Nazi Myth," in *Studies in Eighteenth-Century Culture* 35 (2006): 267–91.

8. Adolf Hitler, "Kunst verpflichtet zur Wahrhaftigkeit" (1934), translated as "Art and Its Commitment to Truth," in *The Third Reich Sourcebook*, eds. Anson Rabinbach and Sander Gilman, trans. Lilian M. Friedberg (Berkeley: University of California, 2013), 491.

9. Eric Michaud, *The Cult of Art in Nazi Germany* (Stanford: Stanford University Press, 2004), 75–84; Richard Steigmann-Gall, *The Holy Reich: Nazi Conceptions of Christianity, 1919–1945* (Cambridge: Cambridge University Press, 2004); Susannah Heschel, *The Aryan Jesus: Christian Theologians and the Bible in Nazi Germany* (Princeton: Princeton University Press, 2010).

10. Hitler, *Mein Kampf*, 290–294.

11. Hitler, "Political Testament," in Rabinbach and Gilman, *Third Reich*

Sourcebook, 872. See also Hugh Trevor-Roper, *The Last Days of Hitler* (New York: Macmillan, 1947), 32.

12. Hans Mommsen, *Der Mythos von der Modernität. Zur Entwicklung der Rüstungsindustrie im Dritten Reich* (Essen: A. Francke Verlag, 1999), discussed in Anson Rabinbach, "Nazi Culture," in *Staging the Third Reich*, eds. Stefanos Geroulanos and Dagmar Herzog (London: Routledge, 2020), 121. On ideology in Nazism see also Jürgen Matthäus, "Antisemitism as an Offer," in *Lessons and Legacies*, vol.7, ed. Dagmar Herzog (Evanston, IL: Northwestern University Press, 2006), 118, 120; Birthe Kundrus, "Kontinuitäten, Parallelen, Rezeptionen," *Werkstatt Geschichte* 43 (2006): 45–62; Eric Michaud, *Cult of Art*.

13. Rabinbach, "Nazi Culture," 121.

14. Stefanos Geroulanos, "The Sovereignty of the New Man after Wagner," in *The Scaffolding of Sovereignty*, eds. Zvi Ben-Dor Benite, Geroulanos, and Nicole Jerr (New York: Columbia University Press, 2017), 440–68.

15. Maurice Olender, *The Languages of Paradise* (Cambridge, MA: Harvard University Press, 1992), chap. 3; Stefan Arvidsson, *Aryan Idols: Indo-European Mythology as Ideology and Science* (Chicago: University of Chicago Press, 2006), 109–18.

16. Cathy Gere, *The Tomb of Agamemnon* (London: Profile, 2006), 89–92, 119–22.

17. Antoine Meillet, *Introduction à l'étude comparative des langues indo-européennes* 5th ed. (Paris, 1922), 375–76; 8th ed. (1937), 418–19. Jean-Paul Demoule, *Mais où sont passés les Indo-Européens?* (Paris: Seuil, 2014), 149; absent from the first four, pre-WWI editions.

18. Meillet, *Introduction à l'étude comparative*, 29. This passage is absent from the earlier editions. Quoted in Demoule, *Mais où sont passés les Indo-Européens?*, 150.

19. Gustaf Kossinna, *Die deutsche Vorgeschichte, eine hervorragend nationale Wissenschaft* (Leipzig: Curt Kabitzsch Verlag, 1912); *Die Indogermanen* (Würzburg, 1920). On Kossinna, see Arvidsson, *Aryan Idols*, 142–43.

20. V. Gordon Childe, *The Aryans* (London: Kegan Paul, 1926), 212.

21. V. Gordon Childe, *The Dawn of European Civilization*, 6th ed. (New York: Vintage, 1958), 172.

22. Henri Hubert, *Les Germains. Cours professé à l'École du Louvre, 1924–1925* (Paris: Albin Michel, 1952), 39–40.

23. Suzanne Marchand, *German Orientalism in the Age of Empire* (Cambridge, UK: Cambridge University Press, 2010), 128.

24. Childe, *Aryans*, 211–12, 200.

25. For the interpretation (and maps) of early cultures in Europe as invented in Germany, as well as for an argument that saw the Nazis as fulfilling scholarly hopes, see the Festschrift for the Indo-Germanist Herman Hirt by his students, including Helmut Arntz (who remained a leading linguist till the end of his career). Helmut Arntz, ed., *Germanen und Indogermanen. Volkstum, Sprache, Heimat, Kultur*, 2 vols. (Heidelberg: Winter, 1936), 1:viii, 207, 217. On Hirt, see Arvidsson, *Aryan Idols*, 196–97.

26. Ingo Haar and Michael Fahlbusch, eds. *German Scholars and Ethnic*

Cleansing, 1919–1945 (New York: Berghahn Books, 2005). (On Erich Keyser and the Nordic type, see Alexander Pinwinkler, "*Volk, Bevölkerung, Rasse, and Raum,*" in this book, 94.) Heather Pringle, *The Master Plan: Himmler's Scholars and the Holocaust* (New York: Hachette Books, 2006). Wolfgang Bialas and Anson Rabinbach, eds. *Nazi Germany and the Humanities* (New York: Oneworld Publications, 2007). On myth, see Michaud, *Cult of Art*, xi.

27. Hans F. K. Günther, *Die Nordische Rasse bei den Indogermanen Asiens* (Munich: Lehmanns, 1934), 237.

28. R. Walther Darré, *Neuadel aus Blut und Boden* (Munich: Lehmanns, 1930), 17.

29. On National Socialist (artificial) visions presented as natural, see especially Tiago Saraiva, *Fascist Pigs: Technoscientific Organisms and the History of Fascism* (Cambridge, MA: MIT Press, 2016), 9.

30. Fritz Todt, "Der Sinn des neuen Bauens," *Die Strasse* 4, no. 21 (November 1, 1937): 616–17.

31. Geroulanos, "Temporal Assemblage," 173–200.

32. On Reinhard Heydrich's views on the Jews, see "The Visible Enemy" in Rabinbach and Gilman, *The Third Reich Sourcebook*, 197–99. On the establishment of the Judenrat, see Heydrich's "Instructions on Policy and Operations Concerning Jews in the Occupied Territories, September 21, 1939" in *Documents on the Holocaust, Selected Sources on the Destruction of the Jews of Germany and Austria, Poland and the Soviet Union* (Jerusalem: Yad Vashem, 1981), Document no. 73.

33. Adam Tooze, *The Wages of Destruction* (London: Allen Lane, 2006), 467.

34. "Aus der Studie Generalplan Ost," in *Europastrategien des deutschen Kapitals 1900–45*, ed. Reinhard Opitz (Bonn: Paul-Rugenstein Nachfolger, 1994), 898–99.

35. "Rede Reinhard Heydrichs über die Grundsätze der nationalsozialistischen, Neuordnung' Europas," in *Vom Generalplan Ost zum Generalsiedlungsplan*, by Czeslaw Madajczyk (Munich: Saur, 1994), 21–22.

36. Joseph Goebbels, "Now, People, Rise Up and Let the Storm Winds Blow!" translated in part in Rabinbach and Gilman, *The Third Reich Sourcebook*, 828–30.

37. For a detailed presentation of this episode, see Christopher Krebs, *A Most Dangerous Book* (New York: Norton, 2012), 15–28. On Nazi looting: Rabinbach, "Nazi Culture," 108–37.

38. Primo Levi, *Survival in Auschwitz*, trans. Stuart Wolfe (1947; New York: Simon and Schuster, 1996), 91.

39. Levi, *Survival in Auschwitz*, 90.

40. Levi, *Survival in Auschwitz*, 41

41. Levi, *Survival in Auschwitz*, 114–15.

42. Levi, *Survival in Auschwitz*, 87.

43. Max Horkheimer and Theodor W. Adorno, *Dialectic of Enlightenment*, trans. Edmund Jephcott (1944/47; Stanford: Stanford University Press, 2002), 1.

44. Horkheimer and Adorno, *Dialectic of Enlightenment*, 187–88.

Chapter 13: Bomb Them Back to the Stone Age!

Epigraph: Aimé Césaire, *Lyric and Dramatic Poetry*, 1946–82 (Charlottesville: University of Virginia Press, 1990), 39.

1. Curtis E. LeMay and MacKinlay Kantor, *Mission with LeMay: My Story* (New York: Doubleday, 1965), 565.
2. *Washington Post* (October 4, 1968), A8.
3. Martti Koskenniemi, *The Gentle Civilizer of Nations* (Cambridge: Cambridge University Press, 2001), 102.
4. James Lorimer, *The Institutes of Law*, rev. ed. (London: Blackwood and Sons, 1883), 1:101.
5. Koskenniemi, *Gentle Civilizer of Nations*, chaps.1–2.
6. John Westlake, *International Law: War*, vol. 2, 2nd ed. (Cambridge: Cambridge University Press, 1913), 59.
7. Westlake, *International Law*, 2:87.
8. Joseph Hornung, "Civilisés et barbares," *Revue de droit international et de législation comparée* 17, no. 1 (1885): 5–7.
9. See, among others, Sven Lindqvist, *A History of Bombing* (London: Granta, 2002), §52.
10. Stefanos Geroulanos and Todd Meyers, *The Human Body in the Age of Catastrophe* (Chicago: University of Chicago Press, 2018), 37. The "brutalization of warfare" and politics is George L. Mosse's thesis—see *The Image of Man: The Creation of Modern Masculinity* (New York: Oxford University Press, 1998), 111, and *Fallen Soldiers* (Oxford: Oxford University Press, 1991), chap. 8.
11. Anthony Bowlby, "Wounds in War," *Lancet* (December 25, 1915), 1385.
12. Henri Barbot, *Paris en feu, Ignis ardens* (Paris: Lettres Françaises, 1914).
13. Lindqvist, *History of Bombing*, §127.
14. Lindqvist, *History of Bombing*, §7.
15. H. G. Wells, *The Outline of History*, 3rd ed. (New York: Macmillan, 1921), 1084.
16. Wells, *Outline of History*, 1089.
17. P. Anderson Graham, *The Collapse of Homo Sapiens* (London: Putnam's Sons, 1923), 33.
18. Peter Fritzsche, *A Nation of Fliers: German Aviation and the Popular Imagination* (Cambridge, MA: Harvard University Press, 1992), 60–61.
19. Giulio Douhet, *Command of the Air* (New York: Arno, 1972), 7.
20. Douhet, *Command of the Air*, 185.
21. Emily Braun, "Shock and Awe" in *Italian Futurism, 1909–1944: Reconstructing the Universe*, ed. Vivien Greene (New York: Guggenheim Museum, 2014), 269–74; Lucia Piccioni, "Aeropittura futurista e colonialismo fascista," *Mitteilungen Des Kunsthistorischen Institutes in Florenz* 59, no. 1 (2017): 108–23.
22. League of Nations Covenant, Article 22.
23. John Bassett Moore, *International Law and Some Current Illusions and Other Essays* (New York: MacMillan, 1924), ix, 7.
24. Moore, *International Law and Some Current Illusions*, 240, 197.
25. Moore, *International Law and Some Current Illusions*, 197.

26. Susan Pedersen, *The Guardians: The League of Nations and the Crisis of Empire* (Oxford: Oxford University Press, 2015), 150.
27. Douhet, *Command of the Air*, 391.
28. Thomas W. Burkman, *Japan and the League of Nations* (Honolulu: University of Hawai'i Press, 2008), 207; Sheldon Garon, "On the Transnational Destruction of Cities: What Japan and the United States Learned from the Bombing of Britain and Germany in the Second World War," *Past and Present* 247, no. 1 (May 2020): 244.
29. Sven Reichardt, "Fascism's Stages: Imperial Violence, Entanglement, and Processualization," in *Journal of the History of Ideas* 82, no. 1 (January 2021): 85–107.
30. Pedersen, *The Guardians*, 297.
31. Quoted in Walter Benjamin, "The Work of Art in the Age of its Technological Reproducibility" (1936/39), in *Selected Writings*, vol. 4: 1938–40 (Cambridge, MA: Belknap Press, 2006), 269.
32. Haile Selassie, Appeal to the League of Nations, June 1936.
33. Bruno Mussolini, quoted in Bertrand Russell, *Power: A New Social Analysis* (1938; London: Routledge, 2004), 18–19.
34. Walter Benjamin, "The Work of Art in the Age of Its Technological Reproducibility: Third Version," in *Selected Writings*, eds. Marcus Bullock and Michael W. Jennings (Cambridge, MA: Belknap Press, 2004), 4:270.
35. States were learning from one another: see Reichardt, "Fascism's Stages," 85–107. Garon, "On the Transnational Destruction of Cities," 242.
36. Walter Benjamin, "On the Concept of History" (1940), thesis 7, in *Selected Writings*, 4:392.
37. Benjamin, "On the Concept of History," thesis 9, in Bullock and Jennings, *Selected Writings*, 4:392. He even invoked prehistory, indirectly, unexpectedly, in his attack on the history written by and for the victors. For a version of the cosmic clock, see Benjamin's thesis 18 of "On the Concept of History," 4:396.
38. The Secretary of State to the Ambassador in the United Kingdom, telegram, *Foreign Relations of the United States Diplomatic Papers*, 1939, General, Volume I, document 564 (740.00116 European War 1939/19a: Telegram).
39. Albert Speer, *Inside the Third Reich* (New York: Macmillan, 1970), 278, 280.
40. United States Strategic Bombing Survey, *The Effects of Strategic Bombing on the German War Economy* (Oct.31, 1945), 14. On the postwar taboo concerning discussion of the allied bombing of Germany, see W. G. Sebald, *On the Natural History of Destruction* (New York: Penguin, 2004) and Alexander Kluge, *The Air Raid* (1979; London: Seagull Books, 2014).
41. Danilyn Rutherford, *Living in the Stone Age* (Chicago: University of Chicago Press, 2018), 1.
42. V. Gordon Childe, *The Prehistory of European Society: How and Why the Prehistoric Barbarian Societies of Europe Behaved in a Distinctively European Way* (Harmondsworth, Middlesex: Penguin Books, 1958) and *Prehistoric Migrations in Europe* (Cambridge, MA: Harvard University Press, 1950).
43. Joanna Curtis, "Nationhood After Infinite Harm: Music, Childhood and the State in West Germany and Austria, c. 1945–1975" (PhD Dissertation, New York University, 2019), chap. 3.

Chapter 14: The Manchurian Catholic and the Future of Humanity

1. Pierre Teilhard de Chardin to Henri Breuil, Dec.16, 1939; Teilhard de Chardin to Max and Simone Begouën, February 8, 1940, in *Lettres de voyage, 1923–55* (Paris: Grasset, 1956), 248, 250.

2. Teilhard de Chardin, "The Moment of Choice" (Dec.1939), in *The Activation of Energy* (1963; New York: Harcourt Brace Jovanovich, 1971), 13, 15.

3. Teilhard de Chardin, "Moment of Choice," 19.

4. Mercè Prats, "Le teilhardisme: Réception, adoption et travestissement de la pensée de Teilhard de Chardin, 1955–1968" (PhD Thesis, Université de Reims Champagne-Ardenne, 2019).

5. In the 1950s, when the prehistorian and Piltdown sleuth Kenneth Oakley wrote to tell him that he had been "hoodwinked" and the fossils were confirmed fakes, Teilhard contributed, somewhat half-heartedly, to an effort to discover the fraud. Kenneth Oakley to Teilhard de Chardin (November 19, 1953), Wenner-Gren Foundation Archive, Gr. 764.

6. Emily Margaret Kern, "Out of Asia: A Global History of the Scientific Search for the Origins of Humankind, 1800–1965" (PhD Dissertation, Princeton University, 2018).

7. Note the awkward treatment in H. G. Wells's *The Outline of History* (New York: Macmillan, 1921). Friends warned Wells about Piltdown. Note also Franz Weidenreich's rejection in *Apes, Giants, and Man* (Chicago: University of Chicago Press, 1945), 22–23.

8. Teilhard de Chardin to his parents (June 16, 1913), quoted in *Teilhard de Chardin Album* (New York: Harper and Row, 1966), 48.

9. Teilhard de Chardin to Claude Aragonnès (September 23, 1917), in *The Making of a Mind* (New York: Harper and Row, 1965), 203–4. Aragonnès was the pen name of Marguerite Teillard-Chambon, a well-known Catholic feminist and a cousin of Teilhard's.

10. Teilhard de Chardin to Aragonnès (October 14, 1916), quoted in *Teilhard de Chardin Album*, 53.

11. On Jersey, theology, and Jesuit politics, see Sarah Shortall, *Soldiers of God in a Secular World* (Cambridge, MA: Harvard University Press, 2021).

12. Teilhard de Chardin, "The Spiritual Power of Matter" (August 8, 1919), in *The Hymn of the Universe* (New York: Harper and Row, 1961), 65–66.

13. Teilhard de Chardin, "Fossil Men" (1921), in *The Appearance of Man* (New York: Harper and Row, 1965), 32.

14. Teilhard de Chardin to Auguste Valensin (May 16, 1925), in *Lettres intimes, 1919–1950*, by Teilhard de Chardin (Paris: Aubier Montaigne, 1972).

15. Teilhard de Chardin to Valensin (May 16, 1925).

16. Henri Begouën to Msgr. Cerette (August 1925), in Muséum national d'Histoire naturelle, Henri Breuil Archive, Br.91, "Vatican et la préhistoire." Teilhard was only one part of this attempt, among Catholic paleontologists, to convince the Vatican to think differently. Stefanos Geroulanos, "Polyschematic Prehistory at the Dusk of Colonialism," *RES: Anthropology and Aesthetics* 69–70 (2018): 138–39.

17. Chris Manias, "*Sinanthropus* in Britain: Human Origins and International Science, 1920–1939," *British Journal for the History of Science* 48, no. 2 (2015): 289–319.

18. Fondation Teilhard de Chardin Archive, "Early Man in China" offprint with handwritten annotations.

19. Teilhard de Chardin, *The Phenomenon of Man* (New York: Harper, 1959), 198. The French title *Le phénomène humain* lacks the gendering that the English translation enforces, so I use *The Human Phenomenon* in what follows, except in citing the English translation.

20. Weidenreich differentiates *sapiens* from *Sinanthropus* in "Six Lectures on Sinanthropus pekinensis and Related Problems" in *Bulletin of the Geological Society of China* 19, no.1 (1939), 75. He believed that "racial differentiations began to be manifested already within the *Sinanthropus* stage" (63) leading to aversion between groups and violence.

21. Teilhard de Chardin, *Phenomenon of Man*, 194.

22. Henri Breuil, "The Bone and Antler Industry of the Choukoutien Sinanthropus Site," in *Palaeontologia Sinica. Series D*, 6:117, trans. Mary Boyle (Beijing: Geological Survey of China, 1939), 1–41.

23. Teilhard de Chardin, *Phenomenon of Man*, 181.

24. Teilhard de Chardin, *Phenomenon of Man*, 193.

25. On Bergson and Teilhard, see Julian Huxley, "Introduction," in Teilhard de Chardin, *Phenomenon of Man*, 22.

26. On latency: Teilhard de Chardin, *Phenomenon of Man*, 67, 77.

27. Sarah Shortall, "Religion and the Divine: A Global History of Twentieth-Century Catholicism through the Eucharist," in *A Cultural History of Ideas in the Modern Age*, ed. Stefanos Geroulanos (London: Bloomsbury, 2022), 113. De Lubac wrote two books on Teilhard after the latter's death. On Lubac, see also Sarah Shortall, *Soldiers of God in a Secular World: Catholic Theology and Twentieth-Century French Politics* (Cambridge, MA: Harvard University Press, 2011).

28. Teilhard de Chardin, *Phenomenon of Man*, 67, 77.

29. Teilhard de Chardin, *Phenomenon of Man*, 35.

30. Teilhard de Chardin, *Phenomenon of Man*, 193, 195.

31. Teilhard de Chardin, *Phenomenon of Man*, 179–80.

32. Teilhard de Chardin, *Phenomenon of Man*, 182–83.

33. Teilhard de Chardin, *Phenomenon of Man*, 261–62. On the pull of the Omega point, see also Teilhard de Chardin, "The Atomism of Spirit" (1941), in *The Activation of Energy: Enlightening Reflections on Spiritual Energy* (Boston: Houghton Mifflin Harcourt, 1976), 50, 55–56.

34. On the rise of antihumanism, see Stefanos Geroulanos, *An Atheism That Is Not Humanist Emerges in French Thought* (Stanford: Stanford University Press, 2010).

35. Teilhard de Chardin, *La place de l'homme dans la nature* (Paris: Albin Michel, 1956), 178n7.

36. Julian Huxley, "Transhumanism" (1949) in *New Bottles for New Wine* (London: Chatto and Windus, 1957), 13.

37. Teilhard de Chardin, *Phenomenon of Man*, 261.

38. Huxley, "Transhumanism," 14.

39. Nasser Zakariya, "Scientific Humanisms and Technological Utopias," in *Perfecting Human Futures*, eds. J. Benjamin Hurlbut and Hava Tirosh-Samuelson (Wiesbaden: Springer, 2016), 275.

40. Wenner-Gren Foundation Archives, Gr.552 and Gr.1033. "Memorandum for Julian Huxley," 2; Huxley to Teilhard, March 27, 1952; Teilhard to Huxley, April 1, 1952; but see also Teilhard's criticism of Huxley's project, which limited its funding by the Wenner-Gren Foundation.

41. Teilhard de Chardin to Paul Féjos, January 21, 1950, in Wenner-Gren Foundation Archives, Gr.119-C, Teilhard de Chardin, Paris-South Africa Conference on Early Man.

42. Pope Pius XII, *Humani Generis* Encyclical, August 12, 1950.

43. Teilhard de Chardin, "South African Archaeological Society, Cape Town, October 1951," in *L'Oeuvre scientifique, Tome X, 1945–1955*, eds. Karl Schmitz-Moormann and Nicole Schmitz-Moormann (Olten: Walter, 1971), 4406; Teilhard de Chardin, "L'Afrique et les origines humaines" (1954), *Revue des questions scientifiques*, January 20, 1955, 5–17.

44. Mercè Prats, "Le Teilhardisme: Réception, adoption et travestissement de la pensée de Teilhard de Chardin, à la croisée des sciences et de la foi, au cœur des 'Trente Glorieuses' en France (1955–1968)" (PhD thesis, Université De Reims–Champagne-Ardenne, 2019).

45. Paul VI, Speech to the Workers of a Chemical Pharmaceutical Factory, February 24, 1966.

46. Léopold Sédar Senghor, "La Négritude, comme culture des peuples noirs, ne saurait être dépassé," in *Liberté 5. Le dialogue des cultures* (Paris: Seuil, 1993); Senghor, "Pierre Teilhard de Chardin et la politique africaine," in *Cahiers Pierre Teilhard de Chardin 3* (Paris: Seuil, 1962); Gary Wilder, *Freedom Time: Negritude, Decolonization, and the Future of the World* (Durham: Duke University Press, 2015), 231–33.

47. Shortall, "Religion and the Divine," 121.

48. See also the arguments of John Sexton, president of NYU and advocate of a global university: Sexton, "Global Network University Reflection" (December 21, 2010); Rachel Aviv, "The Imperial Presidency: John Sexton Has a Vision for N.Y.U.'s Future. His Faculty Aren't Buying It.," *New Yorker*, September 9, 2013.

49. John Paul II to Reverend George V. Coyne, S. J., Director of the Vatican Observatory, June 1, 1988. Benedict XVI, Homily in the Cathedral of Aosta, July 24, 2009.

50. See Francis I's discussion of Teilhard in Pope Francis I, "Laudato Si'," Encyclical of May 24, 2015, footnote 53.

51. Niels de Terra, *Overkill: The Vatican Trial of Pierre Teilhard De Chardin, Sj* (n.p.: Maquisard, 2014).

CHAPTER 15: DARWIN IN THE AGE OF UNESCO

1. Ruth Benedict, *Patterns of Culture* (1934; Boston: Mentor, 1959), 30.

2. Claude Lévi-Strauss, *Race and History* (Paris: UNESCO, 1952), 18.

3. Alfred Métraux to Michel Leiris, August 8, 1945. Collège de France Archives, Laboratoire de l'anthropologie sociale, Fonds Michel Leiris LAS-FML.C.02.04.005.

4. Alfred Döblin, *Destiny's Journey*, trans. Edna McCown (New York: Paragon, 1992), 299-301, 304. Cited in Stefan-Ludwig Hoffmann, "Gazing at Ruins: German Defeat as Visual Experience," in *Journal of Modern European History* 9, no. 3 (2011): 328-50.

5. Hannah Arendt, "The Image of Hell," *Commentary* 2/3(September 1946): 291-95.

6. Paul Betts, *Ruin and Renewal: Civilising Europe after World War II* (New York: Basic Books, 2020).

7. UNESCO Constitution, November 16, 1945, Preamble.

8. The quote is from Julian Huxley, "Transhumanism," in *New Bottles for New Wine* (London: Chatto and Windus, 1946), 13. On Huxley's eugenic dreams, see Julian Huxley, "Is War Instinctive—And Inevitable? A Scientist Answers a Question," *New York Times* (February 10, 1946), 84.

9. Robert Angell to Hadley Cantril, November 2, 1949, in UNESCO 323.12.A.102,81.

10. Ashley-Montagu, *Coming into Being among the Australian Aborigines: A Study of the Procreative Beliefs of the Native Tribes of Australia* (London: Routledge and Sons, 1937), 15n1.

11. A. H. Schultz, review of *"An Introduction to Physical Anthropology,"* in *Quarterly Review of Biology* 21, no. 2 (1946): 197-98. *"An Introduction to Physical Anthropology.* by M. F. Ashley Montagu," *American Naturalist* 80, no. 793 (1946): 485-88. Loren C. Eiseley, review of "Miscellaneous: *An Introduction to Physical Anthropology.* M. F. Ashley Montagu," *American Anthropologist* 48, no. 4 (1946), 647-49. Even Comas, who supported Montagu's politics, had reviewed him severely for relying on personal rather than scientific criteria: Juan Comas, *"An Introduction to Physical Anthropology* by M. F. Ashley Montagu," *Boletín Bibliográfico de Antropología Americana*, 8, no. 1/3 (1945): 152-56.

12. United Nations Educational Scientific and Cultural Organization, Statement of Experts on Race Problems ["Statement on Race"], §1. UNESCO Archives, UNESCO/SS/1 (20 July 1950).

13. UNESCO Statement on Race, §14.

14. Angell to Huxley, January 5, 1950. UNESCO Archives (Paris), 323.12.A.102.

15. Huxley to Montagu, May 24, 1950, UNESCO Archives (Paris) 323.12.A.102, 166-67.

16. Charles Darwin, *The Descent of Man* (London: Murray, 1871), 1:100-1.

17. Montagu to Angell, May 1, 1950, UNESCO Archives (Paris) 323.12.A.102, 165.

18. UNESCO Background Paper 104, July 19, 1950, in UNESCO Archives (Paris) 323.12.A.102, 201-3.

19. UNESCO Archives (Paris) 323.12.A, ODG Memo 5596.

20. William B. Fagg, "32. Note" (editorial note), in *Man* (January 1951), 17-18.

21. National Commission for the United Kingdom, Ministry of Education, to the Director-General of UNESCO (December 12, 1950), UNESCO Archives, 323.12 A 102, p. 297.

22. Alfred Cort Haddon, *The Races of Man and Their Distribution* (New York: Frederick. A. Stokes Company, 1909), 1. All the more intense is Haddon's archive, e.g., Cambridge University Library, Alfred C. Haddon Archive, folders 4040–43, which show his extensive efforts in craniometry and comparative anatomy.

23. Julian Huxley and A. C. Haddon, *We Europeans: A Survey of 'Racial' Problems* (London: Harper, 1936), 92–93.

24. Montagu to Métraux, April 4, 1951, in UNESCO Archives (Paris) 323.A 12, 142.

25. Métraux to Margaret Mead, November 14, 1950, in UNESCO 323.12 A 102, 284.

26. Michelle Brattain, "Race, Racism, and Antiracism: UNESCO and the Politics of Presenting Science to the Postwar Public," *American Historical Review* 112, no.5 (2007): 1386–1413.

27. Alva Myrdal to Métraux, October 22, 1951, SS.262/2115 in UNESCO Archives (Paris) 323.12 A102 (September–December 1951), 131–32.

28. Césaire was at the time a deputy for the Antilles in France's Parliament. On Leiris's pamphlet and UNESCO more broadly, see Stefanos Geroulanos, *Transparency in Postwar France* (Stanford: Stanford University Press, 2017), chap. 6. But the present chapter corrects the account I offered in that book.

29. Juan Comas, *Racial Myths* (Paris: UNESCO, 1951), 8.

30. Scholars, myself included, have treated it as an essay that extended Lévi-Strauss's contribution to the first Statement and defended the Statement. That is wrong.

31. Métraux, "Race and Civilization," UNESCO 323.1/094.4 in UNESCO Archives (Paris) 323.12 A 102, 288–91. Freud used the exact same language about Australia to tie "neurotics" to "primitives": "the most backward and miserable of savages, the aborigines of Australia, the youngest continent, in whose fauna, too, we can still observe much that is archaic and that has perished elsewhere." Sigmund Freud, *Totem and Taboo*, in *The Standard Edition of the Complete Psychological Works of Sigmund Freud*, vol. 13, ed. James Strachey (London: Hogarth, 1955), 1.

32. Lévi-Strauss, *Race and History*, 34.

33. Lévi-Strauss, "Une science révolutionnaire, l'ethnographie," lecture delivered on January 29, 1938 (or 1939), at the Federated Center for Workers' Education: BNF, Fonds Lévi-Strauss, NAF28150: 82/3, 17.

34. Lévi-Strauss, *Race and History*, 12.

35. Lévi-Strauss, *Race and History*, 31.

36. E.g., "Rather than anthropology, we should be writing 'entropology', the name of a discipline devoted to studying . . . this process of disintegration." Claude Lévi-Strauss, *Tristes Tropiques*, trans. John Weightman and Doreen Weightman (1955; New York: Penguin, 1992), 414. See Stefanos Geroulanos and Todd Meyers, *The Human Body in the Age of Catastrophe* (Chicago: University of Chicago Press, 2017), 286–91; Geroulanos, *Transparency in Postwar France*, 243–52.

37. Lévi-Strauss, *Tristes tropiques*, 38.

38. Lévi-Strauss, *Tristes tropiques*, 149.

39. Stefan-Ludwig Hoffmann, "Gazing at Ruins: German Defeat as Visual Experience," *Journal of Modern European History* 9, no. 3 (2011): 328–50.

Chapter 16: A History of Cave Painting

1. Gaston Bachelard, *La Terre et les reveries du répos* (Paris: Corti, 1948), 234.
2. Bachelard, *La terre et les rêveries du répos*, 207.
3. Bachelard, *La terre et les rêveries du répos*, 216.
4. Homer, *Odyssey*, trans. Emily Wilson (New York: W. W. Norton, 2018), 9: 112, 180.
5. Aeschylus, *Eumenides,* in *The Oresteia*, trans. Robert Fagles (New York: Penguin Books, 1984), line 976.
6. Anthony F. D'Elia, *A Sudden Terror: The Plot to Murder the Pope in Renaissance Rome* (Cambridge, MA: Harvard University Press, 2011).
7. Irina Oryshkevich, "Metropolis to Necropolis: The Afterlife of the Roman Catacombs," in *History Takes Place: Rome—Dynamics of Urban Change*, eds. Anna Hoffmann and Martin Zimmermann (Berlin: Jovis Verlag, 2017), 56–70.
8. On engagements with Christian (and Jewish) art in the catacombs, see Ingo Herklotz, *Cassiano dal Pozzo und die Archäologie des 17. Jahrhunderts* (Munich: Hirmer, 1999).
9. Bachelard, *La terre et les rêveries du répos*, 208.
10. The link of Maria di Sautuola to Alice was first proposed by Georges Bataille; see Daniel Fabre, *Bataille à Lascaux* (Paris: L'Echoppe, 2014), 37.
11. There's even a film, *Finding Altamira* (2016), starring Antonio Banderas.
12. Gabriel de Mortillet, *Le Préhistorique. Antiquité de l'homme* (Paris: Renwald, 1883), 476; François Bon, "The Division and Discord of Prehistoric Chronologies," *RES: Anthropology and Aesthetics* 69–70 (2018): 76–84; Noël Coye, *La préhistoire en parole et en acte* (Paris: L'Harmattan, 1997); Marc Groenen, *Pour une histoire de la préhistoire* (Grenoble: Million, 1994); Nathalie Richard, *L'invention de la préhistoire, une anthologie* (Paris: Pocket, 1992); Arnaud Hurel, *La France préhistorienne de 1789 à 1941* (Paris: CNRS, 2007); Nathalie Richard, *Inventer la Préhistoire* (Paris: Vuibert, 2008).
13. Emile Cartailhac and Henri Breuil, "Communication. Les peintures préhistoriques de la grotte d'Altamira à Santillana," *Comptes rendus des séances de l'Académie des Inscriptions et Belles-Lettres* 47, no. 3 (1903): 256–64.
14. Eduardo Palacio-Pérez, "Salomon Reinach and the Religious Interpretation of Paleolithic Art," *Antiquity* 84 (2010): 853–63.
15. Salomon Reinach, "L'art et la magie à propos des peintures et des gravures de l'âge du renne," *L'Anthropologie* 14 (1903): 257–66, 259, 261.
16. Reinach, "L'art et la magie," 260.
17. Louis Capitan and Henri Breuil, "Origines de l'art: Les gravures sur les parois des grottes préhistoriques anciennes," *La Nature* 30, no.1503 (March 1902).
18. James Frazer, *The Magic Art and the Evolution of Kings*, in *The Golden Bough*, 3rd ed. (1911; London: Macmillan, 1976), 1:134. See also Lucien

Lévy-Bruhl, *La Mentalité primitive* (Paris: Presses universitaires de France, 1922), chap. 10: Lévy-Bruhl sought out "mystical" causes of success at war.

19. Reinach, *Apollo. Histoire générale des arts plastiques* (Paris: Picard, 1904); translated as *Apollo* (New York: Scribner, 1910), 7–8.

20. Similarly, in the film *Early Man* (Nick Park, 2018), the invention of soccer is recorded in rock art, inspiring its viewers to return to the game.

21. See also Maria Stavrinaki, *Transfixed by Prehistory: An Inquiry into Modern Art and Time*, trans. Jane Marie Todd (New York: Zone Books, 2022), 161.

22. Reinach, "L'art et la magie," 266.

23. Walter Benjamin would not have been impressed by this; sticking close to the animism point: "Artistic production begins with figures in the service of magic. What is important for these figures is that they are present, not that they are seen. The elk depicted by Stone Age man on the walls of his cave is an instrument of magic, and is exhibited to others only coincidentally; what matters is that the spirits see it." Benjamin, "The Work of Art in the Age of its Technological Reproducibility," and also "The Knowledge That the First Material on which the Mimetic Faculty Tested Itself" in *Selected Works*, eds. Howard Eiland and Michael W. Jennings (Cambridge, MA: Belknap Press, 2006), 3:106, 253.

24. On Africa, the key operator was Leo Frobenius who, together with his institute, was deeply invested (and controversial). More on Frobenius in the next chapter.

25. Dalya Alberge, " 'Sistine Chapel of the Ancients' Rock Art Discovered in Remote Amazon Forest," *The Guardian*, November 29, 2020. Apparently only Christian terms can measure non-Christian art.

26. See Georges Bataille, *Lascaux ou la naissance de l'art* (Geneva: Skira, 1955), 5, for publisher Albert Skira's lament. Bataille, *Lascaux, or The Birth of Art*, trans. Austryn Wainhouse (Geneva: Skira, 1955), 5. Notable photographs were by Belgian photographer Fernand Windels.

27. Henri Breuil, *Quatre Cent Siècles d'art pariétal* (Montignac: Centre d'études et de documentation préhistoriques, 1950), 22–23.

28. Today, some prehistorians debate (among many other things) how stylistic evolution led to particular aesthetic choices. See Emmanuel Guy, *Préhistoire du sentiment artistique: L'invention du style, il y a 20 000 ans* (Paris: Presses du réel, 2012).

29. Henri Breuil, *Beyond the Bounds of History* (London: Gawthorn, 1949), 80–81.

30. Maria Stavrinaki, *Transfixed by Prehistory: An Inquiry into Modern Art and Time* (New York: Zone Books, 2022).

31. Here I follow Stavrinaki's interpretation, published once in an essay we wrote together (Stefanos Geroulanos and Maria Stavrinaki, "Editorial: Writing Prehistory," *RES: Anthropology and Aesthetics* 69–70 (2018): 1–4; (this was her argument, not mine) and more elaborately in her *Transfixed by Prehistory*, esp. 177–212.

32. Alfred H. Barr, Jr. "Preface," in Leo Frobenius and D. C. Fox, *Prehistoric Rock Pictures in Europe and Africa* (New York: Museum of Modern Art, 1937), 8–9.

33. André Malraux, *Anti-Memoirs* (1967; New York: Holt, Rinehart and Winston, 1968), 400, translation amended.
34. Only Max Raphaël, a German-Jewish art historian working in the US, avoided it: "Although the cave paintings appear modern," he declared, "in reality, there is no art more distant and alien to us." Max Raphaël, *Prehistoric Cave Paintings* (New York: Bollingen, 1945), 1.
35. Annette Laming, "Art and Magic," in *The Lascaux Cave Paintings* by Fernand Windels (London: Faber and Faber, 1948), 49.
36. See also the deer on the top right of the cover of Josef Augusta, *U pravěkých lovců*, ill. Zdeněk Burian (Prague: Statní Pedagogické Nakladatelství, 1971). It is clearly a copy from Breuil, and a shaman dominates the rest of the cover.
37. Augusta, *U pravěkých lovců*, 51, 71.
38. For example, in the work of one of the most visible current propagandists of prehistory, Jean Clottes, in his coauthored (with David Lewis-Williams) *The Shamans of Prehistory: Trance and Magic in the Painted Caves* (New York: Abrams Books, 1998).
39. Horst Kirchner, "Ein archäologischer Beitrag zur Urgeschichte des Schamanismus," *Anthropos* 47, no. 1–2 (1952): 244–86.
40. See Johannes Maringer, *The Gods of Prehistoric Man* (1956; New York: Knopf, 1960), 62, 141, 197; Sigfried Giedion, *The Eternal Present: The Beginnings of Art* (New York: Pantheon Books, 1962), 507–8. This interpretation goes much further, from Andreas Lommel's work on the "beginning of art" and on early hunters to John A. Grim's *The Shaman: Patterns of Religious Healing among the Ojibway Indians* (Norman, OK: University of Oklahoma Press, 1988).
41. Mircea Eliade, *Shamanism: Archaic Techniques of Ecstasy* (1951; Princeton, NJ: Princeton University Press, 2004), 503–4.
42. Eliade, *Shamanism*, 11.
43. Eliade, *Shamanism*, 378.
44. Breuil printed this image, over and over, including in *Quatre cents siècles d'art pariétal* (1952), i.e., two years after his drawing of the shamanistic ritual (Figure 16.9).
45. See Stefanos Geroulanos, "Polyschematic Prehistory at the Dusk of Colonialism," *RES: Anthropology and Aesthetics* 69–70 (2018): 136–57.
46. Pierre Teilhard de Chardin, "Some Reflections on the Spiritual Repercussions of the Atom Bomb" [1945], in *The Future of Man* (New York: Image Books, 2004): 133–42, 133. See Jacob Krell, "Genealogies of technology and prehistory in France," *RES: Anthropology and Aesthetics* 69–70 (2018): 158–72.
47. Georges Bataille, "Lecture, January 18, 1955," and "Unliveable Earth?" in *The Cradle of Humanity*, trans. Michelle Kendall and Stuart Kendall (New York: Zone, 2009), 87, 178.
48. Maria Stavrinaki, "Prehistory/Posthistory," *Power and Time*, trans. Lauren Kirk; eds. Dan Edelstein, Stefanos Geroulanos, and Natasha Wheatley (Chicago: University of Chicago Press, 2020), 201.
49. The first book with photographs was Annette Laming and Fernand Windels,

Lascaux, chapelle sixtine de la préhistoire, (Montignac: Centre d'Études et de Documentation Préhistorique, 1948). Breuil's *Four Hundred Centuries of Cave Art* (Montignac: Centre d'etudes et de documentation préhistoriques, 1952) followed a few years thereafter.

50. André Malraux, *La Création artistique* (Paris: Skira, 1948), 120.

51. The caves did not testify to some shamanic ritual, Bataille argued, though he took note of Kirchner's work. He also respectfully criticized Breuil's interpretation.

52. On Greece: Bataille, *Lascaux, ou la naissance de l'art*, 7, 15. See the discussion of this subject in Hal Foster, *Brutal Aesthetics: Dubuffet, Bataille, Jorn, Paolozzi, Oldenburg* (Princeton: Princeton University Press, 2020), chap.2, which I read after completing this book, and with which I generally agree. On Hegel, see Georges Bataille, *Oeuvres completes, Tome* X (Paris: Gallimard, 1987), 725. Also, Stuart Kendall, "The Sediment of the Possible," in *Cradle of Humanity*, 16–17.

53. Bataille, *Lascaux, ou la naissance de l'art*, 115.

54. Bataille, *Lascaux, ou la naissance de l'art*, 121.

55. Bataille, *Théorie de la religion* (Paris: Gallimard, 1973), 32, 25; Bataille, *Theory of Religion* (1948; New York: Zone, 1992), 19, 28; the first to use *Theory of Religion* to explain Bataille's discussion of cave paintings is to my knowledge Yue Zhuo, "Alongside the Animals," *Yale French Studies* 127 (2015): 31.

56. Bataille, "The Cradle of Humanity," in *Cradle of Humanity*, 152.

57. Bataille, "The Passage from Animal to Man and the Birth of Art," in *Cradle of Humanity*, 60.

58. Bataille, "Prehistoric Religion," in *Cradle of Humanity*, 137.

59. Bataille, "A Visit to Lascaux," in *Cradle of Humanity*, 55.

60. Bataille, "The Passage from Animal to Man and the Birth of Art," 78.

61. Jackson Arn, "Inkblots on Stone," in *Lapham's Quarterly*, April 1, 2019. Allan Stoekl, *Bataille's Peak* (Minneapolis: University of Minnesota Press, 2007).

62. Annette Laming, ed., *La Découverte du passé* (Paris: Picard, 1952). Laming-Emperaire was her married name, and I use it throughout for consistency (and also as it's the name by which she is known in France).

63. Annette Laming, "Art and Magic," in *The Lascaux Cave Paintings*, by Fernand Windels, (London: Faber and Faber, 1948), 49.

64. Annette Laming-Emperaire, *La Signification de l'art rupestre paléolithique* (Paris: Picard, 1962).

65. See the even harsher expression of her views by André Leroi-Gourhan in his *Préhistoire de l'art occidental* (Paris: Mazenod, 1965). Translated *Treasures of Prehistoric Art* by Norbert Guterman (New York: Harry N. Abrams, 1967), 174. This is not to cite a man rather than Laming: it is to restore Leroi-Gourhan's ideas to her, especially as (as a junior scholar) she was less blunt in her criticisms.

66. Laming-Emperaire, "Une nouvelle approche des sociétés préhistoriques," *Annales* 24, no.5 (1969): 1261.

67. Laming-Emperaire, *La Signification de l'art rupestre paléolithique*, 275.

68. Laming-Emperaire, *La Signification de l'art rupestre paléolithique*, 287.
69. Leroi-Gourhan, *Treasures of Prehistoric Art*, 211.
70. Laming-Emperaire, *La Signification de l'art rupestre paléolithique*, 276–85.
71. Leroi-Gourhan, *Treasures of Prehistoric Art*, 174.
72. Leroi-Gourhan proceeds in a different interpretive direction: *Treasures of Prehistoric Art*, 315–16. But see the caption to Fig.74.
73. They downplayed the difference and even abandoned the finer details. Laming-Emperaire, "Une nouvelle approche des sociétés préhistoriques," 1261–69. Leroi-Gourhan, *The Dawn of European Art* (Cambridge: Cambridge University Press, 1982).
74. Margaret W. Conkey, "The Structural Analysis of Paleolithic Art," in *Archaeological Thought in America* ed. C. C. Lamberg-Karlovsky (Cambridge: Cambridge University Press, 1991), 135–54.
75. Margaret W. Conkey and Joan M. Gero, eds. "Tensions, Pluralities, and Engendering Archaeology," in *Engendering Archaeology: Women and Prehistory* (Oxford: Blackwell, 1991), 3–30.
76. Margaret W. Conkey, "The Identification of Prehistoric Hunter-Gatherer Aggregation Sites: The Case of Altamira," *Current Anthropology* 21, no. 5 (1980): 609–20.
77. Margaret W. Conkey and Lester B. Rowntree, "Symbolism and the Cultural Landscape," *Annals of the Association of American Geographers* 70, no. 4 (1980): 459.
78. For the criticisms: Randall White, Gerhard Bosinski, Raphaëlle Bourrillon, Jean Clottes, Margaret W. Conkey, Soledad Corchón Rodriguez, Miguel Cortés-Sánchez, et al., "Still No Archaeological Evidence That Neanderthals Created Iberian Cave Art," *Journal of Human Evolution* 144 (2019): 102640.
79. Clottes and Lewis-Williams, *The Shamans of Prehistory*.
80. On Herzog: Richard Baxstrom and Todd Meyers, *Violence's Fabled Experiment* (Berlin: August Verlag, 2018).

CHAPTER 17: KILLER APES FOR AN AGE OF DECOLONIZATION

1. Though I take a quite different approach, this chapter is much informed by the works of Erika L. Milam, *Creatures of Cain: The Hunt for Human Nature in Cold War America* (Princeton: Princeton University Press, 2018) and Nadine Weidman, *Killer Instinct: The Popular Science of Human Nature in Twentieth-Century America* (Cambridge, MA: Harvard University Press, 2021).
2. Raymond A. Dart, "*Australopithecus africanus*: The Man-Ape of South Africa," *Nature* 115 (February 7, 1925): 195–99.
3. Raymond A. Dart, "African Manlike Ape Skull New Link in Man's Ancestry," *The Science News-Letter* 6, no. 202 (1925): 2.
4. Dart, "African Manlike Ape Skull," 2. Hrdlička is today notorious for his racism, particularly his treatment of Indigenous remains.
5. Grafton Elliot Smith, *Human History* (London: Jonathan Cape, 1934), 65.
6. Emily Margaret Kern, "Out of Asia: A Global History of the Scientific

Search for the Origins of Humankind, 1800–1965" (PhD Dissertation, Princeton University, 2018). Ellinor Schweighöfer, *Vom Neandertal nach Afrika* (Göttingen: Wallstein, 2018).

7. See also the map of origins in Roy Chapman Andrews, *Meet Your Ancestors: A Biography of Primitive Man* (New York: Viking, 1945). Andrews reused a 1929 map by Henry Fairfield Osborn and had his own vested interest in prioritizing China, the site of his long-lasting research.

8. Raymond A. Dart, "*Australopithecus africanus*," 199.

9. See Raymond A. Dart, "The Predatory Implemental Technique of *Australopithecus*," *American Journal of Physical Anthropology* 7, no. 1 (1949): 1–38, and Dart, "The Predatory Transition from Ape to Man," *International Anthropological and Linguistic Review* 1, no. 4 (1953): 201–18.

10. Raymond A. Dart, "Recent Discoveries Bearing on Human History in Southern Africa," *Journal of the Royal Anthropological Institute* 70, no. 1 (1940): 13–27.

11. Milam, *Creatures of Cain*, 92.

12. Dart, "The Predatory Transition from Ape to Man," 204.

13. Dart, "The Predatory Transition from Ape to Man," 208. Raymond A. Dart, "The Predatory Implemental Technique of *Australopithecus*," 1–38, 15.

14. Suzanne Marchand, "Leo Frobenius and the Revolt against the West," *Journal of Contemporary History* 32, no. 2 (1997): 153–70. Joseph C. Miller, "History and Africa/Africa and History," *American Historical Review* 104, no.1 (1999): 1–32. Frobenius had also come from a diffusionist tradition, notably thanks to a quasi-mentor, the human-geographer Friedrich Ratzel.

15. Leo Frobenius, *Auf dem Weg nach Atlantis* (Berlin-Charlottenburg: Vita Deutsches Verlag-Haus, 1911).

16. H. A. Waldron, "The Study of the Human Remains from Nubia," *Medical History* 44, no. 3 (2000): 363–88.

17. Allegra Fryxell, "Tutankhamen, Egyptomania, and Temporal Enchantment in Interwar Britain," *Twentieth Century British History* 28, no. 4 (2017): 516–42.

18. In the interwar period, only Nikolai Trubetzkoy curtly denied the need for an original center: "Thoughts on the IndoEuropean Problem" (1939) in *Studies in General Linguistics and Language Structure*, ed. Anatoly Liberman (Durham, NC: Duke University Press, 2001), 87–98. Stefanos Geroulanos and Jamie Phillips, "Eurasianism versus IndoGermanism: Linguistics and Mythology in the 1930s' Controversies over European Prehistory" *History of Science* 56, no. 3 (2018): 343–78.

19. V. Gordon Childe, *Prehistoric Migrations in Europe* (Oslo: Aschehoug, 1950), 2.

20. Emily Martin, "Toward an Ethnography of Experimental Psychology" in *Plasticity and Pathology*, eds. David Bates and Nima Bassiri (New York: Fordham University Press, 2015), 12.

21. Charles G. Seligman, *The Races of Africa* (London: Butterworth, 1930). For Elliot Smith's half-endorsement of such a thesis, see his *The Ancient Egyptians and the Origins of Civilization* (London: Harper, 1911).

22. In this, I follow Saul Dubow, who has tracked Dart's racial thinking and his contributions to white supremacy in South Africa, and who outlines Dart's

frequent invocations of Elliot Smith's diffusionism and his own allegiance to it. See Saul Dubow, "Human Origins, Race Typology and the Other Raymond Dart," *African Studies* 55, no. 1 (1996): 1–30.

23. Grafton Elliot Smith, Bronisław Malinowski, Herbert Joseph Spinden, and Alexander Goldenweiser, *Culture: The Diffusion Controversy* (London: Kegan Paul, 1928).

24. Paul Rivet, *Les origines de l'homme américain* (Paris: Masson, 1925).

25. Henri Breuil, *The White Lady of the Brandberg* (London: Trianon, 1955), 12.

26. Breuil, *White Lady*, 29.

27. It's worth noting who financed Breuil's book: colonial diamond-mining companies, which also had much at stake in theories about Southern Africa. These included the Consolidated Diamond Mines of South-West Africa Limited, the Belgian Société Internationale Forestière et Minière du Congo, the British South Africa Company, and the Portuguese Companhia de Diamantes de Angola.

28. I discuss this episode in detail in "Polyschematic Prehistory at the Dusk of Colonialism" *RES: Anthropology and Aesthetics* 69–70 (2018): 136–157.

29. University of Massachusetts Amherst Libraries, Special Collections and University Archives, W. E. B. Du Bois Papers, series 1A.

30. Léopold Sédar Senghor, *Liberté III: Négritude et civilisation de l'universel* (Paris: Seuil, 1977), 398, 13.

31. See also his later text—Léopold Sédar Senghor, "Les Leçons de Léo Frobenius" *Présence Africaine*, 111 (1979): 142–51—where Senghor describes Frobenius as "the tip of the lance in our struggle for emancipation, for the restitution to us of our truth in our dignity" (150).

32. Aimé Césaire, *Discourse on Colonialism*, trans. Joan Pinkham (1950; London: Monthly Review Press, 2001), 53.

33. Fanon cited Frobenius among others for his own moment of self-recognition: Frantz Fanon, *Black Skin White Masks* (New York: Grove, 1967), 130.

34. Kwame Nkrumah, "Speech by Osagyefo the President at the Opening Session of the First Meeting of the Editorial Board of the *Encyclopaedia Africana*," September 24, 1964.

35. Cheikh Anta Diop, *The African Origin of Civilization* (New York: Lawrence Hill, 1974), 160.

36. Sigrid Schmalzer, *The People's Peking Man: Popular Science and Human Identity in Twentieth-Century China* (Chicago: University of Chicago Press, 2008).

37. Emily Kern, "Alternate Edens: History, Evolution, and the Politics of Universalism in UNESCO's Scientific and Cultural History of Mankind," *Journal of the History of Ideas* 85, no. 1 (January 2024).

38. Milam, in *Creatures of Cain*, concentrates on the paperback as a force of popular science and a major player in postwar scientific education.

39. Robert Ardrey, *African Genesis* (New York: Atheneum, 1961), 9, 12, 13.

40. Ardrey, *African Genesis*, 20.

41. Leakey published two books on the Mau Mau uprising, where he mitigated his strong opposition to them (and his work for the defeat of the uprising) with a support for significant social reforms to benefit the Kikuyu (or Gĩkũyũ): L. S. B. Leakey, *Defeating Mau Mau* (London: Methuen, 1954), and L. S. B.

Leakey, *The Mau Mau and the Kikuyu* (London: Methuen, 1953). Neither work imagined the end of the British colony that ruled over Kenya.

42. Haraway discusses the complex (to put it generously) politics of Goodall's movement to the Gombe Stream in "Donna Haraway Reads 'The National Geographic' on Primates," Paper Tiger Television (1987).

43. Ardrey, *African Genesis*, 20.

44. Konrad Lorenz, *On Aggression* (London: Methuen, 1963). My thanks to Jonas Knatz for his attention to Lorenz.

45. Vernon Mark and Frank Ervin, *Violence and the Brain* (New York: Harper and Row, 1970). I am grateful to Danielle Carr for this line of thought.

46. Robert Ardrey, *The Territorial Imperative* (New York: Atheneum, 1966), 315. Milam mentions his "sympathy" for apartheid but only at the private level: *Creatures of Cain*, 179–80.

47. Ardrey, *African Genesis*, 29.

48. See Weidman, *Killer Instinct*, esp. chap. 5.

49. David A. Hollinger, "How Wide the Circle of the 'We'? American Intellectuals and the Problem of the Ethnos since World War II," *American Historical Review* 98, no. 2 (1993): 319. Cited in Nasser Zakariya, *A Final Story: Science, Myth and Beginnings* (Chicago: University of Chicago Press, 2017), 277.

50. Jacob Bronowski, *The Ascent of Man* (1976; London: BBC, 2011), 25, 28, 29.

51. Paul F. Brain and David Benton, eds., *The Biology of Aggression: Proceedings of the NATO Advanced Study Institute on the Biology of Aggression*, July 21–30, 1980 (Alphen aan den Rijn: Sijthoff and Noordhoff, 1981).

52. Note also political scientist Martin Bernal (son of aforementioned Irish physicist J. D. Bernal), who recuperated the argument of African cultural origins to now cover the Greeks in *Black Athena: The Afroasiatic Roots of Classical Civilization*, 3 vols. (New Brunswick, NJ: Rutgers University Press, 1987–2006).

Chapter 18: Stone-Age Computers

1. Friedrich Engels, "The Part Played by Labor in the Transition from Ape to Man" (1876) in *Marx-Engels Collected Works* by Karl Marx and Friedrich Engels, (New York: International Publishers, 1987), 25:452–65, 453.

2. Many authors made reference to Lucretius as an originator of the idea of tools' importance, but these references were occasional, often even a throwaway.

3. Kenneth Oakley, *Man the Tool-Maker* (1949; London: British Museum, 1975), 1.

4. Oakley, *Man the Tool-Maker*, 1.

5. Oakley, *Man the Tool-Maker*, 2.

6. Grafton Elliot Smith, *Human History* (New York: Norton, 1929), 23, 28. Arthur Keith, *The Antiquity of Man*, second edition (Philadelphia: Lippincott, 1925), 332.

7. For example, in 1959, he credited the survival of generalized anatomical characteristics on mental capacity. Wilfrid Le Gros Clark, *Antecedents of Man* (1959; New York: Harper and Row, 1963), 7, 30.

8. Kenneth Oakley, *Frameworks for Dating Fossil Man* (London: Weidenfeld, 1964).

9. S. L. Washburn, "The Analysis of Primate Evolution with Particular Reference to the Origin of Man," in *Origin and Evolution of Man* (Cold Spring Harbor, LI: Biological Laboratory, 1951), 67.

10. Oakley, "Tools Makyth Man," *Antiquity* 31, no. 124 (1957): 207–8.

11. Oakley, *Man the Tool-Maker*, 9.

12. Oakley, "Tools Makyth Man," 208, 207.

13. Oakley, *Man the Tool-Maker*, 84.

14. Consider Hudson Hoagland and Ralph W. Burhoe, *Evolution and Man's Progress* (New York: Columbia University Press, 1962).

15. The key figure in the agricultural revolution (and, outside the Eastern bloc, the key *communisant* thinker) was V. Gordon Childe, the Australian director of the Institute for Archaeology in London.

16. Oakley, "Tools Makyth Man," 208.

17. Oakley, *Man the Tool-Maker*, 90–91.

18. L. S. B. Leakey, *Adam's Ancestors* (New York: Harper, 1960), 16.

19. Leakey, "A New Fossil Skull from Olduvai," *Nature* 184, no. 4685 (Aug.15, 1959): 491–93, 491, 493. "True man" is in Leakey, *Adam's Ancestors*, x. Oakley immediately brought it into the new, 1960 edition of *Man the Tool-Maker*.

20. Leakey, "New Fossil Skull," 491. For Leakey's mythologization, see *Dr.Leakey and the Dawn of Man*, National Geographic Specials, S1966, Episode 3, directed by Guy Blanchard, written by Nicolas Noxon, aired October 25, 1966.

21. L. S. Leakey, P. V. Tobias, and J. R. Napier, "A New Species of the Genus Homo from Olduvai Gorge," in *Nature* 202 (April 4, 1964): 7–9, 9.

22. Quoted in Erika L. Milam, *Creatures of Cain: The Hunt for Human Nature in Cold War America* (Princeton: Princeton University Press, 2019), 95.

23. Marshall McLuhan, *Understanding Media: The Extensions of Man* (New York: McGraw Hill, 1964). Better still, they were a Derridean supplement: what completes but therefore resets a "thing itself."

24. The title of Le Gros Clark's book, with the teleological triumph reserved for humanity.

25. Huxley and Haddon, in their 1936 *We Europeans*, had privileged a linear development of technological stages, the bulk of which worked in or through Europe. See also Henri Breuil, in *Beyond the Bounds of History* (London: P.R. Gawthorn, 1949), especially regarding *Sinanthropus*.

26. Sherwood L. Washburn, "Tools and Human Evolution," *Scientific American*, 203, no. 3 (September 1, 1960): 62–75, 63.

27. Robert Ardrey discussed Washburn in *The Hunting Hypothesis* (New York: Atheneum, 1976).

28. This showed why race played no role. Washburn, "The New Physical Anthropology," *Transactions of the New York Academy of the Sciences* 13, no. 7 (1951): 300.

29. Washburn, "New Physical Anthropology," 299.

30. Washburn, "New Physical Anthropology," 300.
31. S. L. Washburn, "Speculations on the Interrelations of the History of Tools and Biological Evolution," *Human Biology* 31, no.1 (February 1959): 21–31, especially 25, 25n.1. Like Piltdown, Washburn claimed, the big brain was the fault of English- and French-educated scientists living in long-distance colonial powers. Scientists like Weidenreich or Hrdlička who had trained in the more diverse Central Europe and then left Europe altogether did not care for skull sizes. That claim is specious—to begin with, Weidenreich and Hrdlička had trained in France, not in the Habsburg Empire. It is hence all the more illustrative of Washburn's need to show that evolution did not require the skull. See also his *Anthropology 1* lectures (University of California, Berkeley, Fall 1964). Bancroft Library, S. L. Washburn Papers, BANC MSS 98/132c [henceforth "Washburn Papers"], 3:83, *Anthropology 1* (1964), lecture 15, 18.
32. Franz Weidenreich, *Apes, Giants, and Man* (Chicago: University of Chicago Press, 1946), 21.
33. Weidenreich, *Apes, Giants, and Man*, 23, 100.
34. This was a change from a few years earlier. In a 1941 essay Weidenreich had endorsed seeing the brain's changes present at each stage of the skeleton's evolutionary transformation: Weidenreich, "The Brain and Its Rôle in the Phylogenetic Transformation of the Human Skull," *Transactions of the American Philosophical Society* 31, no. 5 (1941): 320–442.
35. Weidenreich, *Apes, Giants, and Man*, 111, 98; Franz Weidenreich, "The Trend in Human Evolution," in *Evolution: International Journal of Organic Evolution* 1, no. 4 (December 1947): 221–36.
36. Washburn, "The Analysis of Primate Evolution," 76, citing Weidenreich, "The Trend of Human Evolution."
37. Norbert Wiener, *Cybernetics: Control and Communication in the Animal and the Machine* (New York: Wiley, 1947), 107, 169, 36, and elsewhere.
38. Relevant here is Jurgen Ruesch and Gregory Bateson, *Communication: The Social Matrix of Psychiatry* (New York: Norton, 1951).
39. Wilder Penfield and Theodore Rasmussen, *The Cerebral Cortex of Man* (New York: Macmillan, 1950).
40. Washburn, "Speculations . . .," 27–28.
41. Washburn, "Speculations . . .," 28. See also "Tools and Human Evolution," 69.
42. Washburn Papers, 3:83, *Anthropology 1* (1964), lecture 13, 11.
43. Washburn, "Speculations . . .," 26–31. On the mother carrying the baby as an early acquisition, see Washburn Papers 3:83, *Anthropology 1* (1964), lecture 12, 16.
44. Washburn to Oakley, May 12, 1964, in Washburn Papers, cont.1.
45. Washburn, "Tools and Human Evolution," 67.
46. Paul Rivet, ed., *L'Espèce humaine*, in *L'Encyclopédie française*, vol. 7 (Paris: *Encyclopédie française*, 1937).
47. André Leroi-Gourhan, *Gesture and Speech* (1964–66; Cambridge, MA: MIT Press, 1993), 114.

48. Université de Nanterre, Bibliothèque Eric-de-Dampière, Fonds André Leroi-Gourhan, ALG 131/4, "Cours sur le theme: l'évolution humaine," 5, 7.

49. Leroi-Gourhan, "Cours sur le theme: l'évolution humaine," 5–7. See also Leroi-Gourhan, "Libération de la main" *Problèmes: Revue de l'Association des étudiants en médecine de l'Université de Paris* 32 (1956): 6–9.

50. Leroi-Gourhan, *Cours de technologie 1955–1956* (Paris: Institut d'ethnologie, 1956), 19.

51. Leroi-Gourhan, *Gesture and Speech*, 235, 148.

52. Leroi-Gourhan, *Gesture and Speech*, 148.

53. Leroi-Gourhan, *Cours de technologie 1955–1956*, 12.

54. Leroi-Gourhan, *Gesture and Speech*, 290.

55. Leroi-Gourhan, *Gesture and Speech*, 106–7, 114, 145–46.

56. André Leroi-Gourhan, *Mécanique vivante* (Paris: Fayard, 1983), 185.

57. Université de Nanterre, Bibliothèque Eric-de-Dampière, Fonds André Leroi-Gourhan, ALG 130/2: *Cours de technologie 1955–1956*, 7–8.

58. Washburn, "Tools and Human Evolution," 65, also 75.

59. See Leroi-Gourhan's criticism of Teilhard, in *Gesture and Speech*, 361. He planned a further criticism in a "Teilhard lecture" that he apparently never delivered. Collège de France, Fonds Leroi-Gourhan, 76 CdF 17a.

60. Leroi-Gourhan, *Gesture and Speech*, 184–85.

61. Leroi-Gourhan, *Gesture and Speech*, 129.

62. Gregory Bateson, "The Role of Somatic Change in Evolution," in *Evolution* 17 (December 1963): 530. See also Gregory Bateson, *Steps toward an Ecology of Mind* (New York: Ballantine Books, 1972).

63. Clifford Geertz, *The Interpretation of Cultures* (New York: Basic Books, 1973), 44–45.

64. I've described this at length: Stefanos Geroulanos, *Transparency in Postwar France* (Stanford: Stanford University Press, 2017), chap. 19.

65. In *A Thousand Plateaus* (1980) they also followed Leroi-Gourhan in declaring the body *alloplastic*, as constructing itself by engaging its other. Gilles Deleuze and Félix Guattari, *A Thousand Plateaus* (Minneapolis: University of Minnesota Press, 1987), 60.

66. Donna Haraway, "Remodeling the Human Way of Life: Sherwood Washburn and the New Physical Anthropology," in *Bones, Bodies, Behavior*, ed. George W. Stocking, Jr. (Madison: University of Wisconsin Press, 1988), 206–59.

67. Consider even just the concerns expressed in book titles: in the United States, the shift from Daniel Bell's *Work and Its Discontents: The Cult of Efficiency in America* (1956) to *The Coming of Post-Industrial Society* (1973); in France, from Georges Friedmann's *Where Is Human Labor Going?* (1950) to André Gorz's *Goodbye to the proletariat* (1980).

68. Anson Rabinbach, *The Eclipse of the Utopias of Labor* (New York: Fordham University Press, 2018), esp. chap. 7; for some recent discussions, see Phil Jones, *Work without the Worker: Labour in the Age of Platform Capitalism* (London: Verso, 2021) and Aaron Benanav, *Automation and the Future of Work* (London: Verso, 2022).

CHAPTER 19: THE BIRTHS AND ENDS OF PATRIARCHY

1. Stephen Oppenheimer, *The Real Eve: Modern Man's Journey Out of Africa* (New York: Basic Books, 2003). André Leroi-Gourhan, *The Hunters of Prehistory*, trans. Claire Jacobson (New York: Atheneum, 1989), 5.

2. Of this long tradition, see most recently Priscille Touraille, *Hommes grands, femmes petites: Une évolution coûteuse* (Paris: Maison des sciences de l'homme, 2018).

3. Elaine Morgan, *The Descent of Woman* (New York: Stein & Day, 1972), 10.

4. Morgan, *The Descent of Woman*, 4.

5. Claudine Cohen and Lia Marcondes, *Eau et féminismes: Petite histoire croisée de la domination des femmes et de la nature* (Paris: Dispute, 2010). Erika L. Milam, "Old Woman and the Sea: Evolution and the Feminine Aquatic," in *Osiris*, vol. 34: *Presenting Futures Past: Science Fiction and the History of Science*, eds. Iwan Morus and Amanda Rees (Chicago: University of Chicago Press, 2019), 198–215.

6. Jacquetta Hawkes, *Prehistory and the Beginning of Civilization*, vol.1 of *History of Mankind: Cultural and Scientific Development* (London: Allen and Unwin for UNESCO, 1963).

7. Emily Kern, "Alternate Edens: History, Evolution, and the Politics of Universalism in UNESCO's Scientific and Cultural History of Mankind," *Journal of the History of Ideas*, 85.1 (January 2024).

8. Charlene Spretnak, ed., *The Politics of Women's Spirituality: Essays by Founding Mothers of the Movement* (New York: Doubleday, 1982).

9. Sherwood Washburn and Charles S. Lancaster, "The Evolution of Hunting," in *Man the Hunter* eds. Richard Lee and Irven DeVore (Hawthorne, NY: DeGruyter, 1968), 301.

10. V. Gordon Childe, *The Dawn of European Civilization*, 6th ed. (New York: Vintage, 1958), 1–15.

11. Erika Lorraine Milam, "Elaine Morgan and the Aquatic Ape," *The Guardian*, May 13, 2013.

12. Brownmiller, *Against Our Will* (1975; New York: Fawcett Columbine, 1993), 14.

13. Brownmiller, *Against Our Will*, 14–16. Reviewers agreed: "The Sexes: Revolt against RAPE," *Time* (Oct. 13, 1975).

14. Marija Gimbutas, *The Goddesses and Gods of Old Europe* (Berkeley: University of California Press, 1974).

15. Riane Eisler's bestselling *The Chalice and the Blade* (1987) retained the idea of prehistoric Goddess-worship from Gimbutas, while Gerda Lerner's *The Creation of Patriarchy* (1986) dated the creep of male power to the period between 3000 BCE and 600 BCE, the Bronze Age, much as Gimbutas and Stone had argued.

16. Nancy Makepeace Tanner, *On Becoming Human* (Cambridge: Cambridge University Press, 1981).

17. Adrienne L. Zihlman, "Women in Evolution, Part II," *Signs* 4, no. 1 (1978):

4–20; see also the portrait of Zihlman in *Discover* magazine (December 1991). For a related argument, see Jane F. Collier and Michelle Z. Rosaldo, "Politics and Gender in Simple Societies," in *Sexual Meanings*, eds. Sherry B. Ortner and Harriet Whitehead (Cambridge: Cambridge University Press, 1981), 275–329.

18. Richard E. Leakey, *The Making of Mankind* (New York: Dutton, 1981), 52–53. This quote is from the series' accompanying book and it does not quite capture Leakey's tone, which is a hint more suspicious of feminism.

19. Margaret W. Conkey and Janet D. Spector, "Archaeology and the Study of Gender," *Advances in Archaeological Method and Theory* 7 (1984): 5.

20. Conkey and Spector, "Archaeology and the Study of Gender," 9.

21. Conkey and Spector, "Archaeology and the Study of Gender," 12–13.

22. It is worth recalling that Lucy was the subject of a long debate between Don Johanson and Richard E. Leakey. For a counter-position to Johanson's, which proclaimed Lucy's centrality to the human story, see Richard E. Leakey and Roger Lewin, *Origins: What New Discoveries Reveal about the Emergence of Our Species and Its Possible Future* (London: MacDonald and Jane's, 1977), 90–91.

23. The classic paper on African origins out of a single population is L. Vigilant, M. Stoneking, H. Harpending, K. Hawkes, and A. C. Wilson, "African Populations and the Evolution of Human Mitochondrial DNA," *Science* 253, no. 5027 (September 1991): 1503–7.

24. Claudine Cohen, *La femme des origines: Images de la femme dans la préhistoire occidentale* (Paris: Belin-Herscher, 2006).

25. Alice Echols, *Daring to Be Bad: Radical Feminism in America 1967–75*, 30th anniversary ed. (Minneapolis: University of Minnesota Press, 2019), 253, 165.

26. Juliet Mitchell, *Psychoanalysis and Feminism* (London: Allen Lane, 1974), 365.

27. Mitchell, *Psychoanalysis and Feminism*, 362.

28. Mitchell, *Psychoanalysis and Feminism*, 416.

29. Mitchell, *Psychoanalysis and Feminism*, 361.

30. Mitchell, *Psychoanalysis and Feminism*, 364.

Chapter 20: Is Violence Ingrained, and How?

1. Charles Darwin, *The Descent of Man* (London: Murray, 1871), 1:100–1.

2. See also Nadine Weidman, *Killer Instinct: The Popular Science of Human Nature in Twentieth-Century America* (Cambridge, MA: Harvard University Press, 2021), chap. 4.

3. J. D. Bernal, *The Freedom of Necessity* (London: Routledge and Kegan Paul, 1949), 38.

4. Steven Pinker, *The Better Angels of Our Nature* (New York: Penguin, 2011), 671. For a thorough demolition of Pinker's thesis, see Nicolas Guilhot's review of Pinker's *Enlightenment Now*, on H-Diplo, the Robert Jervis International Security Studies Forum on July 4, 2018, and R. Brian Fergu-

son, "Pinker's List" in *War, Peace, and Human Nature*, ed. Douglas P. Fry (Oxford: Oxford University Press, 2013), 112–31.

5. Jared Diamond, *The World Until Yesterday* (New York: Viking, 2012).

6. E. O. Wilson, *The Meaning of Human Existence* (New York: Liveright, 2014), 176–77.

7. See the brilliant discussion by Inga Clendinnen, "'Fierce and Unnatural Cruelty': Cortés and the Conquest of Mexico." *Representations* 33 (Winter 1991): 65–100.

8. John Lubbock, *Pre-historic Times as Illustrated by Ancient Remains and the Manners and Customs of Modern Savages* (1871; New York: Appleton and Company, 1872), 459, 466, 521, 540.

9. "Science: Environmentalist," in *Time*, May 11, 1936.

10. Bronisław Malinowski, "Anthropological Analysis of War," *American Journal of Sociology* 46, no. 4 (1941): 521–50.

11. Raphael Lemkin, "Genocide—A Modern Crime," *Free World* 9, no. 4 (1945): 39–43.

12. Ruth Benedict, *Patterns of Culture* (1934; Boston: Mentor, 1959), 32.

13. Harry Holbert Turney-High, *Primitive War* (Columbia, SC: University of South Carolina Press: 1949). Bronowski, in *The Face of Violence* (1951), presented war as a kind of anarchy and revolt intrinsic to human existence, something "primitive societies" had managed to control. Today it was science's turn to prevent. Jacob Bronowski, *The Face of Violence: An Essay with a Play* (London: Turnstile, 1954).

14. "The Eternal Apprentice," *Time*, Monday, November 8, 1948.

15. Nicolaas Tinbergen, "On War and Peace in Animals and Man," *Science* 160, no. 3835 (June 1968): 1411–18. My thanks to Jonas Knatz for his discussion of Tinbergen.

16. Hannah Arendt, "Reflections on Violence," *Journal of International Affairs* 23, no. 1 (1969): 23.

17. Jane Goodall, *The Chimpanzees of Gombe: Patterns of Behavior* (Cambridge, MA: Harvard University Press, 1986).

18. Jean-Paul Sartre, "Preface" in *The Wretched of the Earth*, by Frantz Fanon, trans. Constance Farrington (New York: Grove Press, 1963), 21. See also Fanon's argument, esp. p. 77.

19. E.g., John Marshall's film, *The Hunters*, about !Kung men hunting (1957) or Timothy Asch and Napoleon Chagnon's series about the Yanomamö (1968–1976).

20. The Hubula and this situation in particular were subjects of books by Jan Broekhuijse, Karl Heider, and Peter Mathiessen (the latter two worked with Gardner). See especially the film by Veronika Kusumaryati and Ernst Karel, "Expedition Content and the Harvard Peabody Expedition to Netherlands New Guinea, 1961," in *MAST: The Journal of Media Art Study and Theory* 2, no.2 (2021): 15–25.

21. Richard Lee and Irven DeVore, eds. *Man the Hunter* (Hawthorne, NY: DeGruyter, 1968), 301.

22. M. I. Budyko, "On the Causes of the Extinction of Some Animals at the End

of the Pleistocene." *Soviet Geography* 8, no. 10 (1967): 783–93. Paul S. Martin, "The Discovery of America," *Science* 179, no. 4077 (March 9, 1973): 969–74. On Martin and overkill, see Melissa Charenko, " 'American Blitzkrieg' or 'Ecological Indian?,' " in *New Earth Histories: Geo-Cosmologies and the Making of the Modern World*, eds. Alison Bashford, Adam Bobbette, and Emily Kern (Chicago: University of Chicago Press, 2023).

23. See Burian's painting on "intertribal clashes" in Josef Wolf and Zdeněk Burian, *The Dawn of Man*, edited by Peter Andrews et al. (New York: Harry Abrams Publishers, 1978), 129.

24. Napoleon Chagnon, *Yanomamö: The Fierce People*, 3rd ed. (1968; New York: Holt, Reinhart and Winston, 1983), 7. I will not replay here the debate over Chagnon's methods.

25. Arendt, "Reflections on Violence," 23n36.

26. André Leroi-Gourhan, *Gesture and Speech* (1964; Cambridge, MA: MIT Press, 1993), 167. He continued: "The outer shell of the farmer is still the same as the mammoth-slayers, but the economic system that made the farmer a producer of resources also made the farmer, by turns, hunter and prey."

27. Claude Lévi-Strauss, *Tristes tropiques* (1955; London: J. Cape, 1973), 38, translation amended.

28. Lévi-Strauss, *Tristes tropiques*, 390.

29. As Jacques Derrida soon pointed out, in an earlier report on his ethnography of the Nambikwara, Lévi-Strauss had offered a rather less pathetic image of the Nambikwara: Geroulanos, *Transparency in Postwar France*, 267–82.

30. Sydney G. Margolin, "A Consideration of Constitutional Factors of the Aggressivity of an Indian Tribe" (1960/62), *American Indian and Alaska Native Mental Health Research* 2, no.2 (1988): 52. I am grateful to Jonas Knatz for pointing out to me Konrad Lorenz's reliance on Margolin.

31. For an early psychoanalytic critique of *Totem and Taboo*, see Géza Róheim, "Introduction," in *Psychoanalysis and the Social Sciences* 1 (1947): 9–33. For a 1969 state-of-the-field publication regarding psychoanalytic and sociological criticisms, see Warner Muensterberger, *Man and His Culture* (New York: Taplinger, 1969).

32. Camille Robcis, "Frantz Fanon, Institutional Psychotherapy, and the Decolonization of Psychiatry," *Journal of the History of Ideas* 81, no. 2 (2020): 303–25. For context, see Robcis, *Disalienation, Politics, Philosophy, and Radical Psychiatry in Postwar France*, (Chicago: University of Chicago Press, 2021).

33. George Devereux, *Reality and Dream* (New York: International Universities Press, 1951); Devereux, *Mohave Ethnopsychiatry and Suicide* (Washington, DC: Government Printing Office, 1961).

34. Marshall Sahlins, "La première société d'abondance," *Les Temps modernes* 268 (October 1968): 239–40.

35. Consider Simone de Beauvoir's discussion of starvation in China and India in her interview with Wilfrid Lemoine for Radio-Canada (1959).

36. Marshall Sahlins, *Stone Age Economics* (Chicago: Aldine-Atherton 1972), 37.

37. Sahlins, *Stone Age Economics*, 182.

38. Clastres, *Society against the State* (1974; New York: Zone Books, 1989), 208–9.

39. Clastres, *Society against the State*, 207.

40. R. Brian Ferguson and Neil L. Whitehead, *War in the Tribal Zone: Expanding States and Indigenous Warfare* (Santa Fe, NM: School of American Research Press, 1992).

41. Lawrence H. Keeley, *War Before Civilization* (Oxford: Oxford University Press, 1996), 22–23.

42. Raymond C. Kelly, *Warless Societies and the Origins of War* (Ann Arbor, MI: University of Michigan, 2000).

43. Raymond C. Kelly, "The Evolution of Lethal Intergroup Violence," *Proceedings of the National Academy of Sciences* 102, no. 43 (2005): 15294–98. This extends much of the argument of his *Warless Societies*.

44. Ruggero Deodato's film *Cannibal Holocaust* (1980), featured the Yanomami as cannibals forever at war against sworn tribal enemies, ready to devour outsiders who offended them. *Cannibal Holocaust*, Deodato's second film in the cannibal exploitation genre—and yes, that was a thing—relished in gory images of rape, murder, and cannibalism, all the while denouncing Western violence for starting it all. Perversely, it is the best, the most indicative of all these documents. Pornographic in its joyful depiction of violence "from both sides," it leaves the viewer feeling superior, like we get the secret joke.

45. James C. Scott, *Against the Grain: A Deep History of the Earliest States* (New Haven, CT: Yale University Press, 2017).

46. Spencer Wells, *Pandora's Seed: Why the Hunter-Gatherer Holds the Key to Our Survival* (New York: Random House, 2011). Kim Sterelny, "How Equality Slipped Away," *Aeon* (June 10, 2021). Through his manifesto, Ted Kaczynski ("the Unabomber") also belongs within this discourse.

47. See the documentary *Cannibal Tours* (Dennis O'Rourke, 1988).

48. Karl G. Heider, *Grand Valley Dani: Peaceful Warriors* (Fort Worth, TX: Harcourt Brace College Publishers, 1997).

49. Moriah Balingit, "The Yanomami Are Dying of Malaria and Malnutrition. Is It Genocide?" in *Washington Post*, April 18, 2023. Jack Nicas, "The Amazon's Largest Isolated Tribe is Dying," *New York Times*, March 25, 2023.

Epilogue: A Storm Blowing from Paradise

1. Jean Clottes and David Lewis-Williams, *The Shamans of Prehistory: Trance and Magic in the Painted Caves* (New York: Harry N. Abrams, 1998).

2. Ewen Callaway, "Oldest *Homo sapiens* Fossil Claim Rewrites our Species' History," *Nature*, June 7, 2017.

3. See some of the discussion in Robert N. Proctor, "Three Roots of Human Recency: Molecular Anthropology, the Refigured Acheulean, and the UNESCO Response to Auschwitz," *Current Anthropology* 44, no. 2 (April 2003): 213–39.

4. Recently, the Mitochondrial Eve seems to have given way to discussions about "ghost" hominid populations that must have existed for DNA to be influenced—even though we know as of yet, nothing about them at present,

nor how this DNA influenced humanity. Carl Zimmer, "Ghost DNA Hints at Africa's Missing Ancient Humans," *New York Times*, February 12, 2020.

5. On neuromarketing and the reptilian brain, see Sebastien Lemerle, *Le cerveau reptilie: Sur la popularité d'une erreur scientifique* (Paris: CNRS, 2021), 8.

6. Joe Klein, "Donald Trump's Lizard Brain," *Time*, February 18, 2016.

7. Team Coco, "Deepak Chopra: Donald Trump 'Thinks with His Penis,'" YouTube: Conan on TBS, October 24, 2016. "Trump's Tweets Are a 'Narnian Wardrobe to His Lizard Brain,'" continued Jeffrey Goldberg in *The Atlantic* two years later (January 2, 2018).

8. Stephen Mithen, *The Prehistory of the Mind* (London: Thames and Hudson, 1996).

9. "If we are dealing with a different type of mentality, as I firmly believe, then the possession of modern language is an obvious candidate for 'what made the difference.' But so too is the possession of a particular creative intelligence." Stephen Mithen, ed, "Introduction to Part II" in *Creativity in Human Evolution and Prehistory* (London: Routledge, 1998), 67.

10. Mithen, "A Creative Explosion?," in *Creativity in Human Evolution and Prehistory*, 135, 127.

11. On creativity and its history, see Jamie Cohen-Cole, *The Open Mind: Cold War Politics and the Sciences of Human Nature* (Chicago: University of Chicago Press, 2014), and Louis Menand, "Inspiration, Inc.: How 'Creativity' Was Created," *New Yorker* (April 24 and May 1, 2023).

12. Richard Wrangham, *Catching Fire: How Cooking Made Us Human* (London: Profile Books, 2009); the original publication was Richard W. Wrangham, James Holland Jones, Greg Laden, David Pilbeam, and NancyLou Conklin-Brittain, "The Raw and the Stolen: Cooking and the Ecology of Human Origins," *Current Anthropology* 40, no. 5 (1999): 567–94.

13. V. S. Ramachandran, *The Tell-Tale Brain* (New York: W. W. Norton, 2011), 242. For a broader criticism of neuroaesthetics see Matthew Rampley, *The Seductions of Darwin: Art, Evolution, Neuroscience* (University Park, PA: Penn State Press, 2017).

14. Ramachandran, *Tell-Tale Brain*, 243.

15. Italo Calvino, *Invisible Cities* (London: Harcourt Brace Jovanovich, 1974), 54.

IMAGE CREDITS

14 Joseph-François Lafitau, *Mœurs des sauvages ameriquains, comparées aux mœurs des premiers temps* (Paris, 1724), frontispiece.

17 *Nouveaux Voyages de Mr. Le Baron de Lahontan dans l'Amerique septentrionale* (Paris, 1703), frontispiece.

22 Thomas Hobbes, *De Cive* (Paris, 1642), frontispiece. Public domain, courtesy of the Houghton Library, Harvard University *EC65 H6525 642e.

28 Philipp Clüver, *Germaniae antiquae libri tres* (Leiden: Lugduni Batauorum/Apud Ludouicum Elzevirium, 1616).

34 Emmanuel, comte de las Cases, *Atlas historique, généalogique, chronologique, et géographique* (Brussels: de Mat, 1827).

41 Gabriel de Mortillet, *Formation de la nation française: Textes, linguistique, palethnologie, anthropologie* (Paris: F. Alcan, 1897).

44 Pierre Boitard, "Fossil Man," in *Le Magasin Universel 5* (April 1838): 27.

48 John Ruskin, *Modern Painters V* (New York: Lovell, 1885), plate opposite p. 321.

50 By permission of the National Museum Cardiff.

51 "Lost animals" from Félix-Edouard Guérin-Méneville, *Dictionnaire pittoresque d'histoire naturelle et des phénomènes de la nature*, vol.1 (Paris: au Bureau de souscription, 1833–1834).

54 Charles Lyell, *Principles of Geology: Being an Attempt to Explain the Former Changes of the Earth's Surface by Reference to Causes Now in Operation* (London: John Murray, 1830).

62 John Lubbock, *Scientific Lectures* (London: Macmillan, 1879), 147.

71 Louis Figuier, *L'homme primitif*, second edition (Paris: Hachette, 1870), plate opposite p. 232.

74 In Charles Pinsard, "La première hache trouvée à St. Acheul. L'ouvrier montre du doigt la hache engagée dans la masse de cailloux" in *Les rues d'Amiens. Notes sur l'histoire et la topographie d'Amiens*, T.43, cote MS_1370_E, p.34. © and courtesy of the Bibliothèques d'Amiens Métropole.

76 Carl von Linné, Gottfried Kiesewetter, and Lars Salvius, *Amoenitates aca-*

demicae, seu dissertationes variae physicae, medicae, botanicae / nunc collectae et auctae . . . Accedit Hypothesis nova de felvium intermittentium causa. (Erlangen: J. J. Palm, 1760), opposite p. 76.

78 Thomas Henry Huxley, *Evidence as to Man's Place in Nature* (London: Williams and Norgate, 1863).

79 Jacques Boucher de Perthes, *De l'homme antidéluvien et de ses œuvres* (Paris: Jung-Treuttel, 1860).

82 August Schleicher, "Die ersten Spaltungen des indogermanischen Urvolkes," in *Allgemeine Monatsschrift für Wissenschaft und Literatur* (September 1853), 787.

84 Ernst Haeckel, *Natürliche Schöpfungsgeschichte* (Berlin: Reimer, 1868), plate VIII.

85 Friedrich Max Müller, *Lectures on the Science of Language* (London: Longman, Green, Longman and Roberts, 1862), 400.

94 Lewis H. Morgan, *Systems of Consanguinity and Affinity of the Human Family* (Washington, DC: Smithsonian Institution, 1871), Plate VII.

109 P. P. Efimenko, *Pervobytnoe obshchestvo: ocherki po istorii paleoliticheskogo vremeni. Vtoroe dopolnennoe i pererabotannoe izdanie* [Primitive Society: Sketches on the History of Paleolithic Time], 2nd ed., expanded and revised, (Leningrad: Gos. sotsial'no-ekonomicheskoe izdatel'stvo, 1938).

111 Arthur Evans, *The Palace of Minos* (London: Macmillan, 1921–35), 3:456.

114 Ernst Haeckel, *Natürliche Schöpfungsgeschichte. Gemeinverständliche wissenschaftliche Vorträge über die Entwickelungslehre im allgemeinen und diejenige von Darwin, Goethe und Lamarck im besonderen*, 8th ed. (1867; Berlin: George Reimer, 1889), plate 20.

130 F. Clark Howell and the Editors of Time-Life Books, *Early Man* (New York: TIME-LIFE Books, 1965), 41–45. © and courtesy of The Zallinger Family, LLC.

132 F. Clark Howell and the Editors of Time-Life Books, *Early Man* (New York: Time-Life Books, 1965), 41–45. © and courtesy of The Zallinger Family, LLC.

133 John Lubbock, *Pre-historic Times* (London: Williams and Norgate, 1865), 331.

134 Hermann Schaafhausen, *Der Neandertaler Fund* (Bonn: Universitäts-Buchdruckerei von Carl Georgi, 1888), 34.

135 Charles Lyell, *Geological Evidences for the Antiquity of Man* (London: John Murray, 1863), 82–83.

136 *Harper's Weekly* 17, no. 864 (July 19, 1873), 1.

138 Paul Jamin, *Fuite devant le mammouth* [Flight from the Mammoth], 1885, oil on canvas. Muséum national d'histoire naturelle, OA 134. © MNHN - JC Domenech, photo reframed.

139 "Femme de la race de Néandertal, in Aimé Rutot, *Un essai de reconstitution plastique des races humaines primitives* (Brussels: Hayez, 1919), plate opposite p. 60.

140 (Top) "An Ancestor: The Man of Twenty Thousand Years Ago. The Man of La Chapelle-Aux-Saints: An Accurate Reconstruction of the Prehistoric Cave-Man Whose Skull Was Found in the Department of Corrèze."

Illustrated London News (27 February 1909): 312–13. (Bottom) Amédée Forestier, "Not in the 'Gorilla' Stage: The Man of 500,000 Years Ago," *Illustrated London News* (May 27, 1911), 779.

141   (Left) Courtesy of the "Cesare Lombroso" Museum of Criminal Anthropology, University of Turin; photograph by R. Goffi. (Right) H. G. Wells, *The Outline of History* (London: Cassell and Company, 1920), 80.

142   Courtesy of the American Museum of Natural History.

143   The sculpture by Frederick Blaschke was made in 1927. Oliver C. Farrington and Henry Field, *Neanderthal (Mousterian) Man* (Chicago: Field Museum of Natural History, 1929), plate VI, opposite p.194.

146   Carleton Coon, *The Races of Europe* (New York: Macmillan, 1939), 24.

148   Zdeněk Burian, *Spears* (1952). Credit: Zdeněk Burian/Shutterstock.com. Shutterstock ID: 12472469a.

149   (Top) Private Collection/Bridgeman Images IL309350. © 2023 Artists Rights Society (ARS), New York / OOA-S, Prague. (Bottom) Private Collection/Bridgeman Images LSE4080457. © 2023 Artists Rights Society (ARS), New York / OOA-S, Prague.

153   Based on the series designed and produced by Albert Barillé, with drawings by Jean Barbaud, colors by Afroula Hadjiyannakis, and backgrounds by Bernard and François Fiévé. www.hellomaestro.fr TM and ©1978Procidis and ©Procidis2023.

154   TM and Copyright © 20th Century Fox Film Corp. All rights reserved. Courtesy: Everett Collection.

155   © Photo Sylvain Entressangle, Reconstructions Elisabeth Daynes/ LookatSciences.

160   Ernst Haeckel, *The Evolution of Man*, 5th ed. (London: Watts and Co., 1910), frontispiece.

164   William K. Gregory, *Our Face from Fish to Man: A Portrait Gallery of Our Ancient Ancestors and Kinsfolk Together with a Concise History of Our Best Features* (London: Putnam's Sons, 1929), frontispiece.

166   Paul Jamin, *Le Rapt à l'âge de pierre* (1888).

171   George W. Crile, *A Mechanistic View of War and Peace* (New York: Macmillan, 1915), frontispiece.

176   Ferdinand Gaillard, engraving after Jean-Auguste-Dominique Ingres, *Oedipe* (1876). National Gallery of Art/Wikimedia Commons.

181   Courtesy of the Freud Museum, London.

187   From *The Red Book* by C. G. Jung, edited by Sonu Shamdasani, translated by Mark Kyburz, John Peck, and Sonu Shamdasani. Copyright © 2009 by the Foundation of the Works of C. G. Jung. Translation copyright © 2009 by Mark Kyburz, John Peck, and Sonu Shamdasani. Used by permission of W. W. Norton & Company, Inc.

194   Alfred Cort Haddon, *The Wanderings of Peoples* (Cambridge: Cambridge University Press, 1911).

199   French postcard (1905).

200   Karl Haushofer, *Raumüberwindende Mächte* (Leipzig: B. G. Teubner, 1934), 42.

202   "The Tracks of Various Transmigrating and Raiding Peoples Between 1AD

and 700AD," in H. G. Wells, *The Outline of History* (London: Macmillan, 1921), 483.

206   Courtesy of Cornell University—PJ Mode Collection of Persuasive Cartography. Republished in Alfred Vogel, *Erblehre und Rassenkunde für die Grund- und Hauptschule* (Stuttgart, 1937).

212   Dietrich Klagges, Geschichtsunterricht als nationalpolitische Erziehung (Frankfurt am Main: Diesterweg, 1937), maps 4 and 5.

222   Walther Gehl, *Nordische Urzeit* (Breslau: Hirt, 1936).

223   Karl Buchholz, "Nordisches Rasse-Schicksal im Altertum" in *Der Schulungsbrief* (Berlin: Reichsschulungsamt der NSDAP, July 1934), 14.

224   From *Olympia* (Leni Riefenstahl, 1936/38). Courtesy of Alamy.

226   Walther Gehl, *Der Deutsche Aufbruch 1918–1938* (Breslau: F. Hirt, 1938), 164.

232   Jules Verne, *Robur-le-Conquérant* (Paris: J. Hetzel et Cie., 1886). Illustration by Léon Bennett.

237   Illustration by Eric Pape for H. G. Wells, *The War in the Air* (New York: Grosset and Dunlap, 1908).

242   Courtesy of the Museo Aeronautico Gianni Caproni, Trento.

245   *Los Angeles Times* (August 7, 1945).

248   Illustration of Peking Man (*Sinanthropus pekinensis*), August 14, 1930. Illustrated London News. Courtesy of Mary Evans Picture Library. 10949240/ILN.

251   *Discussion on the Piltdown Skull* (John Cooke, 1915), Wikimedia Commons.

255   Fondation Teilhard de Chardin (Image 279.2). Courtesy of the Fondation Teilhard de Chardin, Paris.

256   Fondation Teilhard de Chardin (Image 421.2). Courtesy of the Fondation Teilhard de Chardin, Paris.

259   Courtesy of the Fondation Teilhard de Chardin, Paris.

266   Courtesy of UNESCO.

278   Wilfrid Le Gros Clark, *The Antecedents of Man* (New York: Harper and Row, 1959), 45.

285   Peter Paul Rubens, *The Entombment*, about 1612, Oil on canvas, The J. Paul Getty Museum, Los Angeles, 93.PA.9.

286   Wikimedia Commons.

288   Louis Figuier, *L'Homme primitif*, 2nd ed. (Paris: Hachette, 1870).

290   Henri Breuil (1877–1961), [Font-de-Gaume. Grands rennes affrontés polychromes. N°11 et 12 de la bande générale], [1903]. Drawing in red chalk and black pencil on paper. 350 × 550 mm. Muséum National d'Histoire Naturelle, Fonds Henri Breuil, IC BR 541896. Source: MNHN/Abbé Henri Breuil.

291   Wikimedia Commons.

294   (Top) Courtesy of the Muséum National d'Histoire Naturelle/Abbé Henri Breuil. First published in Breuil and Hugo Obermaier, *The Cave of Altamira at Santillana del Mar, Spain* (Madrid: Tipografia de Archivos, 1935). (Bottom) Drawing in red chalk and black pencil on paper. 350 × 480 mm. Muséum National d'Histoire Naturelle, Fonds Henri Breuil, IC BR 541949. Courtesy of the Muséum National d'Histoire Naturelle/Abbé Henri Breuil.

295 Courtesy of the American Museum of Natural History.

296 Henri Breuil, "Scene Twenty-Five: The Sanctuary of Trois-Frères at Montesquieu-Avantès (Ariège)" (detail), in *Beyond the Bounds of History: Scenes from the Old Stone Age* (London: Gawthorn, 1949). Reprinted with permission of the Muséum National d'Histoire Naturelle (Paris) and the rights holders of the Abbé Henri Breuil. The original is not known to have survived.

298 Rudolph Zallinger/The LIFE Picture Collection/Shutterstock.com. First published in Lincoln Barnett and the Editors of Time-Life Books, *The Epic of Man* (New York: Time-Life Books, 1961).

299 Nicolaes Witsen, *Noord en Oost Tartarye* (Amsterdam, 1692).

301 Copyright and courtesy of the Muséum National d'Histoire Naturelle/Abbé Henri Breuil.

306 Annette Laming-Emperaire, "Combarelles superimpositions" in *La Signification de l'art rupestre paléolithique* (Paris: Picard, 1962), 278. Reprinted with permission from Éditions Actes Sud—Errance et Picard.

310 Courtesy Everett Collection.

313 Edwin L. Sundberg, "Mass Migrations of Mankind," *Sunday News* (August 27, 1944), loose page. © David Rumsey Map Collection, Cartography Associates.

316 Grafton Elliot Smith and William James Perry, map of "The Reality of Diffusion," in Grafton Elliot Smith, *Human History* (London: J. Cape, 1934), 489, fig. 67.

320 © FSP.

323 STR/AFP via Getty Images.

327 Deutsche Zentralinstitut für Lehrmittel (DZL), Volk und Wissen Verlag. Collection of the author.

330 Ami Drach and Dov Ganchrow, *MAN MADE; handaxe #5*. Knapped flint and 3D printed polymer, 2014. Photography: Moti Fishbain. Courtesy of the artist.

333 Arthur Keith, *The Antiquity of Man* (London: Williams and Norgate, 1915), 408.

334 Courtesy of UNESCO. Jacquetta Hawkes, Leonard Woolley, *Prehistory and the Beginnings of Civilization*, vol. 1 of *The History of Mankind* (New York: Harper and Row, 1963), 38.

336 Courtesy of the Leakey Foundation Archive.

337 Copyright Jay H. Matternes. Courtesy of Jay H. Matternes and the National Anthropological Archives (NAA MS 381957 Jay H. Matternes Drawings and Paintings of Early Humans). First published in F. Clark Howell and the Editors of Time-Life Books, *Early Man* (New York: Time-Life Books, 1965), 73–74.

338 "Diagram illustrating probable course of Quaternary Prehistory in Western Europe," in Alfred Cort Haddon and Julian Huxley, *We Europeans: A Survey of "Racial" Problems* (London: Harper and Brothers, 1936), 39. By permission of Peters Fraser & Dunlop Ltd, London.

339 Kenneth Oakley, *Man the Tool-Maker* (London: British Museum [Natural History], 1949, 1961), 93. © The Trustees of the Natural History Museum, London.

341 (Top) Arthur Keith, *The Antiquity of Man* (London: Williams and Norgate, 1915). (Bottom) Franz Weidenreich, *Apes, Giants, and Man* (Chicago: University of Chicago Press, 1946), 30 fig. 30. Courtesy of the University of Chicago Press.

343 (Top) Osler Library, McGill University, Wilder Penfield Archives, P142, Box 101, File 6–5/6V—Material concerning Penfield, Baxter, Jasper, & Heller, (1954). Epilepsy and the functional anatomy of the human brain (1954). Courtesy of the Osler Library of the History of Medicine, McGill University. First published in Wilder Penfield and Theodore Rasmussen, *The Cerebral Cortex of Man* (New York: Macmillan, 1953). (Bottom) Charles Sherrington, *The Integrative Action of the Nervous System* (New Haven, CT: Yale University Press, 1906), 274.

346 Collège de France, 76 CdF 24b and 24f (*Mécanique vivante*). Courtesy of the Leroi-Gourhan family and the Library of the Collège de France.

347 André Leroi-Gourhan, "Les différentes étapes de l'évolution du membre antérieur chez le vertébré," *Problèmes: Revue de l'Association des étudiants en médecine de l'Université de Paris*, no. 32 (1956): 6–9. Courtesy of the Leroi-Gourhan family, and with gratitude to Aurélie Montagne-Bôrras of the Maison des sciences de l'Homme Mondes.

352 Valerie Walker, "The Semi-Liberated Woman," *The Feminist Voice* 1, no. 4 (1971), Connie Kiosse Papers, Women and Leadership Archives. Courtesy of Women and Leadership Archives, Loyola University Chicago.

361 Courtesy of UNESCO.

363 Courtesy of Natural History Museum/Mary Evans Picture Library.

365 Ib Ohlsson, "Evolution's Long March," in "The Search for Adam and Eve," *Newsweek* (January 11, 1988), 48–49. By permission of the artist.

368 Louis Figuier, *L'Homme primitif*, 2nd ed. (Paris: Hachette, 1870), plate 143.

377 Courtesy of *Documentary Educational Resources*.

378 artillustratn / Alamy Stock Photo.

388 Walton Ford, *The Flaming Fields*, 2020, watercolor, gouache, and ink on paper; 212.1 × 153 cm. © Walton Ford, courtesy of Private Collection and Galerie Max Hetzler Berlin | Paris | London. Photo: Tom Powel Imaging.

395 Steven Mithen, *The Prehistory of the Mind* (London: Thames and Hudson, 1996), 204. Stephen Mithen and Margaret Mathews © Thames & Hudson Ltd.

INDEX

Page numbers in italics indicate a figure on the corresponding page.